Trace gas exchange in a global perspective

Ecological Bulletins No. 42

Trace gas exchange in a global perspective

Edited by
Dennis S. Ojima
and Bo H. Svensson

Ecological Bulletins

ECOLOGICAL BULLETINS are published in cooperation with the ecological journals Ecography and Oikos. Ecological Bulletins consist of monographs, reports and symposia proceedings on topics of international interest, published on a non-profit making basis. Orders for volumes should be placed with the publisher. Discounts are available for standing orders.

Editor-in-Chief and Editorial Office:
Pehr H. Enckell
Ecology Building
University of Lund
S-223 62 Lund
Sweden

Published and distributed by:
Munksgaard International Booksellers
and Publishers,
P.O. Box 2148, DK-1016 Copenhagen K,
Denmark

Suggested citation:
Author's name. 1992. Ojima, D. S. and Svensson, B. H. (eds), Trace gas exchange in a global perspective. – Ecol. Bull. (Copenhagen) 42: 000–000.

ISBN 87-16-15009-0

Cover photo:
Torbjörn Lilja/N: "Kvällsdimmor". Drängsmark, Västerbotten, Sweden, August 1990.

Contents

Ecological Bulletins still available

9. *Bird Census Work and Environmental Monitoring* (1970). Editor S. Svensson. Price DKK 44.00 (US $ 7.00, £ 4.00, DM 11.00).
11. *Ecology in Semi-arid East Africa* (1971). S. Ulfstrand. Price DKK 66.00 (US $ 10.50, £ 6.00, DM 17.00).
12. *Natural Resources Research in East Africa* (1971). M. Zumer. Price DKK 60.00 (US $ 10.50, £ 6.00, DM 17.00).
15. *Ecology of the Swedish Coniferous Forest landscape – a Research Programme* (1972). Editor F. Andersson. Price DKK 66.00 (US $ 10.50, £ 6.00, DM 17.00).
18. *Scandinavian Aerobiology* (1973). Editor S. Nilsson. Price DKK 66.00 (US $ 10.50, £ 6.00, DM 17.00).
19. *Biocontrol of Rodents* (1975). Editors L. Hansson and B. Nilsson. Price DKK 141.00 (US $ 22.00, £ 13.00, DM 36.50).
21. *Man and the Boreal Forest* (1976). Editor C. O. Tamm. Price DKK 120.00 (US $ 19.00, £ 11.00, DM 31.00).
22. *Nitrogen, Phosphorus and Sulphur – Global Cycles.* SCOPE Report 7 (1979, 2nd reprinted edition). Editors B. H. Svensson and R. Söderlund. Price DKK 141.00 (US $ 22.00, £ 13.00, DM 36.50).
23. *Energetical Significance of the Annelids and Arthropods in a Swedish Grassland Soil* (1977). T. Persson and U. Lohm. Price DKK 130.00 (US $ 20.50, £ 12.00, DM 33.50).
24. *Peut-on-Arrêter l'Extension des Deserts?* (1976). Rédacteurs A. Rapp, H. N. le Houérou et B. Lundholm. Price DKK 184.00 (US $ 29.00, £ 16.50, DM 47.50).
25. *Soil Organisms as Components of Ecosystems* (1977). Editors U. Lohm and T. Persson. Price DKK 271.00 (US $ 43.00, £ 24.00, DM 70.00).
26. *Environmental Role of Nitrogen-fixing Blue-green Algae and Asymbiotic Bacteria* (1978). Editor U. Granhall. Price DKK 206.00 (US $ 32.50, £ 18.50, DM 53.00).
27. *Chlorinated Phenoxy Acids and Their Dioxins. Mode of Action, Health Risks and Environmental Effects* (1978). Editor C. Ramel. Price DKK 184.00 (US $ 29.00, £ 16.50, DM 47.50).
28. *Energy, Economic and Ecological Relationships for Gotland, Sweden* – A Regional Systems Study (1978). A. M. Jansson and J. Zucchetto. Price DKK 141.00 (US $ 22.00, £ 13.00, DM 36.50).
30. *Ecology of a Subarctic Mire* (1980). Editor M. Sonesson. Price DKK 217.00 (US $ 34.00, £ 19.50, DM 56.00).
31. *Environmental Protection and Biological Forms of Control of Pest organisms* (1980). Editors B. Lundholm and M. Stackerud. Price DKK 163.00 (US $ 25.50, £ 14.50, DM 42.00).
32. *Structure and Function of Northern Coniferous Forests – an Ecosystem Study* (1980). Editor T. Persson. Price DKK 325.00 (US $ 51.00, £ 29.00, DM 84.00).
34. *Fish Gene Pools. Preservation of Genetic Resources in Relation to Wild Fish Stocks* (1981). Editor N. Ryman. Price DKK 12.00 (US $ 19.00, £ 11.00, DM 31.0).
35. *Environmental Biogeochemistry* (1983). Editor R. Hallberg. Price DKK 325.00 (US $ 51.00, £ 29.00, DM 84.00).
36. *Ecotoxicology* (1984). Editor L. Rasmussen. DKK 406.50 (US $ 64.00, £ 36.50, DM 105.00).
37. *Lake Gårdsjön. An acid forest lake and its catchment* (1985). Editors F. Andersson and B. Olsson. Price DKK 501.00 (US $ 79.00, £ 45.00, DM 129.50).
38. *Research in Arctic life and earth sciences: present knowledge and future perspectives*. Proceedings of a symposium held 4–6 September, 1985 at Abisko, Sweden (1987). Editor M. Sonesson. Price DKK 174.00 (US $ 27.50, £ 16.00, DM 45.00).
39. *Ecological implications of contemporary agriculture*. Proceedings of a symposium held 7–12 September, 1986 at Wageningen (1988). Editors H. Eijsackers and A. Quespel. Price DKK 271.00 (US $ 43.00, £ 24.00, DM 70.00).
40. *Ecology of arable land organisms, carbon and nitrogen cycling* (1990). Editors O. Andrén, T. Lindberg, K. Paustian and T. Rosswall. Price DKK 312.00 (US $ 49.00, £ 28.00, DM 81.00).
41. *The cultural landscape during 6000 years in southern Sweden – the Ystad Project* (1991). Editor B. E. Berglund. Price DKK 700.00 (US $ 110.00, £ 63.00, DM 181.00).

Ecological Bulletins 42: 7–9. Copenhagen 1992

Introduction

D. S. Ojima and B. H. Svensson

Ojima, D. S. and Svensson, B. H. 1992. Introduction. – Ecol. Bull. (Copenhagen) 42: 7–9.

D. S. Ojima, Natural Resour. Ecol. Lab., Colorado State Univ., Ft. Collins, CO 80523, USA. – B. H. Svensson, Dept of Microbiol., Swedish Univ. of Agric. Sci., Box 7025, S-750 07 Uppsala, Sweden.

Introduction

The rapid change in the atmospheric concentration of CO_2 (0.5% yr^{-1}, cf., Conway et al. 1988, Keeling et al. 1989), CH_4, (1% yr^{-1}, cf., Steele et al. 1987, Blake and Rowland 1988), N_2O (0.25% yr^{-1}, Prinn et al. 1990, Elkins and Rosen 1989), chlorofluorocarbons (CFC, 4% yr^{-1}, cf., Penkett et al. 1985, Houghton et al. 1990), and other radiatively active gases has focused much attention on the factors related to their increase (cf., Bolle et al. 1986, Lashof 1989, Houghton et al. 1990, Matson and Ojima 1990). The atmospheric chemistry community has been instrumental in documenting the changes and identifying the environmental impacts and feedbacks (cf., Crutzen and Gidel 1983, Bolle et al. 1986, Cicerone and Oremland 1988, Crutzen 1988, Rodhe 1990). Through this research effort, a better understanding of the interactions between CO, CH_4, and OH on the cleansing capacity of the atmosphere, of the factors affecting the ozone layer in the atmosphere, and of the fast chemical reactions in the atmosphere involving oxidants and other gases has been developed.

The importance of the radiatively active gases in the atmosphere is now well established (Bolle et al. 1986), however, the factors affecting the processes of production and consumption of these gases are poorly understood. The role of the biosphere in controlling the processes has been the focus of a SCOPE project on Trace Gas Exchange and several workshops by this project, especially the Dahlem Conference on Exchange of Trace Gases between Terrestrial Ecosystems and the Atmosphere (Andreae and Schimel 1989), helped to identify the important issues linking the terrestrial biosphere with trace gas exchange.

The role of biological systems in controlling the production and consumption of important atmospheric constituents, including CO_2, CO, CH_4, volatile organic compounds (VOC), NO_x, N_2O, NH_3, sulfur compounds, and others is considerable (Mooney et al. 1987). The importance of microbial communities on the biogeochemical cycles of C and N were discussed relative to production and consumption of CO, CO_2, CH_4, NO_x, N_2O, and NH_3. These microbial communities are affected by changes in the environment and may act as sources or sinks for trace gases depending upon environmental conditions much as substrate availability and quality, water status, nutrient availability, etc. The vegetation and soils play a major role in controlling the rate of these processes. Vegetation can also play an important role in production and consumption of volatile organic compounds (VOC) that serve as important atmospheric reactants affecting the oxidation capacity of the atmosphere.

In the consideration of these processes, the influence of anthropogenic alterations to the atmospheric composition through intentional and inadvertent activities cannot be ignored. Biomass burning in boreal and tropical regions significantly contribute to the C and N gas input into the atmosphere (Crutzen and Andreae 1990, Andreae 1991, Levine 1991). Agricultural activities alter the flux of N_2O and CH_4 into the atmosphere through alterations in the hydrological cycle and fertilizer management (Mosier et al. 1991, Mosier and Schimel 1992). Fossil fuel production and extraction, as well as consumption, greatly alter the CO_2 and CH_4 composition of the atmosphere.

The Sigtuna workshop was jointly organized by the SCOPE Project On Trace Gas Exchange and the IGBP Coordinating Panel on Terrestrial Biosphere-Atmospheric Chemistry Interactions to facilitate the linkage of the research activities being planned by IGAC and the recommendations from the Dahlem Conference on

Exchange of Trace Gases between Terrestrial Ecosystems and the Atmosphere (Andreae and Schimel 1989) organized by the SCOPE project, so that a more integrated research agenda could be developed which includes the interactions of the terrestrial biosphere with the ongoing atmospheric chemistry research. At the Dahlem conference, the participants identified a number of fundamental gaps in our knowledge relative to the magnitude of trace gas fluxes and the primary controls on the mechanisms of production and consumption in the biotic system that hindered our ability to fully understand what was causing these rapid changes to the atmospheric chemistry.

We designed the workshop to review the existing research plans of IGAC and to discuss the important terrestrial biosphere interfaces with the project plans, and to identify complementary or in some cases new research initiatives to augment the research. In so doing, we accepted the regional organizational structure adopted by the IGAC committee at the Dookie conference (Galbally 1989). The research plans for IGAC were subsequently published in the IGBP Report 12 (1990) and a more detailed account of the recommendations based on this workshop was published in IGBP Report 13 (Matson and Ojima 1990).

The material presented here is based on presentations and discussions held at the 1990 Sigtuna workshop. We have organized the material into five sections, the first containing papers of general relevance to all the developed research plans, followed by three sections divided by geographical regions (high-latitude, mid-latitude, and tropical ecosystem regions), and the final section dealing with rice paddy agriculture, which because of its potential role as a significant source of CH_4 and N_2O was singled out for a specific set of studies. Each section, except the first dealing with general issues, concludes with a working group report developed at the workshop. These reports identify specific research needs to address biosphere atmosphere exchange research for each region.

The first set of papers deals with issues related to determining the source sink strengths of various trace gases, and are of general relevance to all the following research plans. These include discussions of geographic concerns of natural and anthropogenic change, use of multi-isotopic studies in identifying sources and sinks of CH_4, concepts related to large area estimates of fluxes from land surfaces and water bodies, deposition onto the land surface, and finally considerations of the coupled soil-plant system on C fluxes.

The regional sections cover topics relevant to the particular biome or geographical region. In the high-latitude region, for instance, the production and consumption of CH_4 is highlighted in the first paper. This is augmented by the second paper which considers NMHC and other trace gases. This regional focus was expanded from the initial wetland study proposed by IGAC, and now completed by the Canadian research group (Roulet et al. 1992). The boreal forest ecosystem as well as wetlands need to be included in the research effort and encompass boreal and tundra regions around the globe.

In the mid-latitude region, a regional focus not included in the Dookie report, research topics related to C and N compounds were developed in light of the vast amount of extant information on several trace gases and the long history of anthropogenic influence within this region. Papers presented in this section reflect these points relative to the role of landfills and agricultural areas for research in this region.

In the tropical region, two separate working group reports are presented that deal with land use change and biomass burning in the tropics. These research studies augment existing research plans within IGAC. The first paper presents a discussion regarding the application of remote sensing techniques for biomass burning studies. Another paper discusses the need to study the alteration of wet and dry cycles in the tropical region influencing NO and N_2O fluxes.

The last section deals with rice paddy systems, a newly initiated research focus for IGAC, as potential sources of CH_4, N_2O and other trace gas species. The first two papers of this section deal with general aspects of the rice paddy ecosystem from microbial processes to processes mediated by the plant system and modified by agronomic practices related to water management, N inputs and other practices. The following three papers deal with trace gases other than CH_4 emitted from rice paddy systems, including N_2O, NH_3, and dimethyl sulfide (DMS). The working group report identifies a new research focus for the IGAC project and deals with needed trace gas research in the Asian rice paddy region.

All of the research initiatives developed here have been accepted in some manner into the IGBP-IGAC research plans. The outline of the research agenda is provided in the IGBP Reports 12 (IGBP 1990) and 13 (Matson and Ojima 1990). The specific research and experimental activities are evolving within international, regional, and national initiatives within the global change and IGAC research projects. Many scientists have been engaged in the development of the ideas and research initiatives presented in these proceedings, and we are extremely indebted to the participants of the workshop at Sigtuna and to the IGBP Coordinating Panel on Terrestrial Biosphere-Atmospheric Chemistry Interactions, the IAMAP/CACGP/IGAC steering committee, and the SCOPE scientific advisory group.

Acknowledgements – We want to express our gratitude to the staffs of the IGBP Secretariat and the Environmental Committee of the Royal Swedish Academy of Sciences in Stockholm. They took care of adminstration during the conference. Lisa Wanrooy-Cronquist, who led us through the phase of planning and organization, deserves special thanks.

Funding for the SCOPE/IGBP Workshop was provided by grants from the Swedish Natural Science Research Council, the Swedish Council for Forestry and Agricultural Research, the

Andrew W. Mellon Foundation to SCOPE, US National Science Foundation-Division of Atmospheric Sciences, and US Environmental Protection Agency and IGBP all of which are greatly acknowledged.

Thanks are also due to the SCOPE Scientific Advisory Committee on Trace Gas Exchange, who encouraged us to publish the presentations given at the workshop. We are grateful for the patience of the contributing authors and the efforts taken by a number of scientific reviewers. Rieky Stürme took charge of the correspondence with the authors and reviewers and also retyped and adjusted the manuscripts. Finally, we wish to thank the Nordic Publishing Board in Science, who funded the the publication of this volume.

References

Andreae, M. O. 1991. Biomass burning: its history, use and distribution and its impact on environmental quality and global climate. – In: Levine, J. S., (ed.), Global biomass burning: atmospheric, climatic and biospheric implications. MIT Press, Boston, MA., pp. 3–21.

– and Schimel, D. S. (eds). 1989. Exchange of trace gases between terrestrial ecosystems and the atmosphere. Dahlem Workshop Reports. – Life Science Res. Rep. 47.

Blake, D. R. and Rowland, F. S. 1988. Continuing worldwide increase in tropospheric methane, 1978 to 1987. – Science 239: 1129–1131.

Bolle, H. J., Seiler, W., and Bolin, B. 1986. Other greenhouse gases and aerosols. Assessing their role in atmospheric radiative transfer. – In: Bolin, B., and Doos, B. R., (eds), The greenhouse effect, climatic change and ecosystems. SCOPE Vol. 29. – Wiley, New York, pp. 157–203.

Cicerone, R. J. and Oremland, R. S. 1988. Bioeochemical aspects of atmospheric methane. – Global Biogeochem. Cycl. 2: 299–327.

Conway, T. J., Tans, P., Waterman, L. S., Thoning, K. W., Masarie, K. A., and Gammon, R. H. 1988. Atmospheric carbon dioxide measurements in the remote global troposphere, 1981–1984. – Tellus 40b: 81–115.

Crutzen, P. J. 1988. Variability in atmospheric-chemical systems. – In: Rosswall, T., Woodmansee, R. G., and Risser, P.G. (eds), Scales and global change: spatial and temporal variability in biospheric and geospheric processes, SCOPE 35. Wiley, New York, pp. 81–108.

– and Gidel, L. T. 1983. A two-dimensional photo-chemical model of the atmosphere. 2: The tropospheric budgets of the anthropogenic chlorocarbons, CO, CH_4, CH_3Cl and the effect of various NO_x sources on tropospheric ozone. – J. Geophys. Res. 88: 6641–6661.

– and Andreae, M. O. 1990. Biomass burning in the tropics: Impact on atmospheric chemistry and biogeochemical cycles. Science 250: 1669–1678.

Elkins, J. W., and Rossen, R. 1989. Summary report 1988. Geophysical monitoring for climatic change. – NOAA ERL, Boulder, CO.

Galbally, I. E. (ed). 1989. The international global atmospheric chemistry (IGAC) programme: a core project of the international geosphere-biosphere Programme.

Houghton, J. T., Jenkins, G. J., and Ephraums, J. J. (eds). 1990. Climate change: the IPCC scientific assessment. – Univ. Press, Cambridge, UK.

IGBP Report 12. 1990. The international geosphere-biosphere programme: a study of global change. The initial core projects. – IGBP Report Series 12, Stockholm, Sweden.

Keeling, C. D., Bacastow, R. B., Carter, A. F., Piper, S. C., Whorf, T. P., Heimann, M., Mook, W. G., and Roeloffzen, H. 1989. A three dimensional model of atmospheric CO_2 transport based on observed winds: 1. Analysis of observational data. – In: Peterson, D. H. (ed.), Aspects of climate variability in the Pacific and the Western Americas. Geophys. Monograph, 55, AGU, Washington (USA), pp. 165–236.

Lashof, D. A. 1989. The dynamic greenhouse: feedback process that may influence future concentrations of atmospheric trace gases and climate change. – Climatic Change 14: 213–242.

Levine J. S. (ed). 1991. Proceedings of the Chapman conference on global biomass burning: atmospheric, climatic and biospheric implications. – MIT Press, Cambridge, MA.

Matson, P. A. and Ojima, D. S. (eds). 1990. The terrestrial biosphere exchange with global atmospheric chemistry – Terrestrial biosphere perspective of the IGAC project: companion to the Dookie report. – IGBP Report 13, Stockholm, Sweden.

Mooney, H. A., Vitousek, P. M., and Matson, P. A. 1987. Exchange of materials between terrestrial ecosystems and the atmosphere. – Science 238: 926–932.

Mosier, A. R., Schimel, D. S., Valentine, D., Bronson, K., and Parton, W. 1991. Methane and nitrous oxide fluxes in native, fertilized, and cultivated grasslands. – Nature (Lond.) 350: 330–332.

– and Schimel, D. S. 1992. Influence of agricultural nitrogen on atmospheric methane and nitrous oxide. – Chemistry and Industry 23: 874–877.

Penkett, S. A., Jones, M. R., Rycroft, M. J., and Simmons, D.A. 1985. An interhemispheric comparison of the concentration of bromine compounds in the atmosphere. – Nature (Lond.) 318: 550–553.

Prinn, R., Cunnold, D., Rasmussen, R., Simmonds, P., Aleya, F., Crawford, A., Fraser, P., and Rosen, R. 1990. Atmospheric trends and emissions of nitrous oxide deducted from ten years of ALE-GAGE data. – J. Geophys. Res. 95: 18369–18385.

Rodhe, H. 1990. A comparison of the contribution of various gases to the greenhouse effect. Science 248: 1217–1219.

Roulet, N. T., Barrie, L., Glooschenko, W.A., Harris, R.C., MacPherson, J. I., Moore, T. R., Neumann, H., and Schiff, H. I. 1992. Northern Wetlands Study (NOWES): an assessment of the role of northern wetlands in the exchange of atmospheric trace gases. – Ecol. Bull. (Copenhagen) 42: 77–85.

Steele, L. P., Fraser, P. J., Rasmussen, R. A., Khalil, M. A. K., Conway, T. J., Crawford, A. J., Gammon, R. H., and Thoning, K. A. 1987. The global distribution of methane in the troposphere. – J. Atmos. Chem. 5: 125–171.

Participation list

Alperin, M.
12–5 Venable Hall
Univ. of North Carolina
Chapel Hill, NC 27599–3300
USA

Andreae, M. O.
Max-Planck-Inst. für Chemie
Abt. Biogeochemie
Postfach 3060
D-6500 Mainz
Germany

Baer, M.
Inst. für Meteorologie und Klimaforschung
Kernforschungszentrum Karlsruhe
Postfach 3640
D-7500 Karlsruhe
Germany
Present address:
BASF AG
Emissionsüberwachung Luft und Lär
D-6700 Ludwigshaven
Germany

Baldocchi, D.
NOAA/ERL/Atmospheric Turbulence and Diffusion Div.
P.O. Box 2456
Oak Ridge, TN 37831
USA

Bogner, J. E.
Argonne Nat. Lab.
EES-362–2A 9700 S Cass Ave.
Argonne, IL 60439
USA

Bolin, B
Arrhenius Lab.
Univ. of Stockholm
S-106 91 Stockholm
Sweden

Bouwman, A. F.
ISRIC P.O. Box 353
6700 AJ Wageningen
The Netherlands

Conrad, R.
Fak. für Biologie
Univ. Konstanz
Postfach 5560
D-7750 Konstanz
Germany

Crutzen, P.
Max-Planck-Inst. für Chemie
Abt. Luftchemie
Postfach 3060
D-6500 Mainz
Germany

Davidson, E. A.
Nat. Res. Counc. Ass.
NASA Ames Res. Cent.
Mail Stop 239–12
Moffett Field, CA 94035
USA

Denmead, O. T.
CSIRO
Cent. for Environ. Mechanics
GPO Box 821
Canberra, ACT, 2601
Australia

Desjardin, R.
Agrometeor. Sec.
Land and Res. Inst.
Cent. Exp. Farm, Bldg. 74
Ottawa, Ontario
Canada KIA 0C6

Duyzer, J. M.
MT TNO P.O. Box 217
2600 AE Delft
The Netherlands

Ehhalt, D. H.
Inst. für Atmos. Chemie
Kernforschungsanlage Jülich
D-5170 Jülich
Germany

Fitzjarrald, D.
Atmos. Sci. Res. Cent.
State Univ. of New York at Albany
100 Fuller Road
Albany, NY 12205
USA

Fowler, D.
Inst. of Terrestrial Ecology
Edinburgh Res. St.
Bush Estate
Penicuik
Midlothian EH26 OQB
Scotland
UK

Freney, J.
CSIRO,
Plant Industry
P.O. Box 1600
Canberra, ACT 2601
Australia

Galbally, I.
CSIRO
Div. of Atmos. Physics
Private Bag No. 1
Mordialloc, Victoria 3195
Australia

Galchenko, V.
Inst. of Microbiol.
Acad. of Sci. USSR
117811 Prosp. 60-letija
Octjabrja 7
Moscow
Russia

Glooschenko, W. A.
CIRAC, York Univ.
4700 Keele St. North York,
Ontario
Canada M3J IP3

Groffman, P.
Univ. of Rhode Island
Dept Natural Resour. Sci.
Kingston, RI 02881
USA

Haider, K.
Inst. für Pflanzenernährung und Bodenkunde
Bundesforschungsanstalt für Landwirtschaft (FAL)
Bundesallee 50
D-3300 Braunschweig
Germany

Harriss, G.
Max Planck Inst. Chem.
Air Chem. Dept
Saarstrasse, 23
D-6500 Mainz
Germany

Isidorov, V. A.
Chem. Dept.
St. Petersburg State Univ.
198904 St. Petersburg
Russia

Jacob, D. J.
Pierce Hall
29 Oxford St.
Harvard Univ.
Cambridge, MA 02138
USA

Johansson, C.
Dept of Meteorol.
Univ. of Stockholm
S-106 91 Stockholm
Sweden

Keller, M.
Nat. Cent. for Atmos. Res.
P.O. Box 3000
Boulder, CO 80307
USA

Kennedy, J. P.
Inst. for Remote Sensing Appl.
Joint Res. Cent.
Comm. of the European Communities
21020 Ispra, Varese
Italy

Klemedtsson, L.
Dept of Microbiol.
Swedish Univ. of Agric. Sci.
Box 7025
S-750 07 Uppsala
Sweden

Klug, M. J.
Kellog Biol. Stn
Hickory Corners, MI 49060
USA

Lassey, K.
DSIR
Inst. of Nuclear Sci.
P.O. Box 31312
Lower Hutt
New Zealand

Lassiter, R.
EPA
Environ. Res. Lab.
Athens, GA 30605
USA

Levine, J.
Atmos. Sci. Div.
NASA Langley Res. Cent.
Hampton, V 23665–5225
USA

Lundberg, H.
Environ. Secretary
Royal Swedish Acad. of Sci.
Box 50005
S-104 05 Stockholm
Sweden

Matson, P.
NASA-Ames Res. Cent.
Mail Stop 239–12
Moffett Field, CA 94035
USA

Melillo, J. M.
Ecosystem Cent.
Marine Biol. Lab.
Woods Hole, MA 02593
USA

Minami, K.
Nat. Inst. of Agro-Environ. Sci.
Tsukuba Norin Kenkyu Danchi-Nai
P.O. Box 2, Ibaraki 305
Japan

Moore, T.
Dept of Geography
McGill Univ.
Burnside Hall
805 Sherbrooke St West
Montreal, Quebec
Canada H3A 2K6

Mosier, A.
USDA-ARS
P.O. Box E
Ft Collins, CO 80523
USA

Neue, H. U.
Int. Rice Res. Inst.
P.O. Box 933
Manila
Philippines

Nicholson S.
Dep. of Meteorology
Florida State Univ.
Tallahassee, FL 32306
USA

Nobre, C.
Dept of Meteorol.
Univ. of Maryland
College Park, MD 20742
USA

Ojima, D. S.
IGBP Secretariat
Royal Swedish Acad. of Sci.
Box 50005
S-104 05 Stockholm
Sweden

Panikov, N. S.
Inst. of Microbiol.
Acad. of Sci.
117811 Prosp. 60-letija
Octjabrja 7
Moscow
Russia

Penkett, S. A.
School of Environ. Sci.
Univ. of East Anglia
Norwich NR47TJ
UK

Reeburgh, W. S.
Inst. of Marine Sci.
Univ. of Alaska
Fairbanks, AL 99775–1080
USA

Rennenberg, H.
Fraunhofer-Inst. für Atmosphärische
Umweltforschung
Kreuzeckbahnstrasse 19
D-8100 Garmsich-Partenkirchen
Germany

Robertson, G. P.
W.K. Kellog Biol. Stn
Michigan State Univ.
Hickory Corners, MI 49060
USA

Rodhe, H.
Dept of Meteorol.
Univ. of Stockholm
S-106 91 Stockholm
Sweden

Rosswall, T.
IGBP Secretariat
Royal Swedish Acad. of Sci.
Box 50005
S-104 05 Stockholm
Sweden

Roulet, N. T.
Dept of Geogr.
York Univ.
4700 Keele St. North York,
Ontario
Canada M3J 1P3

Sanhueza, E.
I.V.I.C.
Apartado 21827
Caracas 1020-A
Venezuela

Sass, R. L.
Biol. Dept.
Rice Univ.
Houston, TX 77251
USA

Schiff, H.
CIRAC
York Campus
4700 Keele St, North York
Ontario
Canada M3J IP3

Schimel, D. S.
Nat. Resour. Ecol. Lab.
Colorado State Univ.
Ft Collins, CO 80523
USA

Schimel, J. P.
Inst. of Arctic Biol.
Univ. of Alaska
Fairbanks, AK 99775
USA

Smith, K. A.
The Edinburgh School of Agric.
West Mains Rd,
Edinburgh EH9 3JG
UK

Stewart, J.
Inst. of Pedol.
Univ. of Saskatchewan
Saskatoon
Canada SN7 OWO

Su, W. H.
Dir. of Lab. of Atmos. Environ. RCEES
P.O. Box 934
Beijing
China

Svensson, B. H.
Dept of Microbiol.
Univ. of Agric. Sci.
Box 7025
S-750 07 Uppsala
Sweden

Vitousek, P. M.
Dept of Biol. Sci.
Stanford Univ.
Stanford, CA 94305
USA

Wahlen, M.
Wadsworth Cent. for Lab. and Res.
New York State Dept of Health
Alabany, NY 12201
USA

Wang, M.-X.
Inst. of Atmos. Physics
Acad. Sinca
De Sheng Men Wai Street
Beijing
China

Wieprecht, W.
Heinrich-Hertz-Inst.
Atmos. Res.
Rudower Strasse 5
1199 Berlin
Germany

Zavarzin, G.
Inst. of Microbiol.
Acad. of Sci.
117312 Prosp. 60-letja
Octjabrja 7
Moscow
Russia

Zepp, R. G.
EPA
Environ. Res. Lab.
Athens, GA 30605
USA

Zimmermann, P.
Nat. Cent. for Atmos. Res.
P.O. Box 3000
Boulder, CO 80307
USA

Ecological Bulletins 42: 12–23. Copenhagen 1992

Prediction of trace gas emissions and their climatic impacts: some geographical considerations

Sharon E. Nicholson

Nicholson, S. E. 1992. Prediction of trace gas emissions and their climatic impacts: some geographical considerations – Ecol. Bull. (Copenhagen) 42: 12–23.

This paper examines two major areas of uncertainty in the prediction of the impact of trace gas emissions on climate. The first is socioeconomic factors which determine the rate of such processes as resource use, industrial production or land conversion. The second is the feedback between the earth's land surface and climate. Since the land surface is the source of trace gas emissions, both natural and anthropogenic changes of vegetation will affect the nature and quantity of emissions. This paper demonstrates large-scale land surface changes which have taken place naturally or from human activities, either intentionally or inadvertently, and describes the dwindling availability of natural resources, using water as an example. Vegetation is also examined as both a response to and a determining factor in climate. Hence, the intricate feedback between vegetation and climate complicates any attempt to predict climatic change. Better quantitative assessment of all relationships and processes is required to achieve realistic forecasts of global change.

S. E. Nicholson, Dept of Meteorology, Florida State Univ., Tallahassee, FL 32306-3034, USA.

Introduction

Emissions of trace gases into the atmosphere are of concern because many are radiatively active "greenhouse" gases and their increased concentrations may lead to a global warming and other worldwide climatic changes. The trace gases include carbon dioxide, carbon monoxide, ozone, methane, nitrous oxide and other nitrogen compounds, and numerous other constituents (Graedel and Crutzen 1989). Two pressing scientific issues are the prediction of the rate of emissions and the prediction of their climatic impact, especially on a regional scale. The former requires future projections of emission sources, which include various industrial processes, human resource use and microbiological and biogeochemical processes at the earth's surface. These biological processes are a function of the nature of the land surface and human land use. Since the global distribution of ecosystems – and their potential for use – is climatically determined, climatic change is potentially both a factor in and a response to trace gas emissions. This climate vegetation feedback thus must enter into any attempt to predict future trace gas emissions or their climatic impact.

Some of the major surface sources of radiatively active trace gases are summarized in Fig. 1. The by-products of industrial processes, such as the production of aerosol sprays, refrigerants and plastics, are a variety of chemicals, such as chlorofluorocarbons, which ultimately alter the ozone chemistry of the stratosphere. When land is cleared of vegetation, the biogeochemical cycles are altered both by the loss of that component of the ecosystem and by the decomposition of the removed biologic debris. The decomposition is often hastened by agricultural burning or consumption of the residue for fuel. Deforestation and other forms of land clearing are prime factors in the increased concentration of carbon dioxide in the atmosphere. Methane, ozone, carbon monoxide, nitrous oxide, carbon dioxide and other gases are all produced from biomass burning. The combustion of fossil fuels is likewise a significant source of trace gas emissions, particularly carbon dioxide and carbon monoxide. Natural ecosystems, such as tundra, swamp and other wetlands, emit methane and other gases. Agriculture also provides a number of sources: ruminant grazing animals and rice paddies produce methane and crops fertilized with nitrogen compounds emit nitrous oxide and other nitrogen-containing gases.

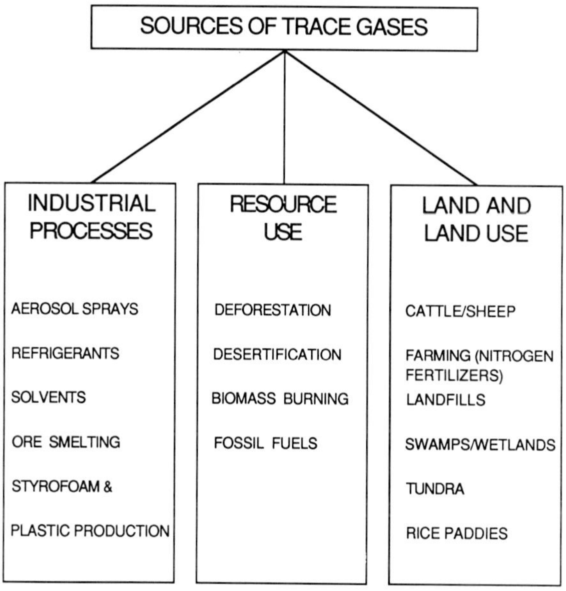

Fig. 1. Schematic of the major sources of trace gas emissions and carbon dioxide.

Urbanization concentrates waste products, necessitating landfills, which produce methane. It passively affects emissions by replacing natural landscapes or agricultural fields, which are likewise sources of trace gases.

The geographical distribution and intensity of these sources is changing as a consequence of human activities. This paper puts forth some considerations which must be made in attempts to predict future emissions and their climatic consequences. The first is the geographical distribution of natural ecosystems and the impact human activities have had upon them. The second is the constraints and uncertainties involving the availability of resources and other socioeconomic factors. Finally, the relationship between climate and vegetation is considered.

Table 2. Major plant communities of the earth (from Woodwell 1978).

	Area (10^6 km^2)	Net primary production (10^{15} grams of carbon yr^{-1})	Planet mass (10^{15} grams of carbon)
Tropical rain forest	17.0	16.8	344.0
Tropical seasonal forest	7.5	5.4	117.0
Temperate evergreen forest	5.0	2.9	79.0
Temperate deciduous forest	7.0	3.8	95.0
Boreal forest (conifers)	12.0	4.3	108.0
Woodland and shrubland	8.5	2.7	22.0
Savanna	15.0	6.1	27.0
Temperate grassland	9.0	2.4	6.3
Tundra and alpine meadow	8.0	0.5	2.3
Desert scrub	18.0	0.7	5.9
Rock, ice and sand	24.0	0.0	30.2
Cultivated land	14.0	4.1	6.3
Swamp and marsh	2.0	2.7	13.5
Lake and stream	2.0	0.4	0.02
Total continental	149.0	52.8	826.5
World total	510.0	77.6	828.0

Human-induced changes of the global land surface

The land areas of the globe are the prime sources of the natural atmospheric greenhouse gases and the nature and amount of emissions vary with the land cover type and with a number of environmental variables, such as soil type. The natural land surface has been altered by human activities for centuries. The processes of transformation include deforestation, urbanization, dam-building, irrigation, and wetland drainage. In the future, anthropogenic modification of the earth's surface will continue but at unpredictable rates. If major climatic changes occur, they will also alter the patterns of natural vegetation and this in turn will impose new limits on the human modification of the landscape.

The development and expansion of agriculture moti-

Table 1. Comparison of areal estimates of major ecosystems. Ecosystem reduction figures indicate the extent to which ecosystems have been modified, through agricultural activities, relative to each other, and relative to their respective original extents (from Matthews 1983).

	Pre-agricultural vegetation data Area (10^6 km^2)	Present land cover data Area (10^6 km^2)	Ecosystem reduction Area (10^6 km^2)	Ecosystem reduction Total ecosystem (%)
Total forest	46.28	39.27	7.01	15.15
Tropical rainforest	12.77	12.29	0.48	3.75
Other forest	33.51	26.98	6.53	19.50
Woodland	15.23	13.10	2.13	13.8
Shrubland	12.99	12.12	0.87	6.7
Grassland	33.90	27.43	6.47	19.1
Tundra	7.34	7.34	0.00	0.0
Desert	15.82	15.57	0.25	1.6
Cultivation	0.93	17.56	−16.63	–

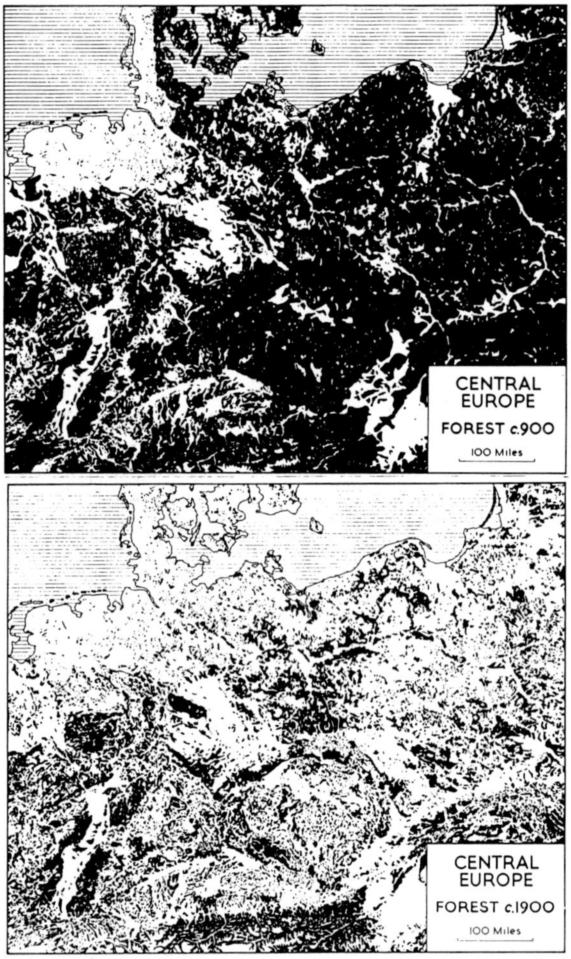

Fig. 2. The changing distribution of forest in central Europe between a) AD 900 and b) 1900 (from Darby 1956).

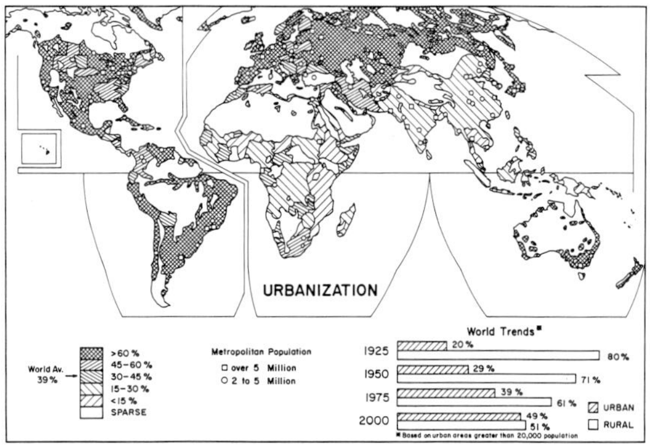

Fig. 3. World map of the percent of urbanization and a comparison of urbanization in 1925, 1950 and 1975 with projections for 2000 (from Espenshade 1982).

vated much of the change. Matthews (1983) has attempted to quantify land conversion for agriculture. Table 1 summarizes her estimates of the areal extent of major ecosystems, both now and prior to intensive agriculture. In pre-agricultural times, forests covered the largest area, c. 46×10^6 km^2. Grasslands, the second most extensive, covered c. 15×10^6 km^2. Large-scale changes have taken place: crops have replaced 15% of the forests, 19% of the grasslands, 14% of the woodlands and 7% of the shrublands. This has altered the release of carbon to the atmosphere, since each ecosystem differs with respect to carbon production and storage (Table 2). Similar changes have been effected in other trace gases.

Some the most extensive land surface change has been in forest cover. Deforestation has gone on for millennia in parts of South America and India and it has been widespread for centuries (Goudie 1984). Precise quantitative assessment is lacking, but one estimate is available from Darby (1956), who appraised the changes which took place from 900 to 1900 A.D. Within that time, forest cover in western Europe was reduced from 90 to 20% (Fig. 2), and similar changes occurred in the Mediterranean. In New England there was a continual reduction in forest cover from the arrival of the first European settlers until about 1900. In the tropics deforestation continues at a rate of ~0.6% yr^{-1} (Lanly 1982), a rate which implies complete destruction in less than two centuries. By some estimates, 72% of the rain forest has vanished from West Africa, 64% from southern Asia (Goudie 1984). The largest existing rain forest, the Amazon, is likewise rapidly disappearing. Until recently deforestation could be largely attributed to agricultural expansion. Now, however, much deforestation takes place for urbanization, fuel wood and lumber, charcoal production and cattle ranching.

Increased efficiency in agricultural production has fa-

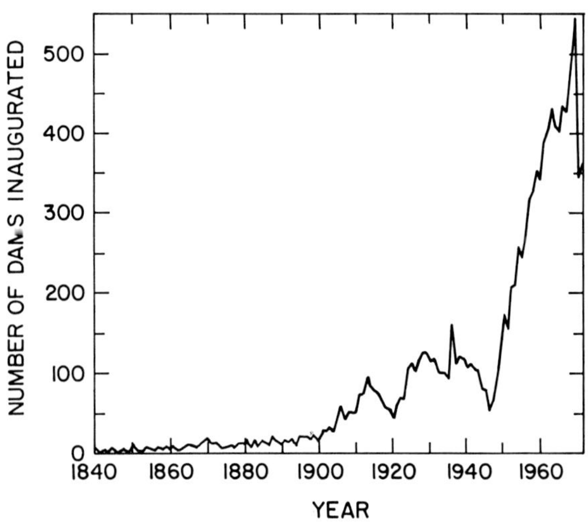

Fig. 4. The construction of dams worldwide (from Goudie 1984).

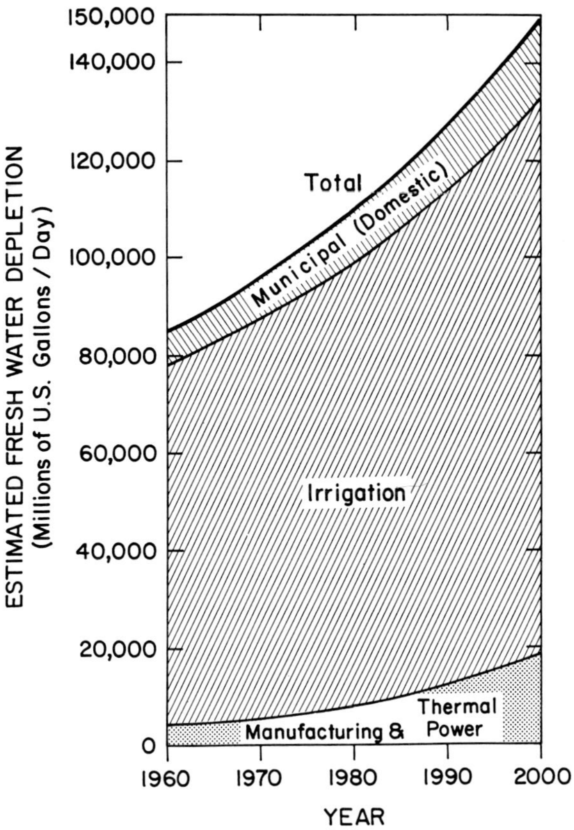

Fig. 5. Estimated fresh-water depletion for the United States to A.D. 2000. These totals do not include water used for electric generation which is immediately returned to the surface (from Chorley 1978).

cilitated urbanization. It has been most intense in Europe, Soviet Asia, and North and South America (Fig. 3). In large regions of these continents >60% of the land is considered urban. In 1925, c. 80% of the global land surface was rural, while now this figure is down to c. 60%. Urban areas are projected to approach the size of rural ones early in the 21st century. Urbanization influences not only trace gas emissions; it also alters the surface heat and water balance. This likewise could have an impact on at least regional-scale climate.

Agricultural efficiency has in part been achieved through control of the hydrologic cycle via irrigation and dam building. Dam construction has been widespread since the 1940s (Fig. 4) and currently c. 15 to 20% of the world's streamflow is regulated by dams. Two of the world's largest dams, the Akosombo in Ghana and the Aswan in Egypt, created reservoirs several hundred miles long (Goudie 1984). Natural irrigation, i.e., that produced by river floods, has been practiced for millennia, but during the 20th century the demand for irrigation has grown with population, particularly in developed countries. In the western U.S., 50–60% of agricultural land is cultivated through irrigation, which creates the country's biggest demand for freshwater (Adler et al. 1981, Pillsbury 1981). Nationally, this demand has increased by c. 50% since 1960 and continues to grow (Fig. 5).

Another process of landscape change has been the drainage of wetlands, a natural source of methane. Wetlands are drained for both agriculture and cities. The state of Florida provides an excellent example of the extent of this change. While swamps once covered most of the state, they are now limited in area and confined mostly to coastal areas and along rivers. The everglades have been reduced to one tenth of their original size (Fernald and Patton 1984). Wetland drainage is common in Europe, where in many countries >60% of the agricultural lands had previously been wetlands (Fig. 6). There are ambitious plans to drain the vast swamps of the Sudd in the southern Sudan by constructing a canal to the Nile river. This swampland now covers >200,000 km^2 and comprises c. 4% of the global wetland area.

The processes described so far have been active or intentional. An extensive passive, or inadvertent, change is that of desertification. It results from many of the other processes: overcultivation, deforestation, overgrazing, improper irrigation, slash/burn agriculture and other forms of land mismanagement (United Nations 1977). Desertification reduces the productivity of the land and leads to erosion. The United Nations esti-

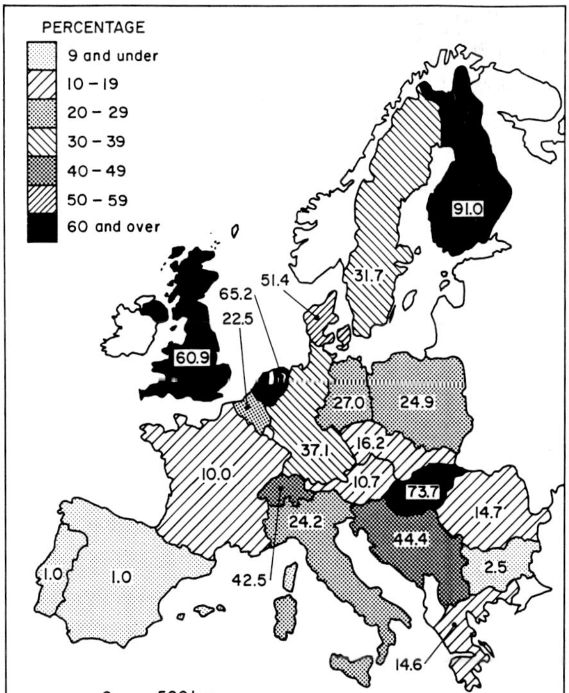

Fig. 6. Percentage of drained agricultural land in Europe; there are no data for blank areas (after Green 1976, Goudie 1984).

Table 3. Global surface changes due to anthropogenic modifications (based on Henderson-Sellers and Gornitz 1984).

Process	Change in area (over 30 yr) relative to global land surface area
Tropical deforestation	0.0259
Temperate deforestation	small
Dam-building	0.0024
Irrigation	0.0029
Salinization	0.0003
Urbanization	0.0048
Desertification	0.0254

mated in 1977 that 9 mill km^2 of primarily semi-arid land has been desertified. Although this figure is often disputed, desertification will definitely be a factor to reckon with in the future.

In summary, the earth's natural landscape has been extensively altered by deforestation, desertification, urbanization, irrigation and dam construction, agriculture and wetland drainage. The effect of these processes on the surface emission of trace gases and hence on atmospheric concentrations is widely recognized. However, projection of future trace gas emissions and their climatic effect requires a quantification of both the rate of these processes and of their impact on biogeochemical cycles. Therein lies the problem, as there is little agreement concerning either the areal extent of natural ecosystems or the extent of anthropogenic change. As an example, Matthews' (1983) estimate of total forest cover is 39.3×10^6 km^2, while that of Lieth (1975) is 48.5×10^6, or c. 25% greater. Similar disparities exist for woodlands, shrublands and other ecosystems. Estimates of deforestation from Myers (1980) and FAO (Woodwell et al. 1983), although in agreement for the period 1860 to 1980, differ by nearly a factor of three for the year 1980. The United Nations' (1977) published estimate of global desertification, 5.4 mill km^2 yr^{-1}, is even more controversial. The figures for Africa, for example, have been greatly exaggerated (Olsson 1985) but the precise extent of such lands is unknown.

Henderson-Sellers and Gornitz (1984) have attempted to summarize the extent of anthropogenic land surface modification (Table 3). Their work amply illustrates the discrepancies between various estimates of the processes of change, and their published rates of change represent "best guesses". They conclude that over the last 30 years anthropogenic changes have been small relative to the total global land surface. However, the extent of surface modification has been regionally significant. This is especially true for tropical deforestation, which has affected the largest area. Moreover, these processes are continuing at an accelerated pace. More accurate quantification of these processes and their effects on climate and trace gas emissions must be taken into account to achieve realistic predictions of global change.

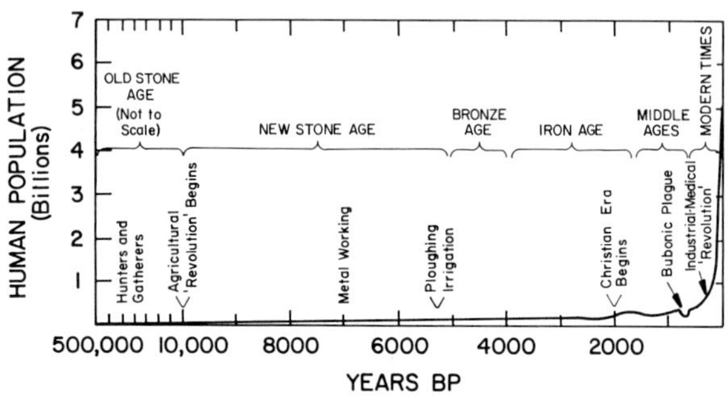

Fig. 7. The growth of population for the past half-million years (from Ehrlich et al. 1977, Goudie 1984).

Socioeconomic factors determining future trace gas emissions

The simplest projection of future trace gas emissions is the assumption of a continuation of current rates of release. This fails to take into account socioeconomic changes such as population growth and decreased availability of resources, as supplies (e.g. forests) are depleted. The impact of socioeconomic factors on trace gas emissions is illustrated by the long-term trends of carbon release into the atmosphere. Water supply is an example of the potential for reduced availability of resources in the near future.

Population growth is perhaps the single most important factor which has contributed to anthropogenic alteration of the natural land surface change and to increased trace gas emission to the atmosphere. World population is rising exponentially (Fig. 7); since the industrial-medical revolution in the 19th century it has increased seven-fold. This increase parallels such trends as dam construction (Fig. 4), extent of irrigation (Fig.

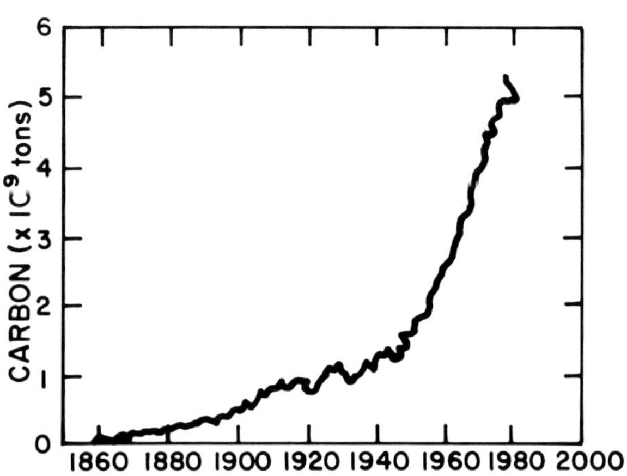

Fig. 8. Fossil fuel emissions of carbon dioxide 1860–1982 (from Marland and Rothy 1983).

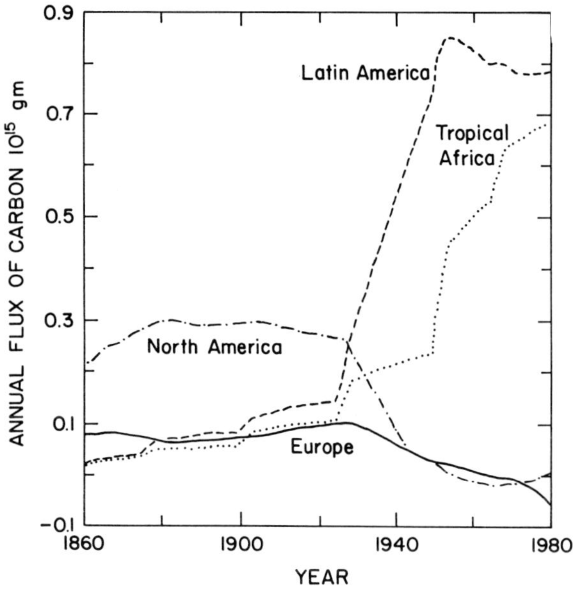

Fig. 9. Net annual release of carbon from various regions of the earth between 1860 and 1980. This is an indicator of land cover change, particularly deforestation (from Woodwell et al. 1983).

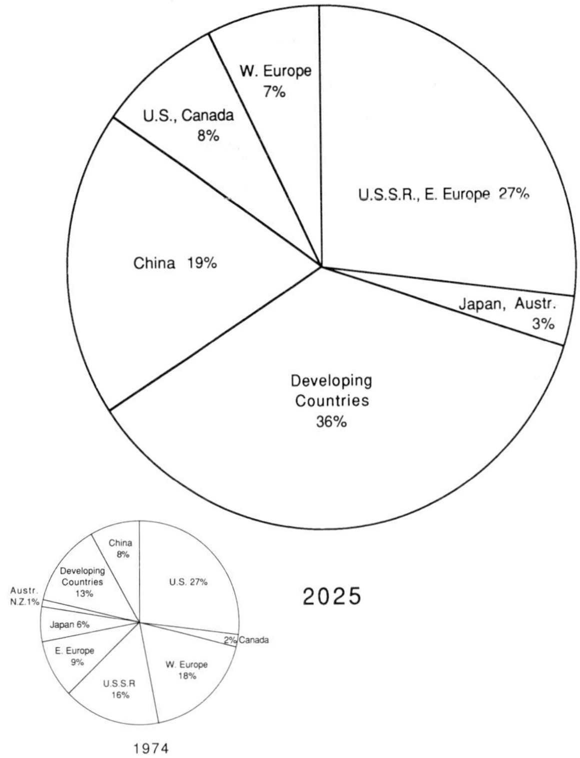

Fig. 10. The changing patterns of global CO_2 emissions (from U.S. Dept of Energy 1979). Area is proportional to CO_2 emission.

5), deforestation, and fossil fuel consumption (Fig. 8). Many estimates of future trace gas emissions and their effects are based on extrapolation of this population curve. That approach, however, has several short comings. Medical epidemics can drastically reduce human population, as the bubonic plague did many centuries ago (Fig. 7). Moreover, the world may not be able to sustain the projected population.

Although such socioeconomic factors as resource use and agricultural production roughly parallel population growth, they are also influenced by such factors as standard of living or industrial development. Thus, the release of carbon into the atmosphere has not shown a steady increase, as population growth would suggest, but has fluctuated as societies changed from agricultural to industrial or nations developed (Woodwell et al. 1983). In the 19th century, technology and the industrial revolution enhanced the efficiency of agricultural production in the developed world. Agricultural land was increasingly abandoned in North America and Europe as industries developed, and much of the forest cover regenerated (Fig. 9) and carbon release decreased. In New England, remnants of stone fences, which once marked pastures and fields, are scattered throughout the woods.

At the same time, population and agriculture continued to grow at an accelerated pace in the developing world. As a consequence, forest cover has declined exponentially in Latin America and tropical Africa since the 1920s (Fig. 9). The net result globally has been a sharp reduction the extent of forests and a rapid increase in carbon release to the atmosphere.

Since population is growing most rapidly in developing countries, estimates of the impact of trace gas emissions must be based on geographically specific growth estimates. A likely scenario is that in the future, developing countries will be responsible for an increased proportion of the global carbon release (Fig. 10). The greatest population growth is in areas which are now predominantly rural (Fig. 3). A likely result is increased conversion of natural landscapes to cities. Urban area is expected to increase c. 20% by the end of the century. The gaseous emissions from the "paved" landscape will differ in composition and quantity from those emitted by the natural ecosystems and agricultural

Table 4. Per capita carbon dioxide emissions (10 kg of carbon) for ten countries (based on data in Marland and Rotty 1983).

United States	4.3
Canada	3.7
U.S.S.R.	3.0
West Germany	2.7
United Kingdom	2.2
Japan	1.7
Italy	1.3
Mexico	.8
China	.4
India	.1

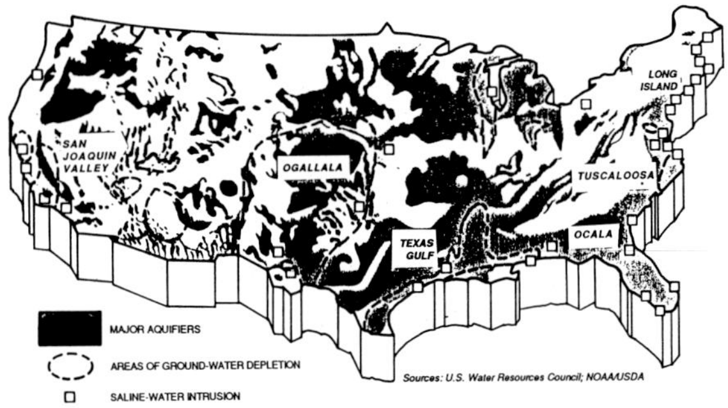

Fig. 11. Major aquifers in the U.S. and areas of groundwater depletion and saltwater intrusion (from Adler et al. 1981).

lands which the cities replace. On the other hand, more agricultural land may be required to support the world's population, although technological advances and improved productivity could alter that.

Economics, policies and practices will also influence the per capita use of resources and the environmental impacts of the world's increasing population. The effect of two world wars and the depression were apparent in the trends of carbon production from fossil fuel in the U.S. (Fig. 8). As the economies of the developing nations improve, they are expected to be responsible for increasingly large emissions of carbon dioxide. This is evident from per capita emissions of carbon dioxide for various countries (Table 4). For the U.S. and Canada, per capita emissions are about twice that of European countries and c. 40 times that of India. These same considerations may be true for other trace gases affected by resource use, industrial processes, and fossil fuel consumption. Yet, the potential development of policies to restrict such practices, such as the Montreal protocol aimed at reducing ozone destruction, make future emissions very hard to predict.

Another potential limitation is simply the availability of resources. After the forests vanish, deforestation will cease. So will fossil fuel consumption once fuel supplies have been depleted. Such scenarios sound extreme, but the case of irrigation will show they are, in fact, plausible. The tremendous amount of water consumed worldwide and the extent of our dependence on irrigation to supply this water are generally not realized. All in all, each km^2 in the western U.S. is irrigated annually with the equivalent of 1 m of rainfall, or in all c. 150 km^3. Starting with the water required to grow corn, it takes 13 m^3 to produce one steak; 55 m^3 are required for a bushel of wheat. Much of the water for irrigation is supplied by the Ogallala aquifer of the Great Plains; as much as one meter of water is withdrawn in many locations each year. This rate exceeds replenishment by seepage of fresh rain water by 79 mill m^3 d^{-1} (Adler et al. 1981). Consequentially, the Ogallala will probably dry up within the next 40 years.

Much evidence indicates that irrigation may be only a temporary augmentation to water supply (Adler et al.

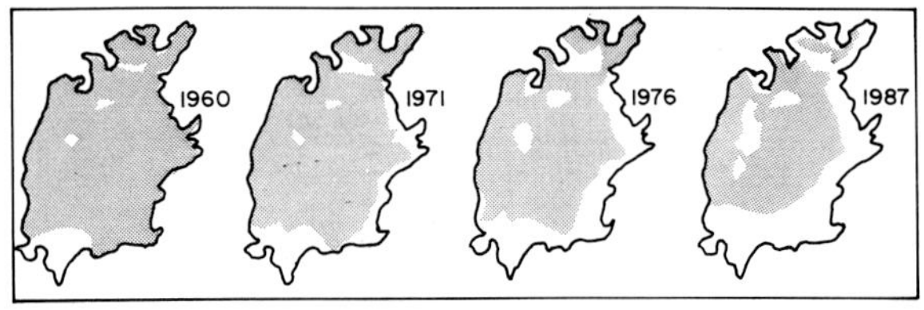

Fig. 12. The changing profile of the Aral Sea, 1960–2000 (from Micklin 1988).

1981, Pillsbury 1981). In the U.S., the water table is falling by up to 1 m yr^{-1} in many regions, demonstrating rapid depletion of many aquifers (Fig. 11). In the fertile San Joachin Valley of California, the drop of the water table has caused land subsidence of 10 m in 50 years. Irrigation has also tremendously reduced the levels of the Caspian Sea, the world's largest lake, and the Aral Sea, once the fourth largest lake (Chorley 1978, Goudie 1984, Micklin 1988). The level of the Aral has dropped c. 3 m since extensive irrigation began, reducing its surface area by 40% and its volume by 66% between 1960 and 1987 (Fig. 12). The reduced availability of fresh water must be taken into account in any estimate of future trace gas emissions from agricultural lands.

An additional problem associated with irrigation is the salinization of the water supply and soils (Chorley 1978, Pillsbury 1981). Since c. 75% of the water used for irrigation evaporates in the fields, the original salt residues are concentrated into 25% of the original volume. Other chemicals are added when runoff traverses fertilized fields. The result is salinization of both soils and the runoff which re-enters streams. Worldwide, c. 10% of the total arable lands suffers from soil salinization. The problem is so severe in the Colorado River that the salty river water is essentially useless as it crosses the border to Mexico unless desalinized. The water draining from California's Imperial Valley into the Salton Sea is saltier than the oceans.

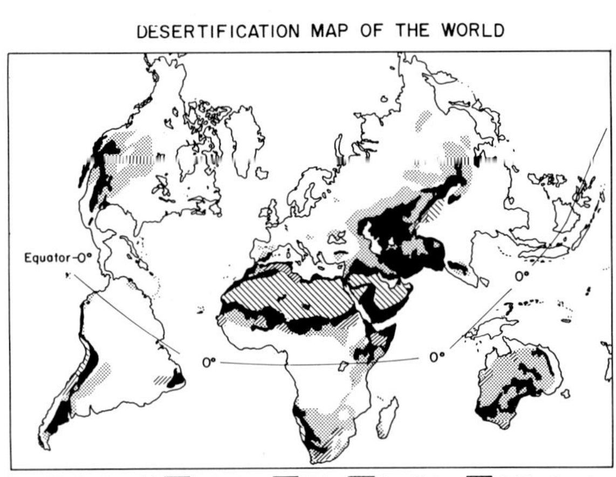

Fig. 13. Desertification map of the world (from Ibrahim 1978).

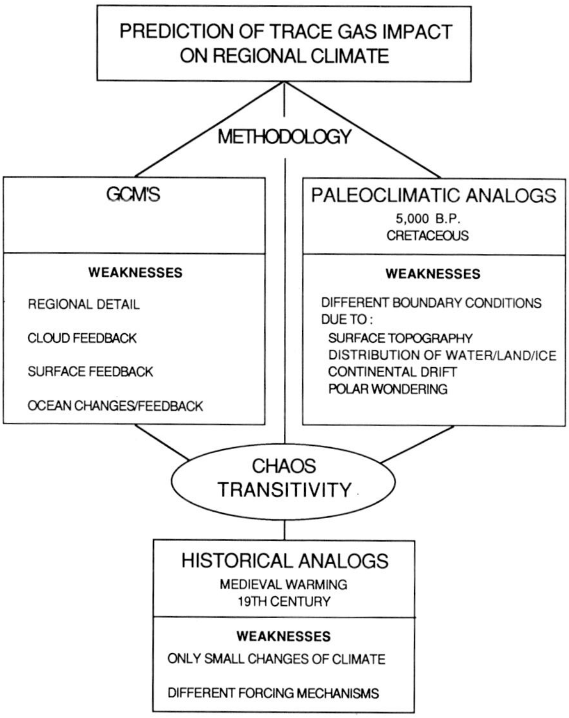

Fig. 14. Schematic of possible methodological approaches for predicting the climatic impact of increased greenhouse gas emissions and the uncertainty in them produced by chaos and transitivity in the climate system.

Another common result of irrigation, particularly in arid lands, is that the soils become much more prone to erosion. This problem also results from agricultural burning, urbanization, deforestation and agriculture. On cultivated land in the U.S. (most of which is irrigated), c. 2.7 mill kg of topsoil are eroded annually from each km^2; this is eight times quicker than the soil is reformed (Pimentel 1976).

Thus, the result of intensive land use is reduced availability of fresh water and declining land productivity due to such processes as salinization and erosion. These are the processes of desertification. The United Nations estimated in 1977 that 9 mill km^2 had been desertified worldwide (Fig. 13); another 48 mill km^2 were deemed "at risk". These estimates are probably too high and they include land surface changes both climatologically and anthropogenically induced. Nevertheless, it is clear that current land use practices may have to be greatly altered in the future and that this will affect the surface sources of atmospheric trace gases.

In summary, of the various socioeconomic factors related to trace gas emissions, there are two major unknowns: population growth and per capita utilization of land and resources. The latter generally increases with development and economic growth. However, this increase could be curtailed by dwindling natural resources, especially fuel, forests, and water for irrigation. These factors create large, unavoidable uncertainties in projections of future trace gas emissions and their atmospheric concentrations.

Predicting the impact of trace gas emissions on regional climate

Prediction of an overall global warming trend due to increased concentrations of carbon dioxide and other trace gases is relatively straightforward and there is strong consensus that a global warming will occur (Schneider 1989). The associated regional patterns of climatic change are much more difficult to extrapolate. The suggested methods for regional prediction (Fig. 14) are general circulation models, paleoclimatic analogs and historical analogs; all have serious shortcomings. The problem is compounded by two other factors. The first is the chaotic behavior of climate: several states are possible for the same set of forcing factors and a transition from one state to another with no change of forcing is plausible. The second is vegetation. Surface vegetation cover is an important factor in both global and regional climate, but it is also a response to climate and it is being altered by human activities. Hence an understanding of the interaction between the land surface and atmosphere is critical for predicting climatic change. This section examines this interaction and illustrates the degree to which vegetation patterns might be expected to vary in response to climate.

The characteristics of the earth's surface are prime factors in global climate because the atmosphere is highly transparent to solar radiation and is most directly heated by energy from the earth's surface. The surface transfers energy to the atmosphere in three ways: the processes of conduction and convection (termed "sensible" heat), latent heat released in the atmosphere as water vapor evaporated at the surface condenses, and emission of radiation to the atmosphere. Latent heat exchange is particularly important because it is released high in the atmosphere and can have a regional impact. The properties of the surface determine both the amount of energy available to the atmosphere (for a given amount of solar isolation) and its partitioning into sensible and latent heat and thermal radiation. The surface also exchanges momentum, moisture, particulates and gases with the atmosphere. The most important characteristics determining the rate of exchange are surface temperature, albedo or reflectively, soil moisture, and surface roughness, all of which are markedly influenced by vegetation type.

The energy exchange processes are described by the surface energy balance equation:

$$R_{SW\downarrow}(1 - a_s) = R_{LW\downarrow} - R_{LW\uparrow} + H + LE \quad (1)$$

where $R_{SW\downarrow}$ is solar insolation, a_s is surface albedo, $R_{LW\downarrow}$

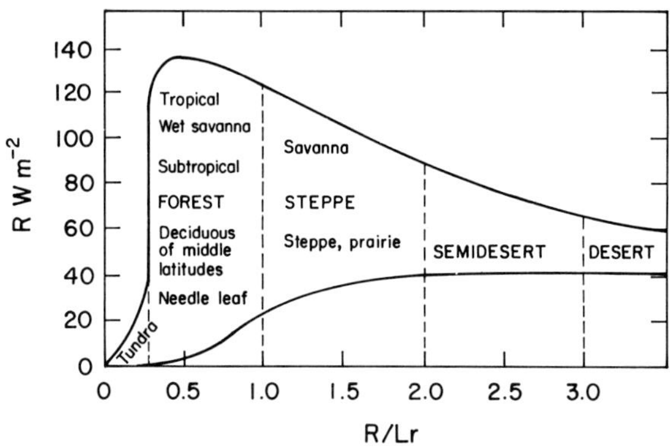

Fig. 15. Budyko's (1986) concept of geographical zonality, defining vegetation limits in terms of net radiation received at the surface (y-axis) and the dryness ratio (x-axis). The dryness ratio is the ratio of net radiation to the product of the latent heat of vaporization and mean annual precipitation and it represents the degree to which the net radiation suffices to evaporate the mean precipitation.

is longwave atmospheric radiation, $R_{LW\uparrow}$ is longwave terrestrial radiation, H is sensible heat transfer, and LE is latent heat transfer. The terms a_s, $R_{LW\uparrow}$, H and LE are all determined principally by either vegetation cover directly or by factors which it influences, such as soil moisture, surface temperature and surface roughness. Even the longwave radiation emitted by the atmosphere $R_{LW\downarrow}$ can be influenced by the land surface, since particulates derived from the surface affect longwave emission.

Thus, the earth's surface converts and redistributes solar radiation into the energy balance components identified by Eq. 1. The characteristics of the surface, including vegetation cover or absence of it, prescribe the specifics of the redistribution. Conversely, the surface energy balance of a given region, in turn, is a prime determinant of the dominant regional vegetation. This is illustrated with a diagram from Budyko (1986), using two parameters referred to as net radiation and the dryness ratio, which essentially represent insolation and rainfall (Fig. 15). Net radiation is a measure of energy availability and is defined as:

$$R_{net} = R_{SW\downarrow}(1-a_s) + R_{LW\downarrow} - R_{LW\uparrow} \quad (2)$$

Over land, the terms a_s and $R_{LW\uparrow}$ are largely a function of surface vegetation cover. The dryness ratio is an inverse indicator of "usable" annual rainfall (i.e. that which is not evaporated and lost to the ecosystem). It is the ratio of net radiation to annual rainfall (divided by the latent heat of condensation, a conversion factor which tranforms rainfall into the units of energy required to evaporate it).

Budyko (1986) shows that the major biomes (Fig. 15) are prescribed to a large degree by net radiation and the dryness ratio. The border between forests and dryland steppe vegetation types, for example, is roughly delineated by a dryness ratio of one, a quantity which indicates an equality between available energy and energy

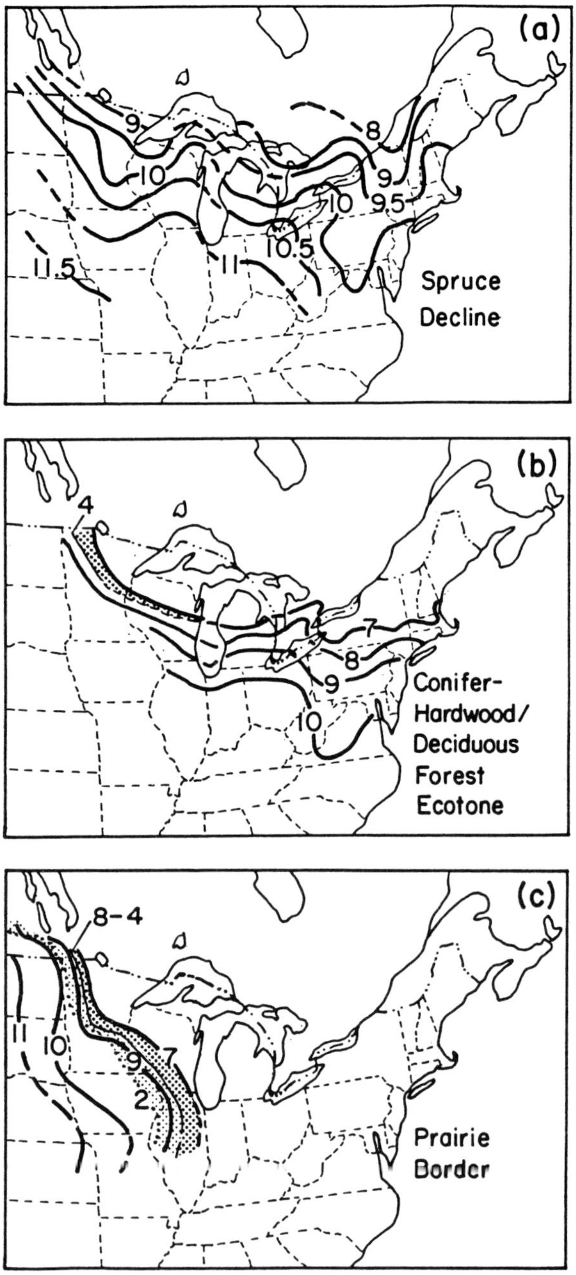

Fig. 16. Isochrones (in thousands of years) on the migration of various vegetation ecotones, or boundaries, based on pollen analysis. In the top diagram isochrones indicate the position of the southern margin of the boreal forest as it migrated northward. In the center diagram the conifer/deciduous forest ecotone is indicated. In the bottom diagram the "prairie border" is delimited. Shading indicates the area across which the prairie border first expanded (to ~7000 years BP) then retreated (westward) as conditions became more moist in the region in the mid to late Holocene (after Bernabo and Webb 1977).

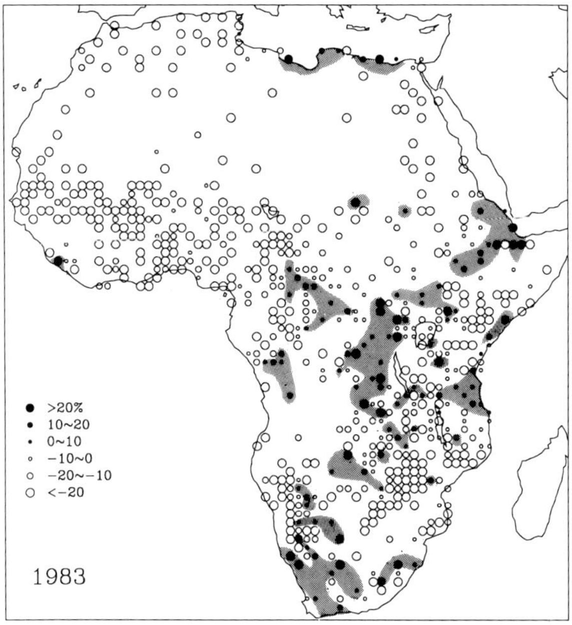

Fig. 17. Annual rainfall in 1983 expressed as a percent departure from the long-term mean (from Nicholson et al. 1988). Positive anomalies are shaded.

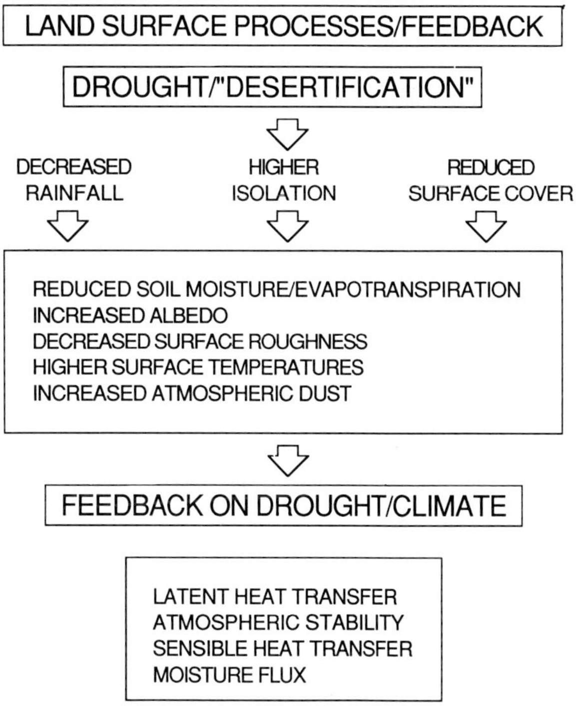

Fig. 18. Schematic of the land surface-atmosphere feedback resulting from large-scale drought and/or desertification (i.e. anthropogenic denuding of the surface).

required to evaporate the mean annual rainfall' (Fig. 15).

The above discussion demonstrates an inextricable feedback relationship between regional climate and regional vegetation cover: vegetation cover is prescribed by rainfall and net radiation, but the latter parameter is to a significant degree a function of vegetation cover. This concept becomes particularly important in view of the large changes of surface vegetation which do occur in response to climate.

One example of large-scale vegetation change is that of North America during the Pleistocene and Holocene. As the Laurentide ice sheet retreated about 11,000 years ago, the borders of the boreal forest, the deciduous and conifer forests, and the prairie moved with it (Fig. 16). The impact of such a transformation on surface energy balance is clear from simple consideration of the albedo differences between snow, ice and forest. While the forest reflects only 5 to 15% of incident solar radiation back to space, ice reflects c. 50% and snow up to 95%. This feedback helps to stabilize polar climates, once developed.

The Ice Age example represents an exaggeration of the extent of change which might be expected in the near future, but even in recent times significant changes have occurred. This is illustrated by the recent situation in Africa. Rainfall in the semi-arid Sahel has steadily declined >50%, causing the Saharan border to "advance" perhaps several hundred kilometers (Nicholson 1989). Moreover, on longer time scales much of the continent becomes uniformly wetter or drier; such was the case, for example, in 1983 (Fig. 17). Evidence suggests that as recently as 700 years the Saharan border was displaced as much as 500 km, northward (Nicholson 1988). This was during the "medieval warming", which is one possible analog to the anticipated global warming. Such long-term climatic changes not only redistribute vegetation, but they can also alter the areal expanse of some ecosystems, especially those in semi-arid regions.

The effect of a large-scale climate or vegetation change in semi-arid regions is illustrated in Fig. 18. If a prolonged drought or human-induced desertification occurs, the likely effects on the surface include reduced soil moisture, higher temperature, reduced vegetation

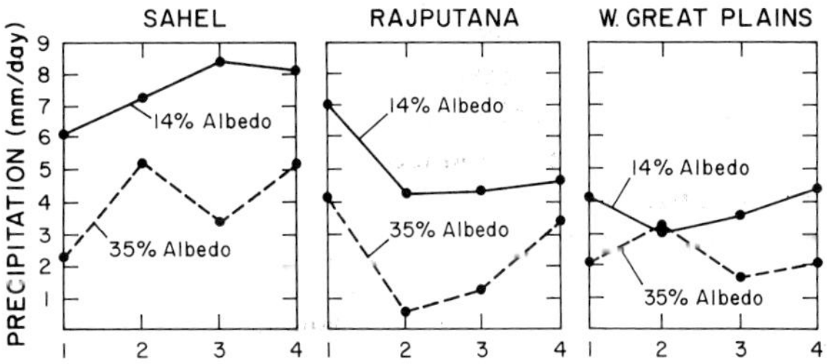

Fig. 19. General circulation model simulation of the effect of changing surface albedo from 14 to 35% in three geographic regions (Charney et al. 1977).

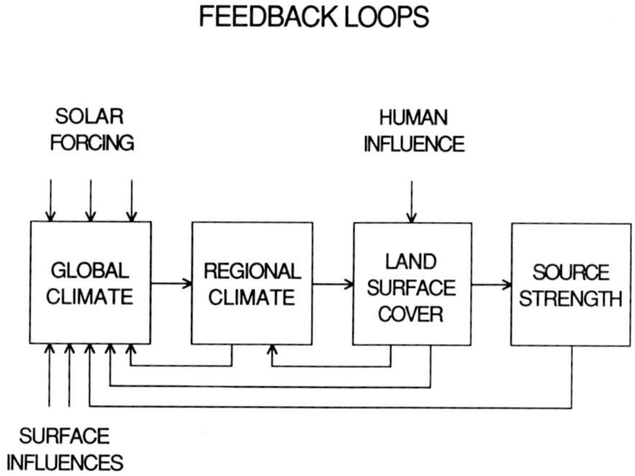

Fig. 20. Schematic of feedback loops involving climate, land surface and surface sources of trace gases and carbon dioxide.

cover, and decreased surface roughness. Rough estimates of the impact on surface energy balance can be gleaned from calculations for dry and wet years in the Sahel. A model from Lare and Nicholson (unpubl. data) indicates that in the July-to-Sept wet season, latent heat transfer to the atmosphere is about twice as great in wet years as in dry ones. Surface temperature is 4 or 5°C higher in the dry years, representing a change of radiative emission to the atmosphere of c. 6 to 8%.

The impact of land surface changes on climate has recently become a major focus of research (see review in Nicholson 1988). Most of the research has involved numerical models, such as the classic study of Charney et al. (1977) which suggested that increased surface albedo (a likely consequence of human activities) could reduce rainfall in semi-arid regions (Fig. 19). The albedo hypothesis has been largely discounted and, admittedly, observational proof of land surface effects on climatic change is lacking. For example, there is no evidence that centuries of deforestation directly altered the climate of Europe or North America (although this may have contributed to global warming at the end of the 19th century). Nevertheless, there is much indirect evidence to indicate that land surface changes can alter climate and that a common result of projected changes is increased aridity in many locations.

This brings us back to the question of predicting regional climatic impacts of trace gas emissions. Summarizing the above, any significant climatic change affects surface vegetation, which in turn influences climate both regionally and locally. A series of feedback loops exist, as illustrated in Fig. 20. The most direct impact of increased greenhouse gas concentrations would be on "global" climate, e.g., on the mean global temperature and temperature gradients. This would alter the global-scale pressure, wind and rainfall patterns which, in turn, are the factors which produce the regional-scale characteristics of climate. As global and regional climate changes, so do surface vegetation patterns. These in turn influence the source strength and geographic distribution of the sources of atmospheric greenhouse gases, factors which exert feedback on both global and regional climate. Thus, any attempts to predict the climatic impact of trace gases and carbon dioxide must account for such feedback loops. The realistic and interactive representation of land surface processes in general circulation models is imperative, if accurate predictions of trace gas impact on global change are to be realized.

Summary and conclusions

The anthropogenic increase of greenhouse gases in the atmosphere is likely to induce a net warming of the earth, i.e., an increase in its mean temperature. Although there is little scientific debate as to whether a warming will take place, considerable controversy surrounds such questions as when, how much and how the climatic changes will be manifested on a regional scale. Questions as to the magnitude of the greenhouse warming and when it will become apparent require, at the very least, projections of future surface emissions of trace gases. They also require a better understanding and quantitative assessment of the feedback relationships between climate and vegetation.

A number of socioeconomic variables introduce uncertainties in prediction of trace gas emissions. The location and intensity of surface sources or sinks depend on such factors as industrial development, resource use, and land cover change. These processes are, in turn, influenced by population growth, national or international policy, and resource availability. These must be quantified in order to forecast future surface emissions of trace gases.

The vegetation cover is a prime determinant of trace gas emissions. Through this and through its impact on surface energy balance, vegetation cover influences climate and, most probably, climatic change. It is also a response to climate and climatic change will alter the surface vegetation patterns. This climate-vegetation feedback is further complicated by the possibility that higher levels of carbon dioxide in the atmosphere will alter such biophysical parameters as the efficiences of growth, evapotranspiration or photosynthesis. Quantification of the vegetation-climate feedback relationships must be a high priority for research if predictions of the regional consequences of global change are to be achieved.

Acknowledgements – The author would like to thank J. Kim, of FSU, and L. Fortier, of NCAR, for their preparation of many of the figures. The support of NASA grant NAG 5–764 is also acknowledged.

References

Adler, J., Cook, W. J., McGuire, S., Lubenow, G. C., Kasindorf, M., Maier, F. and Morris, H. 1981. The browning of America. – Newsweek, Feb 23: 26–37.

Bernabo, J. C. and Webb, T. III. 1977. Changing patterns in the Holocene pollen record of northeastern North America: a mapped summary. – Quaternary Res., 8: 64–96.

Budyko, M. I. 1986. The evolution of the biosphere. – D. Reidel Publ. Comp., Dordrecht.

Charney, J. G., Quirk, W. J., Chow, S. H. and Kornfield J. 1977. A comparative study of the effects of albedo change on drought in semi-arid regions. – J. Atmos. Sci. 34: 1366–1385.

Chorley, R. J. (ed.). 1978. Introduction to geographical hydrology. – The Chaucer Press, Bungay, Suffolk.

Darby, H. C. 1956. The clearing of the woodland in Europe. – In: Thomas, W. L. (ed.), Man's role in changing the face of the earth. Univ. of Chicago Press, Chicago, pp. 183–216.

Ehrlich, P. R., Ehrlich, A. H. and Holdren, J. P. 1977. Ecoscience: population, resources, environment. – Freeman, San Francisco.

Espenshade, E. B. (ed.). 1982. Goode's world atlas. – Rand McNally & Co., Chicago.

Fernald, E. A. and Patton, D. J. (eds). 1984. Water resources atlas of Florida. – Florida State Univ. Inst. of Sci. and Public Affairs, Tallahassee, Florida.

Goudie, A. 1984. The human impact. Man's role in environmental change. – The MIT Press, Cambridge, MA.

Graedel, T. E. and Crutzen, P. J. 1989. The changing atmosphere. – Sci. Am. 261: 58–68.

Green, F. H. W. 1976. Field drainage in Europe. – Geogr. J. 144: 171–174.

Henderson-Sellers, A. and Gornitz, V. 1984. Possible climatic impacts of land cover transformations, with particular emphasis on tropical deforestations. – Climatic Change 6: 231–258.

Ibrahim, F. 1978. Desertification, ein weltweites problem. Geogr. Rundschau 30: 104–107.

Lanly, J. P. 1982. Tropical forest resources, FAO Forestry Paper 30. – FAO, Rome.

Lieth, H. 1975. Primary production of the major vegetation units of the world. – In: Lieth H. and Whittaker, R. H. (eds), Primary productivity of the biosphere. Ecol. Studies 14: Springer, Berlin, pp. 203–205.

Marland, G. and Rotty, R. M. 1983. Carbon dioxide emissions from fossil fuels: a procedure for estimation and results for 1950–81 (DOE/NBB-0036). – U.S. Dept of Energy, Washington, D.C.

Matthews, E. 1983. Global vegetation and land use: new high resolution data bases for climate studies. – J. Climate and Appl. Meteorol. 22: 474–487.

Micklin, P. P. 1988. Desiccation of the Aral Sea: a water management disaster in the Soviet Union. – Science 241: 1170–1176.

Myers, N. 1980. Conversion of tropical moist forests. – Nat. Acad. of Sci., Washington, D.C.

Nicholson, S. E. 1988. Land surface-atmosphere interaction: physical processes, surface changes and their impact, and potential for remote monitoring. – Progress in Phys. Geogr. 12: 36–65.

– 1989. African drought: characteristics, causal theories and global teleconnections. – In: Berger, A., Dickinson, R. E. and Kidson, J. W. (eds), Understanding climate change. Am. Geophys. Union, Washington, D.C., pp. 79–100.

– , Kim, J. and Hoopingarner, J. 1988. Atlas of African rainfall and its interannual variability. – Florida State Univ., Tallahassee, Florida.

Olsson, L. 1985. An integrated study of desertification – Lunds Univ. Geogr. Inst. Avhandl., XCVIII.

Pillsbury, A. F. 1981. The salinity of rivers. – Sci. Am. 245: 54–65.

Pimentel, D. 1976. Land degradation: effects on food and energy resources. – Science 194: 149–155.

Schneider, S. H. 1989. The changing climate. – Sci. Am. 261: 70–79.

United Nations Conference on Desertification. 1977. Desertification: its causes and consequences. – Pergamon Press, Oxford.

U.S. Dept of Energy. 1979. Carbon dioxide effects research and assessment program. – Workshop on the global effects of carbon dioxide from fossil fuels, Miami Beach, Florida, March 7–11, 1977.

Woodwell, G. M. 1978. The carbon dioxide question. – Sci. Am. 238: 34–43.

– , Hobbie, J. E., Houghton, R. A., Melillo, J. M., Moore, B., Peterson, B. J. and Shaver, G. R. 1983. Global deforestation: contribution to atmospheric carbon dioxide. – Science 222: 1081–1086.

Ecological Bulletins 42: 24–30. Copenhagen 1992

Sampling strategies to quantify gas exchange for complex ecosystems

R. L. Desjardins, J. I. MacPherson and P. H. Schuepp

Desjardins, R. L., MacPherson, J. I. and Schuepp, P. H. 1992. Sampling strategies to quantify gas exchange for complex ecosystems. – Ecol. Bull. (Copenhagen) 42: 24–30.

Carbon dioxide, water vapor and sensible heat flux values from several flight patterns are analysed to examine several techniques of estimating large area fluxes of trace gases. The relative contributions of various length scales to flux estimates are calculated using high-pass filters with different cut-off. Covariance analysis is examined as a technique to account for the effects of different altitudes and changing radiation conditions on flux measurements. Repeated passes over a 4×4 km wheat growing area during changing radiation conditions provide the data to relate CO_2 fluxes to incident solar radiation. Carbon dioxide fluxes obtained from two successive grid flights over a 15×15 km grassland ecosystem are presented as an example of the differences in CO_2 flux measurements due to intermittency of gas exchange and the effect of wind direction. Vertical flux profiles upwind and downwind of the same area are presented as a sample of the data required to estimate the main exchange processes within the atmospheric boundary layer above the surface under investigation. These results demonstrate that, in order to obtain large area flux estimates using aircraft-based systems, many factors need to be taken into account and that the optimum measuring strategy is dependent upon the information desired.

R. L. Desjardins, Agric. Canada, Res. Branch, Land Resource Res. Cent., Ottawa, Ontario, Canada, K1A OC6. – J. I. MacPherson, Nat. Res. Council, Flight Res. Lab., Ottawa, Ontario, Canada, K1A OR6. – P. H. Schuepp, MacDonald Coll. of McGill Univ., Dept of Renewable Resources, Ste-Anne-de-Bellevue, Quebec, Canada, H9X 1CO.

Introduction

The atmospheric concentrations of radiative gases contributing to the greenhouse effect, such as CO_2, N_2O, O_3 and CH_4, are increasing at a unprecedented rate (Mooney et al. 1987). In order to predict the rates of increase of these gases and, where possible, take some remedial action, it is important to quantify both their anthropogenic and non-anthropogenic sources and sinks (Schimel et al. 1989). Changes in land use, such as flooding of large areas for a hydro dam, and climate induced response of natural ecosystems can lead to processes that are very difficult to reverse and also are not well understood.

Biomes such as wetlands, boreal forests, tropical forests and agricultural lands are complex composites of numerous species, all of which have a unique relationship with their environment. Even though sophisticated enclosure and tower-based systems have been developed to quantify exchange processes, it is usually very difficult to obtain representative measurements, due to the heterogeneity of most ecosystems (Stewart et al. 1989). For this reason, aircraft-based systems are starting to be used extensively, even though an optimum sampling strategy for each particular situation being studied has yet to be developed.

Airborne flux measurements of latent and sensible heat have been made for almost twenty years (McBean and Paterson 1975, Bean et al. 1976). Flux measurements of O_3, CO_2 and CH_4, determined primarily by long transects at several altitudes, have also been obtained more recently (Lenschow et al. 1982, Alvo et al.

1984, Harriss et al. 1988). Little of this work was focused on relating these measurements to surface processes. However, there have been recent efforts to compare low level airborne flux values to results from tower-based measurements (Desjardins et al. 1989). Although an optimum observational program might combine tower and aircraft-based measurements for the study of temporal and spatial variability of flux, the discussion will be restricted to airborne sampling.

This paper presents several techniques for deriving large areal flux estimates. It demonstrates that flight patterns and analysis procedures must be tailored to the physical and environmental conditions. Flux variations caused by intermittency of transfer and changing environmental and physical conditions are discussed, as well as a means to verify the accuracy of flux estimates. The techniques presented here are based on large area measurements of the fluxes of sensible and latent heat and CO_2, but should nevertheless be equally applicable for any trace gases for which fast response sensors are available.

Instruments and methods

Aircraft are excellent platforms for measuring the flux of gases, since their mobility permits the sampling of large areas in a short period of time. They are particularly well suited to measure fluxes in areas which are not easily accessible. This is advantageous because very little is known about the reaction of these ecosystems to environmental changes.

The sensors used on aircraft to measure fluxes have been extensively described by Lenschow (1986) and Desjardins and MacPherson (1989). Flux measuring aircraft, flying at 50–120 m s^{-1}, require turbulence and chemical sensors with a frequency response of 10–20 Hz if the eddy correlation technique is to be used. Several kinds of systems have been found to give satisfactory turbulence measurements (Greenhut and Gilmer 1985). Fast response chemical sensors have been the main limiting factor, despite recent technical progress in the measurement of trace chemical species such as O_3, CO_2, SO_2, CH_4, N_2O and NO_2 (Lenschow and Hicks 1989). When sensors are located at different positions on the aircraft, great care must be taken to adjust for time delays between measurements from different sensors (MacPherson et al. 1987). It is also essential to take density variations into consideration for the case of gases for which density rather than mixing ratios are measured (Webb et al. 1980).

It is generally agreed that for gases for which fast responding sensors are available, the limitations in making large area flux estimates are due to sampling difficulties rather than instrumentation problems (Mahrt 1987).

Turbulent fluxes depend on the scale of atmospheric motion included in the measurements, which is usually dictated by the sampling length. In this paper, a third order high-pass filter is used to reduce the flux variability resulting from poorly defined long wavelength contributions. It is a convenient method to remove trends in the data and it permits the comparison of data collected in runs of different lengths (MacPherson et al. 1987). Spectral analysis is used to characterize the wavenumber of the primary fluctuations in wind (w′) and admixture concentration (s′). The wavelengths where these fluctuations are significantly correlated in "coherent events" that dominate the flux contributions, are shown by cospectral analysis.

This paper analyzes flux measurements obtained from repeated passes over extensive flat wheat fields in Manitoba, Canada in order to demonstrate the kind of data required to relate the gas fluxes to environmental factors considered important for a particular ecosystem. These data will also be used to assess the relative contribution from different turbulent scales to the flux estimates. It also presents data from a grid flight pattern of eight 15 km parallel runs with a lateral spacing of one nautical mile (1.8 km), which was flown at constant pressure altitude, over grasslands of relatively complex topography (Sellers et al. 1988). These data will be used to determine the mean flux over an area, as well as to examine the spatial variation of the fluxes. Temporal effects are investigated by repeating the grid pattern after a few hours.

Covariance analysis, which combines the features of analysis of variance and regression analysis, is used to adjust the latent heat flux estimates for biases due to altitude and solar radiation (Searle 1971). Each grid run (1 to 8) for one grid pattern is divided into four segments (A, B, C, D) resulting in a data set of 32 flux values, Fij, with corresponding terrain elevation, Zij, and incident solar radiation, Sij. Assuming that both variables are linearly related to the measured flux, a model of the type shown in Eq. (1) is then developed:

$$Fij = \mu + \alpha\,(Zij - \bar{Z}) + \beta\,(Sij - \bar{S}) \tag{1}$$

where μ is the mean flux for a whole grid flight and α and β are the regression coefficients for the model.

Whatever the sampling scenario, airborne flux observations must deduce surface flux densities from measurements made at some distance above the surface. Since negligible flux divergence and absence of horizontal advection cannot generally be assumed, a double stack flight pattern, with vertical flux profiles upwind and downwind of the test area, may be used for "budget studies" (Betts et al. 1990). The budget equation for a passive admixture s, in the absence of sources and sinks and for negligible vertical advection, can be written as:

$$\frac{\partial \bar{s}}{\partial t} + \bar{u}\frac{\partial \bar{s}}{\partial x} + \bar{v}\frac{\partial \bar{s}}{\partial y} + \partial\frac{\overline{w's'}}{\partial z} = 0 \tag{2}$$

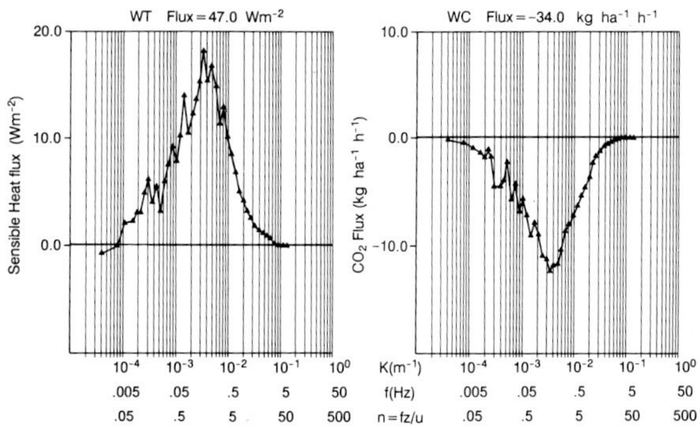

Fig. 1. Average cospectra of vertical wind and air temperature (WT) and vertical wind and CO_2 concentration (WC) as a function of wavenumber (K), frequency (f), and normalized frequency (n), both based on six 45 km runs at 50 m above a homogeneous wheat growing area.

The vertical flux profiles provide the flux divergence term, $\frac{\partial \overline{w's'}}{\partial z}$, which can be extrapolated to estimate the flux at the surface. The other two important terms are the rate of change term, $\frac{\partial \bar{s}}{\partial t}$ and the horizontal advection terms $\bar{u}\frac{\partial \bar{s}}{\partial x} + \bar{v}\frac{\partial \bar{s}}{\partial y}$, where $\bar{u}$ and $\bar{v}$ are the mean wind speed in the x and y direction, respectively. These can also be computed using data from the double-stack flight pattern.

Results and discussion

Cospectral estimates quantify the relative contributions of different frequencies (turbulent scales) to the fluxes of sensible heat and CO_2 for six 45 km runs at 50 m above a homogeneous wheat-growing area (Fig. 1). The average cospectra show that most of the contribution to the flux is found between wavenumber 2×10^{-2} to 2×10^{-4} m^{-1}, equivalent to wavelengths from 50 m to 5 km. The cospectral peaks, which are at wavelengths of c. 300 m, appear identical for heat (WT) and mass (WC) transfer. This peak contribution shifts to longer wavelengths as the sampling altitude increases (Kaimal et al. 1976). This increases the need for longer runs at higher altitude.

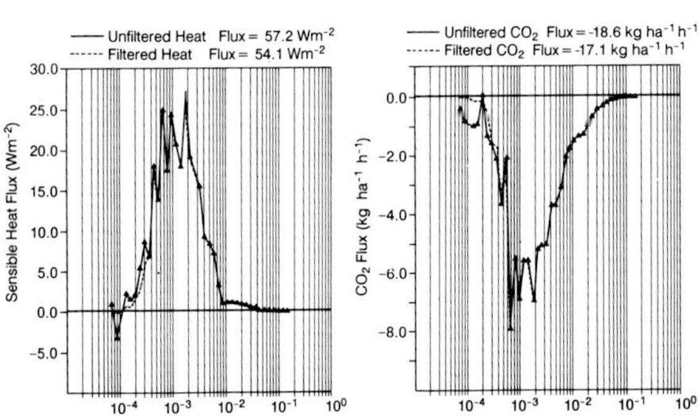

Fig. 2. Filtered and unfiltered sensible and CO_2 fluxes as a function of wavenumber.

Table 1. Unfiltered and filtered fluxes estimated of sensible heat (WT), latent heat (Wq) and CO_2 (WC) for six sequential 15 km runs at 100 m above a grassland ecosystem in Kansas, USA, on 9 July 1987.

Unfiltered				Filtered		
WT Wm^{-2}	Wq Wm^{-2}	WC kg $ha^{-1}h^{-1}$		WT Wm^{-2}	Wq Wm^{-2}	WC kg $ha^{-1}h^{-1}$
64	223	–40.3		63	208	–36.9
38	260	–38.4		54	229	–34.6
75	224	–22.6		42	148	–22.8
68	202	–37.5		64	209	–35.8
75	160	–23.8		70	215	–37.2
91	207	–31.4		70	215	–37.7
69	213	–32.3	X	60	204	–34.2
18	33	7.7	±S.D.	11	28	5.7

High-pass filtering has been found useful to reduce the variability of flux measurements. It is primarily used to remove larger scale motions which are poorly defined by an inadequate sample size. Average cospectra from sixteen runs at c. 100 m above the ground are shown in Fig. 2, with and without high-pass filtering by a third-order algorithm with a breakpoint of 0.012 Hz. This corresponds to a wavelength of c. 5 km which represents c. 1/3 of the run length. The effect of the filter is not very significant in this situation because the cospectra are based on the average of several runs and because the long wavelength contribution is limited at low altitudes. However, when dealing with individual runs even at these low altitudes, it is useful to high-pass filter the data before comparing results from one run to another because the long wavelength contribution can be very variable. Table 1 shows a series of data from individual runs, with and without a 0.02 Hz high-pass filter. The variations in the flux estimates are substantially reduced by the filter, as illustrated by the two data sets.

Flux estimates have also been calculated for eight grid runs 15 Aug 1987, using several high-pass filter cut off frequencies to quantify the relative contribution of various length scales (Fig. 3). In this case, practically identical results were obtained for CO_2, sensible and latent heat fluxes.

In order to evaluate the representativeness of the case studies shown in Figs 2 and 3, the ratio of unfiltered/filtered latent and sensible heat fluxes are presented for several series of grid patterns (Table 2). All the data presented were obtained at a mean altitude of 100 m over a grassland ecosystem. Two filter breakpoints were used, 0.02 Hz in 1987 and 0.012 Hz in 1989, which

Fig. 3. Ratio of filtered to unfiltered fluxes of CO_2, sensible and latent heat over a grassland ecosystem during eight 15 km grid runs at a mean altitude of 100 m on 15 Aug 1987.

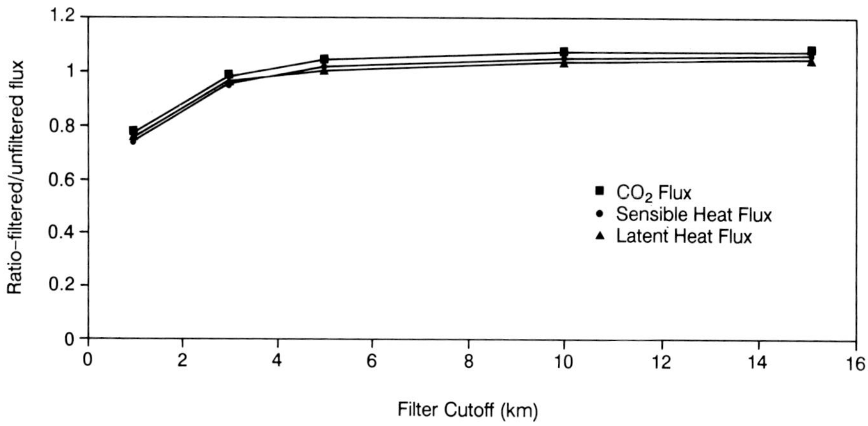

correspond to 3 and 5 km wavelengths, respectively. Results show that, contrary to the results presented in Fig. 3, there are several cases in which the long-wavelength contribution for sensible heat is considerably different than that for latent heat. Results are relatively consistent from one hour to the next, but are substantially different from day to day. In general, a slightly larger underestimation is observed with a higher instability.

Flux estimates are strongly susceptible to changing environmental conditions. This source of variability may be minimized if a functional relationship can be found between fluxes over an ecosystem and environmental conditions. Fig. 4 shows a correlation of 0.85 between CO_2 and incident solar radiation obtained from repeated low altitude passes over a wheat growing area during a two-hour period of changing radiation conditions. In order to minimize intermittency effects of transfer of CO_2 over short distances, successive runs with relatively constant radiation conditions were averaged. Each data point in Fig. 4 represents three to five successive 4 km runs. The high correlation suggests that it should be possible to estimate the functional relationship between CO_2 flux and solar radiation for areas as small as 4×4 km using low level flights.

Fig. 5 shows the measured latent heat flux, the predicted values based on Eq. (1), and the adjusted values based on covariance analysis. The latter is the measured

Table 2. Mean ratio of unfiltered/filtered fluxes of latent heat (RWq) and sensible heat (RWT) for eight 15 km runs. High-pass filter breakpoint of 0.02 Hz was used in 1987 and 0.012 Hz in 1989. The mean time of each grid run is also given, as well as the stability conditions (Z/L).

Year	Date	Time (CDT)	RWq	RWT	Z/L
1987	11 July	1120	1.04	0.79	–0.2
	11 July	1205	1.05	0.97	–0.2
	11 July	1505	1.09	1.00	–0.2
	11 July	1550	1.15	1.07	–0.1
	15 Aug	1130	1.03	1.08	–0.5
	15 Aug	1220	1.05	0.98	–0.6
	15 Aug	1530	1.07	1.09	–0.3
	15 Aug	1620	1.20	1.16	–0.2
	7 Oct	1210	1.31	1.20	–1.8
	7 Oct	1305	1.46	1.22	–1.3
	7 Oct	1530	1.18	1.11	–2.3
	7 Oct	1625	1.21	1.14	–1.4
1989	28 July	1110	0.89	1.21	–5.6
	28 July	1200	0.90	1.29	–5.1
	4 Aug	1020	1.01	1.13	–0.9
	4 Aug	1130	1.09	1.02	–1.2
	4 Aug	1400	1.10	1.14	–1.4
	4 Aug	1505	0.98	2.44	–0.9
	6 Aug	1515	1.08	1.14	–0.7
	6 Aug	1630	1.08	1.16	–0.5
	8 Aug	1130	1.19	1.13	–4.0
	8 Aug	1245	1.26	1.23	–9.0
	12 Aug	1215	1.07	0.86	–2.2
	12 Aug	1330	1.02	0.89	–2.9

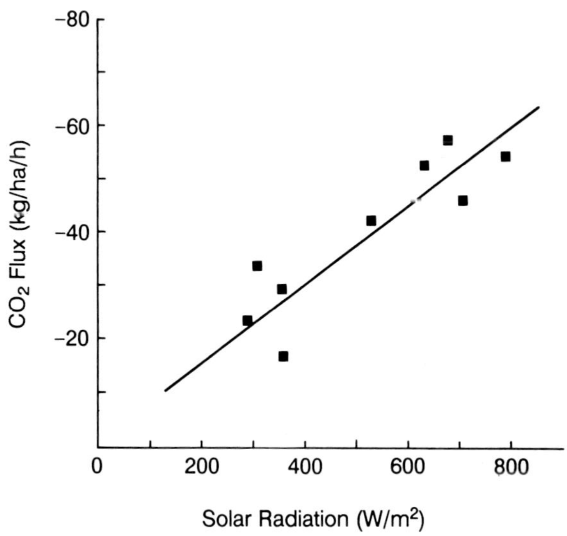

Fig. 4. CO_2 flux versus incident solar radiation measured over a 4×4 km agroecosystem during a 2 h period with changing radiation conditions, from repeated passes at a mean altitude of 50 m.

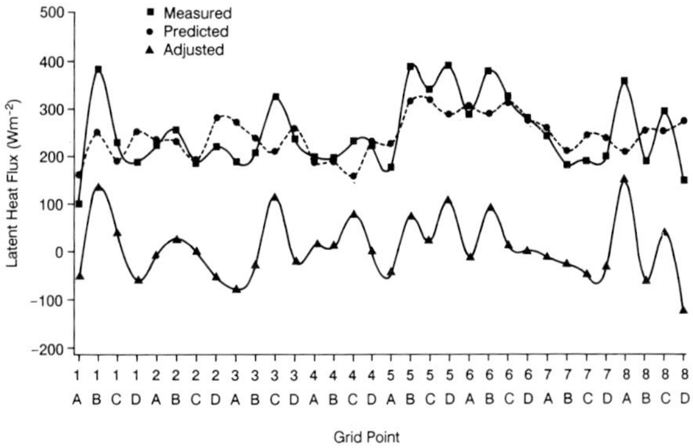

Fig. 5. Measured, predicted and adjusted latent heat fluxes obtained over a 15×15 km grassland ecosystem during a grid flight. The adjusted flux values have the mean removed. The numbers on the x axis represent the grid run and the letters represent the segment.

difference between the measured and the predicted flux. It represents the actual latent heat flux with the effects of changing elevation and radiation removed. The mean of the adjusted flux has been removed for purpose of clarity on the graph. It is presented as an example of possible ways to account for complex terrain and changing environmental conditions. More development of this kind of technique remains to be done. A term to account for the intermittency of turbulence transfer would also significantly improve the flux estimates over short distances.

Temporal effects on flux estimates are illustrated in Fig. 6. Shown are CO_2 flux contours over a grassland ecosystem for two sets of grid patterns, the first recorded between 1030–1204 CDT and the second between 1338–1536 CDT 4 Aug 1989. Both periods were completely clear with a south westerly wind in the morning of 6–7 m s^{-1} and a southerly wind, i.e. cross-track of 7–8 m s^{-1} in the afternoon. A mean flux of -17.7 kg ha^{-1} h^{-1} with a standard deviation of 6.2 was obtained for the morning grid and -15.5 ± 6.4 for the afternoon grid.

Each data point represents the average of two passes over each 3.8 km segment of the pattern. In order to minimize the variability due to long wavelength contributions, data used in the flux calculations were high-pass filtered at a breakpoint of 5 km. No other corrections of any kind have been performed to these data. It is presented as an indication of the differences we have observed between morning and afternoon passes. In spite of some similarity in the patterns, marked differences can be observed. Some of these differences may be real because moisture stress on vegetation is usually higher in the afternoon and the wind direction shifted by 39° between the morning and the afternoon flight. The development of an improved covariance analysis (Searle 1971) and application of the so-called "footprint correction" to allow for downwind displacement from flux sources (Schuepp et al. 1990) should result in more comparable flux patterns from one time to another and improve the correlation between the observed fluxes and the underlying terrain and vegetation.

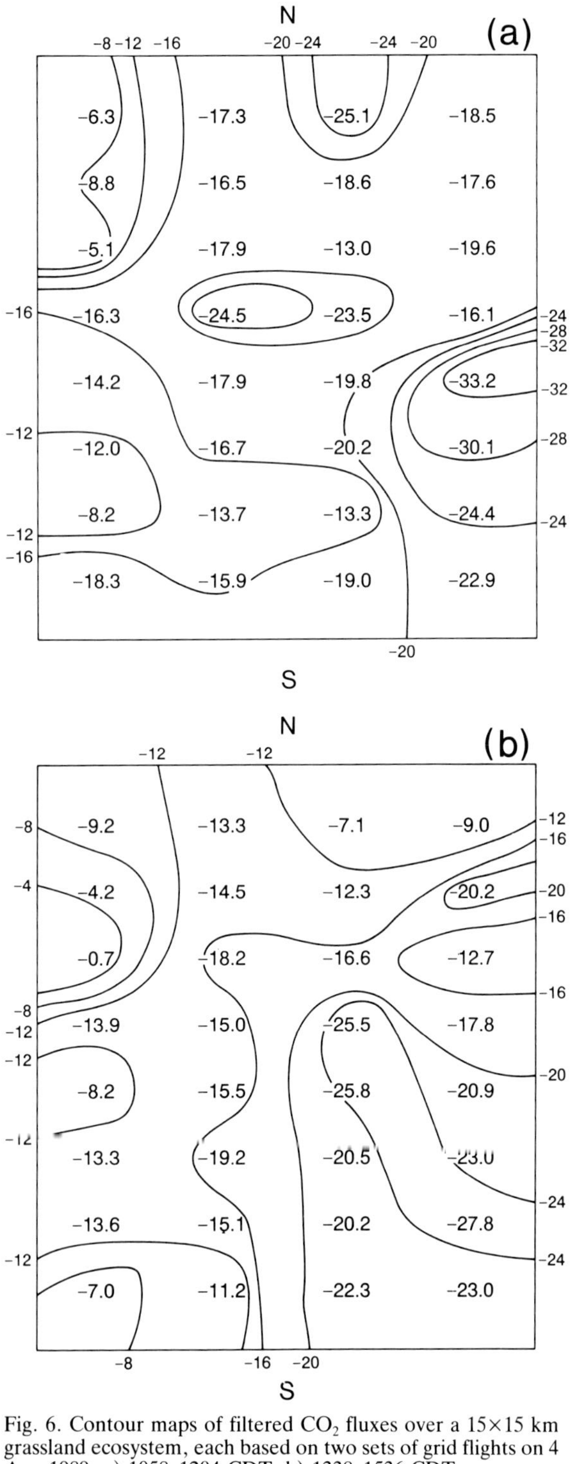

Fig. 6. Contour maps of filtered CO_2 fluxes over a 15×15 km grassland ecosystem, each based on two sets of grid flights on 4 Aug 1989: a) 1058–1204 CDT, b) 1338–1536 CDT.

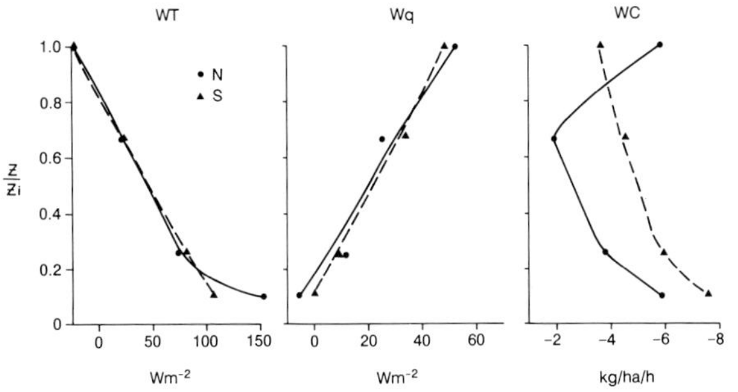

Fig. 7. Measurements of filtered flux profiles of sensible heat (WT), latent heat (Wq) and CO_2 (WC) at the upwind (south) and downwind (north) edge of a grassland test area on 8 Oct. 1987.

Budget studies are recommended in cases of non-negligible horizontal advection. Examples of some of the data required to estimate the flux divergence terms in Eq. (2) are shown in Fig. 7. These data were obtained by flying a repeated double-stack pattern upwind (south end) and downwind (north end) of the area of interest at four levels ($0.1Z_i$, $0.2Z_i$, $0.6Z_i$ and Z_i), where Z_i is the height of the mixed layer. As can be seen, the flux divergence terms are reasonably well defined. Once the advection terms and the rate of change term of the scalar of interest are calculated, we have all the required data to assess the accuracy of the areal flux estimates. This can be done using the data from the same flight pattern (Betts et al. 1990). The accuracy of flux measurements of sensible and latent heat, based on the inclusion of all these corrections, is now estimated to be c. 10% (unpubl. data). Such data also provide information on the representativeness of the site compared with the surrounding area. It shows that, based on the similarity of the sensible heat and latent heat flux profiles upwind and downwind of the site, the moisture conditions upwind of the area must have been very similar to those over the site. However, the CO_2 profiles indicate that the net photosynthesis was substantially different.

Concluding remarks

This paper shows that considerable progress has been achieved in obtaining large area flux estimates for CO_2 and H_2O from airborne measurements. It refers to several sampling strategies such as repeated low level passes, grid runs and budget studies, and analysis techniques such as cospectral analysis, high pass filtering and covariance analysis. More accurate definition of the area sampled, as well as improved estimates of the long wavelength contributions to the fluxes are required to improve the usefulness of airborne flux estimates for developing satellite-based algorithms to estimate ecosystem processes.

Acknowledgments – The financial support of NASA, and the technical assistance of the FIFE organizers and the staffs of the Flight Res. Lab. of the Nat. Res. Counc. (NRC) and the Res. Branch of Agric. Canada are gratefully acknowledged.

References

Alvo, P., Desjardins, R. L., Schuep, P. H. and MacPherson, J. I. 1984. Measurements of CO_2 exchange over various ecosystems. – Boundary-Layer Meteorol. J. 29: 167–183.

Bean, B. R., Gilmer, R. O., Hartman, R. F., McGavin, R. E. and Reinking, R. F. 1976. Airborne measurement of vertical boundary-layer fluxes of water vapor, sensible heat and momentum during GATE. – In: NOAA Tech. Memo., ERL WMPO-36 – NOAA Environ. Res. Lab., Boulder, CO.

Betts, A. K., Desjardins, R. L., MacPherson, J. I. and Kelly, R. D. 1990. Boundary-layer heat and moisture budgets from FIFE. – Boundary-Layer Meteorol. J. 50: 109–138.

Desjardins, R. L. and MacPherson, J. I. 1989. Aircraft-based measurements of trace gas fluxes. – In: Andreae, M. O. and Schimel, D. S. (eds), Exchange of trace gas between terrestrial ecosystem and the biosphere. Berlin, pp. 135–152.

– , MacPherson, J. I., Schuepp, P. H. and Karanja, F. 1989. An evaluation of airborne eddy flux measurements of CO_2, water vapor and sensible heat. – Boundary-Layer Meteorol. J. 47: 55–69.

Greenhut, G. K. and Gilmer, R. O. 1985. Calibration and accuracy of the NOAA/ERL gust probe system and intercomparison with other systems. – NOAA Tech. Memo. ERL ESG-22. Boulder, CO.

Harriss, R. C., Wofsy, S. C., Garstang, M., Molion, L. C. B., McNeal, R. S., Hoell, J. M., Bendura, R. J., Beck, S. M., Navaro, R. L., Riley, J. T. and Shell, R. C. 1988. The Amazon boundary-layer experiment (ABLE 2A): dry season 1985. – J. Geophys. Res. 93: 1351–1360.

Kaimal, J. C., Wyngaard, J. C., Haringen, D. A., Côté, O. R., Izumi, Y., Caughey, S. J. and Readings, C. J. 1976. Turbulence structure in the convective boundary layer. – J. Atmos. Sci. 33: 2152–2169.

Lenschow, D. H., Pearson, R. Jr. and Stankow, B. B. 1982. Measurements of ozone vertical flux to ocean and forest. – J. Geophys. Res. 87: 8833–8837.

– 1986. Aircraft measurements in the boundary-layer. – In: Lenschow, D. H. (ed.), Probing the atmospheric boundary-layer. Amer. Meteorol. Soc. pp. 39–55.

– and Hicks, B. B. 1989. Global tropospheric chemistry chemical fluxes in the global atmosphere. – In: Report of workshop on measurements of surface exchange and flux divergence of chemical species in the global atmosphere. Columbia Univ., New York, NY, p. 107.

MacPherson, J. I., Desjardins, R. L. and Schuepp, P. H. 1987. Gaseous exchange measurements using aircraft-mounted sensors. – In: Proceedings of sixth symposium meteorological observations and instrumentation. New Orleans, LA, pp. 128–131.

Mahrt, L. 1987. Boundary-layer eddies and surface inhomogeneity and their impact in measurement strategy and data analysis for FIFE, Report. – Oregon State Univ. for NASA-Goddard Space Flight Center.

McBean, G. A. and Paterson, R. D. 1975. Variations of the turbulent fluxes of momentum, heat and moisture over Lake Ontario. – J. Phys. Oceanography 5: 523–531.

Mooney, H. A., Vitousek, P. M. and Matson, P. A. 1987. Exchange of materials between terrestrial ecosystems and the atmosphere. – Science 238: 926–932.

Schimel, D. S., Andreae, M. O., Fowler, D., Galbally, I. E., Harriss, R. C., Ojima, D., Rodhe, H., Rosswall, T., Svensson, B. H. and Zavarzen, G. A. 1989. Priorities for an international research program on trace gas exchange.– In: Andreae, M. O. and Schimel, D. S. (eds), Exchange of trace gases between terrestrial ecosystem and the biosphere. Berlin, pp. 321–331.

Schuepp, P. H., Leclerc, M. Y., MacPherson, J. I. and Desjardins, R. L. 1990. Footprint prediction of scalar fluxes from analytical solutions of the diffusion equation. – Boundary-Layer Meteorol. J. 50: 355–374.

Searle, S. R. 1971. Linear models. – Wiley Series in Probability and Mathematical Statistics.

Sellers, P. J., Hall, F. G., Asrar, G., Strebel, D. E. and Murphy, R. E. 1988. The First ISLSCP Field Experiment (FIFE). – Bull. of the Amer. Meteorol. Soc. 69: 22–27.

Stewart, J. W. B., Aselmann, I., Bouwman, A. F., Desjardins, R. L., Hicks, B. B., Matson, P. A., Rodhe, H., Schimel, D. S., Svensson, B. H., Wassmann, R., Whiticar, M. J., Yang, W.-X. 1989. Extrapolation of flux measurement to regional and global scales. – In: Andreae, M. O. and Schimel, D. S. (eds), Exchange of trace gases between terrestrial ecosystems and the biosphere. Berlin, pp. 155–174.

Webb, E. K., Pearman G. I. and Leuning, R. 1980. Correction of flux measurements for density effects due to heat and vapor transfer. – Quart. J. Meteorol. Soc. 106: 85–100.

Ecological Bulletins 42: 31–41. Copenhagen 1992

Transfer coefficients for water–air exchange of ammonia, carbon dioxide and methane

Owen T. Denmead and John R. Freney

Denmead, O. T. and Freney, J. R. 1992. Transfer coefficients for water–air exchange of ammonia, carbon dioxide and methane. – Ecol. Bull. (Copenhagen) 42: 31–41.

We have examined the ability of a stagnant film model of gas transfer to and across an air–water interface to predict emissions of ammonia, carbon dioxide and methane from a large, shallow, freshwater pond. The model gives quite different transfer coefficients for the three gases in their air and water phases.
The data come from an experiment in which, because of fertilization or gas production from decomposing plant residues, supersaturation of the water with respect to the overlying air occurred for all three gases. Micrometeorological measurements using aerodynamic approaches were used to calculate the atmospheric flux densities of water vapour and the gases. The water vapour density at the surface was inferred from the surface temperature. Wet-trapping was used to measure atmospheric gradients of NH_3 concentration, and knowledge of the water chemistry to calculate aqueous NH concentrations. The CH_4 and CO_2 concentrations in air and water were measured by infrared gas analysis. Transfer coefficients were calculated from the fluxes and the water–air concentration differences, and compared with those predicted by Deacon's model. These comparisons can be summarised as follows: 1) For water vapour, the only transfer coefficient to be considered is that in the air. Deacon's model overestimates atmospheric transfer coefficients by c. 25% in low winds, but this may not be so serious for predicting the exchange of dissolved trace gases where the gas phase is only part of the overall pathway. 2) For NH_3, c. ⅔ of the resistance to transport is in the air, c. ⅓ in the water. Again, the model overestimates the fluxes in low winds, presumably because it overestimates transfer coefficients for the gas phase. 3) For CO_2, almost all resistance to transport resides in the water. In contrast to the micrometeorological investigations of air–sea exchange which have been published so far, our measurements show that Deacon's model would provide a very good description of CO_2 exchange. 4) For CH_4, most of the transport appears to be by ebullition rather than diffusion, a result deduced in a number of other studies, and Deacon's model would have underestimated the exchange by two orders of magnitude. Some escaping bubbles were ⅓ methane.

O. T. Denmead, Centre for Environ. Mechanics, CSIRO, GPO Box 821, Canberra, ACT 2601, Australia. – J. R. Freney, Div. of Plant Industry, CSIRO, GPO Box 1600, Canberra, ACT 2601, Australia.

Introduction

Exchanges of trace gases between water bodies and the atmosphere are important in many contexts: assessing the extent of ammonia volatilization and the efficiency of use of nitrogen fertilizers in irrigated agriculture, for instance, or predicting the role of air–sea exchange in the global carbon dioxide balance, or quantifying emissions of methane from lakes, rivers and wetlands. It is common to describe these exchanges in terms of transfer coefficients in the water and the air as e.g. in Liss and Slater (1974). If we denote the vertical flux density of the gas at the water surface as F, we can write for transport in the water,

$$F = V_L (C_w - C_0) \qquad (1)$$

where V_L is a transfer coefficient (in units of velocity), C_w is the concentration of the dissolved gas in the liquid phase in the main, well-mixed body of the water (in units of density, mass per unit volume), and C_0 is the corresponding concentration at the surface (in the same

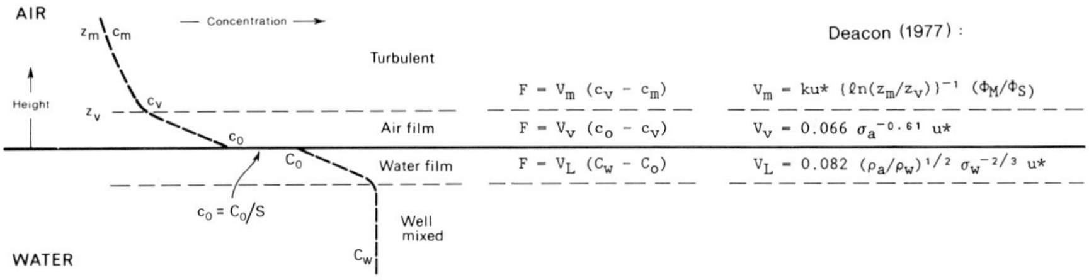

Fig. 1. Concentrations, fluxes and transfer coefficients in air and water, and parameterisations in Deacon (1977).

units as C_w). There, the liquid phase is in equilibrium with air with a gas phase concentration of c_0. The two concentrations are related through the solubility S, the concentration ratio liquid/gas.

$$c_0 = C_0/S. \tag{2}$$

In the air, it is usual to invoke a viscous sublayer of depth z_v next to the water surface, where transport is dominated by molecular processes, and a fully turbulent layer between z_v and a reference measuring height z_m. The gas concentrations at z_v and z_m are, respectively, c_v and c_m and the transfer coefficients for each layer are V_v and V_m. Thus,

$$F = V_v(c_0 - c_v) = V_m(c_v - c_m). \tag{3}$$

We can describe transport for the whole layer between the surface and z_m in terms of an atmospheric transfer coefficient, V_a, such that

$$F = V_a(c_0 - c_m), \tag{4}$$

and

$$V_a = (1/V_v + 1/V_m)^{-1}. \tag{5}$$

In a similar fashion, we can describe gas transfer between the main body of the water and z_m in terms of a bulk transfer coefficient, V_b, such that

$$F = V_b(C_w/S - c_m), \tag{6}$$

$$V_b = (1/SV_L + 1/V_a)^{-1}. \tag{7}$$

The concentrations, fluxes and transfer coefficients in the various domains are shown in Fig. 1.

All the transfer coefficients can be expressed as functions of wind speed (and atmospheric stability). The purpose of formulations like Eq. (6) is to enable predictions of gas fluxes from relatively simple observations: wind speed and concentrations of the gas in the bulk water and the air.

Much of our knowledge of the transport processes comes indirectly: from wind-water tunnels, or enclosures placed over the water, or tracer releases, or fluid-mechanical models. There are few direct measurements of the water–air exchange of trace gases in the undisturbed natural environment, from which transfer coefficients could be derived or the predictive ability of existing models tested. This contribution reports such measurements for ammonia, carbon dioxide and methane, three gases important in the biosphere, which have quite different transfer coefficients in the air and water phases, and it examines the ability of a much-invoked stagnant-film model, due to Deacon (1977), to predict the gas transfers.

Deacon's model

This model of gas transfer to and across an air–water interface is best suited to conditions of light winds and smooth or nearly smooth water (Deacon 1977). It uses Reichardt's formulation of the velocity profile in turbulent flow over a smooth plane surface to calculate transfer coefficients for near-surface viscous sublayers in the water and the air. It combines this with an eddy-diffusivity description of transport in the turbulent air-layer above. The model's parameterisations of the transfer coefficients in water and air are shown in Fig. 1.

For the water, Deacon has

$$V_L = 0.082\ (\varrho_a/\varrho_w)^{1/2}\ \sigma_w^{-2/3}\ u_*, \tag{8}$$

where ϱ_a is the density of the air, ϱ_w is the density of the water, σ_w is the Schmidt number, the ratio of the water viscosity, μ, to the molecular diffusivity of the gas in water, D_w, and u_* is the friction velocity of the surface air flow, which can be measured by eddy-correlation or calculated from the wind profile, as, for example, in Thom (1975). For the viscous sublayer in the air

$$V_v = 0.066\ \sigma_a^{-0.61}\ u_* \tag{9}$$

and for the turbulent layer,

$$V_m = ku_*/[\ln(z_m/z_v) - \{\psi_2(z_m) - \psi_2(z_v)\}] \tag{10}$$

In Eq. (9) and (10), σ_a is the ratio of the kinematic viscosity of the air, ν, to the molecular diffusivity of the gas in air, D_a, k is von Karman's constant, ψ_2 is an integrated form of the stability function for scalars, which can be calculated from the temperature, humidity and wind profiles (Thom 1975), and

$$z_v = 50\ \nu/u_*. \quad (11)$$

Substitution of typical values into Eq. (8) to (10) indicates that for ammonia, approximately two-thirds of the resistance to transport is in the air, only one-third in the water. For CO_2, $SV_L \ll V_a$, so that Eq. (7) becomes $V_b \simeq SV_L$ and we need only consider transport in the liquid phases. Using Deacon's formulation, Wesely et al. (1982) approximate V_b by $7.7\times10^{-5}\ u_*$. The situation for CH_4 is similar to that for CO_2; virtually all the resistance to transport is in the liquid phase.

Previous work

Ammonia

Processes of ammonia volatilization in shallow ponds have been studied by Leuning et al. (1984), Denmead et al. (1987) and Trevitt et al. (1988). Leuning et al. (1984) and Trevitt et al. (1988) found that the rates of NH_3 loss from water to air were often considerably smaller than those predicted by models of gas transfer across the air–water interface, specifically by a surface-renewal model developed by Bouwmeester and Vlek (1981) and the stagnant-film model of Deacon (1977). They attributed the discrepancies to the development of stable temperature stratifications in the water, caused in one case by the near-surface absorption of solar radiation in turbid water, and in the other by the failure of weak solar radiation to penetrate to the pond bottom. By day, the surface temperatures of the ponds were often 3° higher than the bulk water temperatures. The stable stratification effectively increased the resistance to transport (in the aqueous phase) by suppressing the mechanical mixing induced by the wind. In the case of NH_3, however, the effects of the slower rate of transport may be partially offset by the decreased solubility of the gas at higher temperatures (Eq. 20).

Denmead et al. (1987), working with clear water and high incident radiation, found that unstable temperature stratifications usually developed in their experimental ponds. By day, surface temperatures were typically 1° to 3° less than in the bulk water. The unstable stratification could be expected to enhance the mechanically induced mixing, but the lower surface temperature would reduce c_0. They found generally good agreement between the measured fluxes and those predicted by Deacon's model, provided the solubility S was evaluated at the surface water temperature rather than the bulk temperature. These three investigations were conducted in relatively small ponds (50 m ∅). Although the reference height was set at 0.2 m above the water in the centre of the pond so as to ensure a fetch to height ratio of >100:1, one might expect some influence of pond size on the results through effects of fetch on wave action and other parameters such as surface temperature and atmospheric NH_3 concentration. This paper reports results of an experiment conducted to examine NH_3 volatilization in a situation closer to equilibrium, with fetches of 100 m or more. As described later, the NH_3 source was created by spreading urea fertilizer into a large pond.

Carbon dioxide

Experiments in wind-water tunnels and lakes, radon-evasion experiments in ocean water, and theoretical calculations all indicate a very small transfer coefficient for the exchange of CO_2 between water and air (Deacon 1977, Wesely et al. 1982, Broecker et al. 1986). Yet, micrometeorological measurements of the air–sea exchange of CO_2 by eddy-correlation suggest transfer coefficients which are 4 to 60 times higher (Jones and Smith 1977, Wesely et al. 1982). Whichever estimates are correct is of vital importance both for modelling the role of air–sea exchange in the global CO_2 balance, and for determining the feasibility of making direct measurements of the atmospheric CO_2 flux over the oceans (or other water bodies). To enlarge on this last point, we note that if the smaller coefficients are correct, likely flux densities of CO_2 over the oceans will be just on the present limits of detection by micrometeorological techniques. Background to the controversy is provided by four papers published by way of a review (Broecker et al. 1986, O'Brien 1986, Smith and Jones 1986, and Wesely 1986).

Our ammonia experiment provided an opportunity to address this question. Our experience has been that the addition of fertilizer nitrogen to the water promotes intense algal growth which depletes dissolved CO_2. It was anticipated that this would provide an opportunity to study CO_2 transfer from the air into the water in a situation where there was a large CO_2 concentration difference between the two: 100 or 200 ppmv. This situation was realised initially, but decomposition of plant residues in the soil underneath the floodwater, and the accompanying release of CO_2 into it, led eventually to dissolved CO_2 concentrations up to 1600 ppmv higher than those in the air. We were thus able to examine CO_2 exchange in a much more favourable experimental situation than could be achieved over the oceans where CO_2 differences between water and air are usually in the range ± 100 ppmv (Broecker et al. 1986). A more detailed account of the work is given in Denmead (1989).

Methane

So far as we are aware, there are no published micrometeorological measurements of methane fluxes over water bodies; present knowledge comes entirely from investigations employing enclosures. The evidence from these is that measured bulk transfer coefficients agree with those observed for other gases such as O_2 and CO_2 in wind-water tunnels (Sebacher et al. 1983), but in many field situations, much of the transport through the water appears to have occurred by ebullition rather than the processes of solution, diffusion and dissolution underlying Eq. (1) and (4) (Holzpapfel-Pschorn et al. 1986, Bartlett et al. 1988, Crill et al. 1988, Devol et al. 1988). In such circumstances, the transfer coefficient description has limited value. Further, the unpredictable pattern of bubble release means that large sample numbers are required to establish an average flux if enclosures are employed. This points to the need for micrometeorological techniques which can integrate the emissions occurring over large areas and smooth out local variations caused by differences in the mechanisms of gas transport through the water.

Methods

Site

The main data come from an experiment designed primarily to study ammonia loss from a large, shallow, freshwater pond 300×100 m with an average depth of 0.22 m. The pond was established at Whitton, N.S.W., Australia, on bare soil in a field where the residues from a previous wheat crop had been recently incorporated. Urea fertilizer was spread into the water by aircraft. The hydrolysed urea provided a strong source of NH_3, but gas production from the decomposing plant residues led to supersaturation of the water with respect to the overlying air for CO_2 and CH_4 as well.

Micrometeorological

Nondisturbing micrometeorological techniques were employed to measure the water–air exchanges of all three gases as well as the friction velocity u_* and the exchanges of heat and water vapour. The last two fluxes provided checks on the validity of the micrometeorological approach, through comparisons with the energy balance, while the fluxes of water vapour also permitted direct calculation of the atmospheric transfer coefficients V_a, which were used to test Deacon's parameterisation of surface to air exchanges, Eq. (4), (5), (9) and (10).

Many investigators have shown that in the air layers above aerodynamically smooth land surfaces, the transport of scalars can be represented adequately by an eddy-diffusion description in which the vertical flux density, F_s, of a scalar whose atmospheric concentration is s, is given by the product of a height-dependent eddy diffusivity, K_s, calculable from the wind profile, and the vertical concentration gradient, i.e.

$$F_s = -K_s \, \partial s/\partial z. \tag{12}$$

Miyake et al. (1970) have shown that this description is adequate for water bodies as well.

Instruments were deployed to calculate the exchanges of heat, water vapour, NH_3, CO_2, CH_4, and momentum between the water and the air from aerodynamic relationships of the form of (12).

Heat and water vapour

Following Thom (1975), the vertical flux density of sensible heat, H, was calculated from a finite difference formulation of (12):

$$H = \frac{\varrho_a c_p k u_*(\theta_1 - \theta_2)}{\ln(z_2/z_1) - \{\psi_2(z_2) - \psi_2(z_1)\}}, \tag{13}$$

where c_p is the specific heat of air at constant pressure, u is horizontal wind speed, θ is potential temperature, Φ_H is the stability function for heat, and the subscripts refer to particular measuring heights. In the same fashion, the flux density of latent heat is obtained from:

$$\lambda E = \frac{\varrho_a c_p k u_*(e_1 - e_2)/\gamma}{\ln(z_2/z_1) - \{\psi_2(z_2) - \psi_2(z_1)\}}, \tag{14}$$

where λ is the latent heat of vaporization of water, E is the flux density of water vapour, e is vapour pressure and γ is the psychrometric constant.

Gases

Depending on whether concentrations are expressed in density or volumetric units, the aerodynamic formulations for gases are:

$$F = \frac{k u_*(c_1 - c_2)}{\ln(z_2/z_1) - \{\psi_2(z_2) - \psi_2(z_1)\}}, \tag{15}$$

or

$$F = \frac{\varrho_a(m_s/m_a) k u_*(c_1 - c_2)}{\ln(z_2/z_1) - \{\psi_2(z_2) - \psi_2(z_1)\}}, \tag{16}$$

where c_1 and c_2 are measured relative to dry air, ϱ_d is the density of the dry air, m_s and m_a are the molecular masses of the gas and dry air respectively, and s is the atmospheric, volumetric content of the gas, again relative to dry air. Note that when gas concentrations are measured relative to dry air, and temperature differences between the air-streams are removed, no corrections are needed for density effects associated with simultaneous heat and vapour transfer (Webb et al. 1980).

Momentum

For our purpose, it is convenient to work with the friction velocity u_* rather than the momentum flux τ; $u_* = (\tau/\varrho_a)^{\frac{1}{2}}$.

$$u_* = \frac{k(u_2 - u_1)}{\ln(z_2/z_1) - \{\psi_2(z_2) - \psi_2(z_1)\}}, \quad (17)$$

where the ψ_1 is an integrated form of the stability function for momentum.

Energy balance

The energy balance of the pond can be written

$$R_n = W + G + H + \lambda E, \quad (18)$$

where R_n is net radiation, W is the rate of change of heat storage in the water, and G is the flux of heat into the underlying soil. For water bodies of this type, W is by far the largest term on the right hand side of Eq. (18) and G is also relatively large (Fig. 2). The difficulties of measuring them precisely often preclude the use of the energy balance for estimating H and λE. Nevertheless, the term $R_n - W - G$ provides a useful, if somewhat imprecise, check on the accuracies of the aerodynamic formulae.

Solubilities and diffusivities

For ammonia, we use relationships given by Emerson et al. (1975) and Hales and Drewes (1979), first to partition the dissolved ammoniacal nitrogen, A_w, between NH_3 and NH_4^+ and then to calculate c_0. From these relationships,

$$C_w = A_w[1 + 10^{(0.09018 + 2729.92/T_w - pH)}]^{-1} \quad (19)$$

and

$$S = 10^{(1477.7/T_w - 1.69)} \quad (20)$$

In the above, T_w is the water temperature in K. Note that c_w (= C_w/S) increases with both the temperature and the pH of the water.

The diffusivity of NH_3 in solution has been taken as 2.5×10^{-9} m^2 s^{-1} at 25°C and adjusted for temperature on the assumption that $\mu D_w/T_w$ is constant (Bouwmeester and Vlek 1981). In the air, D_a was taken as 2.16×10^{-5} m^2 s^{-1} at 25°C (Moeller and Vlek 1982) and adjusted for absolute temperature on the basis that $D/D_0 = (T/T_0)^{1.81}$, where D_0 is the diffusivity at T_0 (List 1957).

For CO_2, we use the values and temperature dependence for S and D_w adopted by Deacon (1977), viz. S = 0.815 at 20°C, and over the range 5 to 30°C, $S \propto T^{-7.5}$; $D_w = 1.62\times10^{-9}$ m^2 s^{-1} at 20°C and varies with temperature in the same way as for NH_3. For CH_4 we have used a value of 0.024 for S derived from Table 1 of Liss and Slater (1974) and a value of 2.32×10^{-9} m^2 s^{-1} for D_w at 23.5°C (Ledwell 1984), with adjustment for temperature as for NH_3.

Instrumentation

Measurements of the atmospheric temperature, humidity, CO_2 and CH_4 gradients were facilitated by a reversing apparatus which interchanged psychrometers and airlines between heights of 0.3 and 1.1 m above the water surface every 7.5 min. A wet trapping technique described by Denmead et al. (1976) was used to measure the differences in atmospheric NH_3 concentration between the same two heights. The upper height was set at 1.1 m in order to have a minimum fetch-to-height ratio of 100:1, thereby ensuring near-equilibrium conditions for flux measurement.

The psychrometers consisted of radiation-shielded, aspirated, wet- and dry-bulb, platinum resistance thermometers. For CO_2 and CH_4 analysis, air-streams were dried with magnesium perchlorate before being pumped down 150 m of 2.4 mm radius polypropylene tube to a mobile laboratory where they passed through a heat exchanger to remove temperature differences and then through Infra-Red Gas Analysers, IRGA, (Analytical Development Company Series 225) operating in differential mode. For CH_4 analysis, the air streams were also passed through Ascarite to remove CO_2 before passage through the IRGA.

The wind profile was measured with small, sensitive cup anemometers mounted at 0.3, 0.7, 1.1 and 2 m above the water surface.

Net radiation was measured with a hemispherical net radiometer and the water storage term W was calculated from measurements of temperature change in the water made with platinum resistance thermometers.

$$W = \varrho_w C_W \int_0^Z (\partial T_w/\partial t)dz, \quad (21)$$

where C_W is the specific heat of water, t denotes time, and Z is the water depth. To evaluate $\partial T_w/\partial t$, water temperatures were measured continuously with platinum resistance thermometers at two depths at each of two locations, situated 25 m either side of the psychrometer mast. At one location, the thermometers were mounted at 0.02 and 0.10 m below the water surface and at the other, at 0.02 and 0.20 m below the surface. The temperature time series were first smoothed by fitting 3-harmonic Fourier series, which were then differentiated to yield analytic expressions for $\partial T_w/\partial t$, in the manner described by Trevitt et al. (1988).

The soil heat flux G was calculated from the relationship

$$G = -K_S \partial T_S/\partial z, \quad (22)$$

where K_S is the soil thermal conductivity and $\partial T_S/\partial z$ is the temperature gradient at the soil surface. The conductivity was calculated from the temperature waves at 0.01 and 0.04 m in the soil, measured at each of two locations, following procedures described by van Wijk and de Vries (1963), and the temperature gradient was obtained from the time course of the mean of the water temperature registered by the lower water thermometers and knowledge of the soil thermal properties. That procedure is described by Trevitt et al. (1988). The data were logged in 15 min runs.

To calculate V_a for water vapour from Eq. (4), we needed measurements of the vapour density at the water surface. This was taken to be the density of water vapour at saturation at the temperature of the water surface, T_0, which can be obtained from tabulations as in List (1957). The surface temperature itself was calculated from the measured temperature and humidity gradients. If we make the usual assumptions that the V_a for heat and water vapour are identical and that $\Phi_H = \Phi_v$, then application of Eq. (4) to the simultaneous transfers of sensible and latent heat gives

$$\beta = H/\lambda E = \gamma(T_0 - T_m)/(e_s(T_0) - e_m), \tag{23}$$

where β is the well-known Bowen ratio and e_s is the saturation vapour pressure, and Eq. (13) and (14) give

$$\beta = \gamma(T_1 - T_2)/(e_1 - e_2). \tag{24}$$

As outlined by Leuning et al. (1984), the value of β given by Eq. (24) can be used to solve Eq. (23) for T_0 by iteration. With T_0 known, e_s follows.

Direct measurements of the apparent surface temperature were also made with an infrared thermometer, but like Leuning et al. (1984), we found that these radiative temperatures were usually lower than those calculated from β by up to 2°C, presumably because of uncertainty in the emissivity of the water surface and the background radiation (Fuchs and Tanner 1966).

The concentrations of dissolved NH_3 and the corresponding solubilities were calculated from measurements of the ammoniacal nitrogen concentration, temperature and pH of the bulk water using Eq. (19) and (20). To measure the concentration of dissolved CO_2 and CH_4, we used a manual equilibration technique similar to the automated one described by Takahashi (1961). Five hundred ml of water were taken into a 1000 ml flask. A Suba-Seal was inserted into the neck of the flask and 60 ml of air were injected by syringe through the seal into the flask head space, resulting in a slight excess pressure. The flask was then shaken vigorously for 1 min to bring the air to equilibrium with the water and 60 ml of air were then withdrawn from the head space (thereby returning it to normal pressure) and injected, through a connecting plastic tube and a magnesium perchlorate drier (and a tube of Ascarite in the case of CH_4), into the analysis cell of an ADC Series 225 IRGA operating in absolute mode. The change in signal was noted. This procedure was repeated twice, when there was no more signal change. The water temperatures at the times of sampling and gas analysis were determined and corrections made to the dissolved gas concentration for any temperature-induced changes in solubility using the formula presented earlier. A similar injection procedure was followed for measuring the concentrations of CO_2 and CH_4 in samples of ambient air collected from a height of 1.1 m. Transfer coefficients were calculated from the gas fluxes and the appropriate concentration differences, Eq. (6) and (7).

Results and discussion

Energy balance and fluxes of sensible and latent heat

Earlier, we discussed the use of the energy balance to check the accuracy of the aerodynamic flux calculations. Fig. 2 shows the net radiation R_n and the storage terms W and G for a typical day. As pointed out previously, both storage terms are large, particularly W, so that estimation of the surface heat fluxes, H and λE, as residuals in the energy balance is prone to large experimental error. Fig. 2 compares the residual energy available for conversion to sensible and latent heat, i.e. $R_n - W - G$, with the sum of H and λE calculated from Eq. (13) and (14).

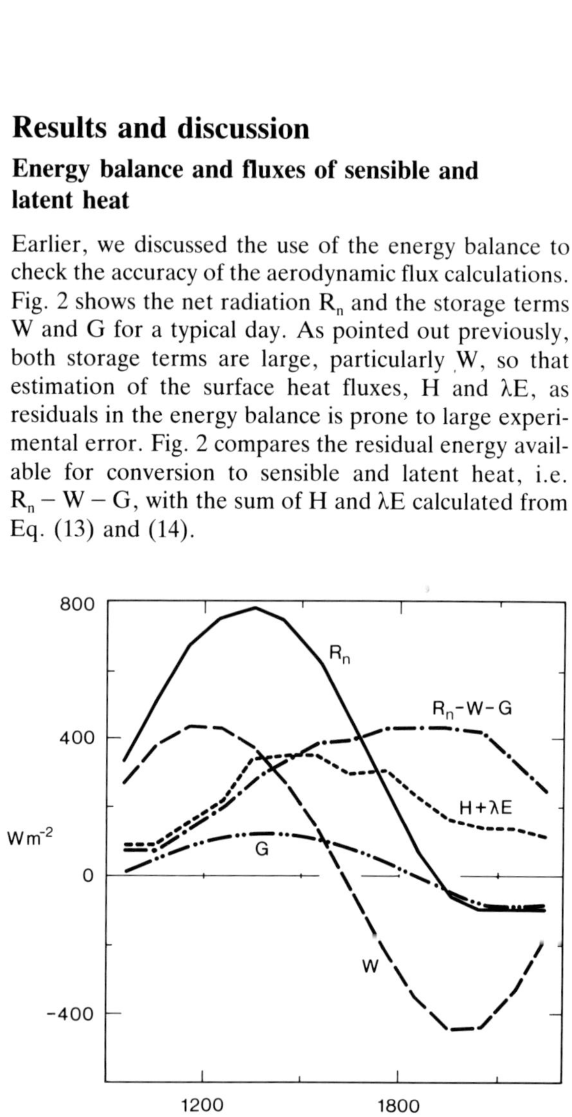

Fig. 2. Components of the energy balance for experimental pond 0.22 m deep: net radiation R_n, changes in heat storage in the water W, and the soil underneath G, and the sum of the surface flux densities of sensible heat H and latent heat λE.

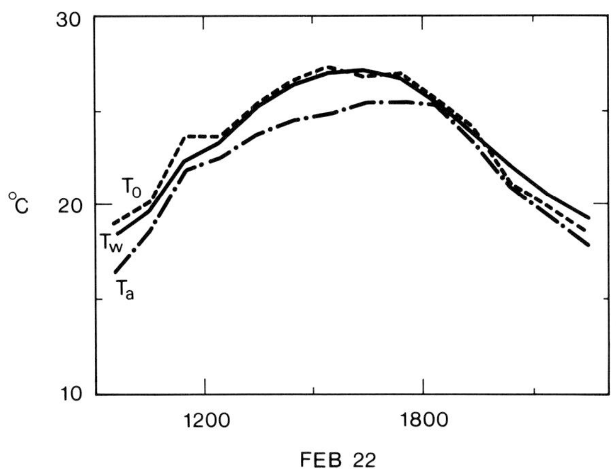

Fig. 3. Temperatures of the water surface T_0, the bulk water T_w, and the air at a height of 1.1 m T_a, for the same day as in Fig. 2.

Although the energy balance and the aerodynamic estimates of the total surface heat flux diverged markedly over night when R_n was small and W large, the agreement in the core period between 0900 and 1800 when the gas fluxes were measured was reasonably good, giving credence to our calculated transfer coefficients.

V_a for water vapour

Figure 3 shows typical diurnal variations in the temperatures of the air at 1.1 m, at the water surface, and at a depth of 0.02 m within the water. The bulk water was usually well mixed; temperatures at depths of 0.1 and 0.2 m were within hundreds of a degree of those at 0.02 m for a good part of the time, and the mixing appeared to extend right to the surface. Between mid-morning and dusk, the calculated T_0 seldom differed by more than a few tenths of a degree from T_w. Thermal stratification was thus very small.

Figure 4 shows determinations of V_a for water vapour, from Eq. (4), (23) and (24), plotted against friction velocity u_*. All values are averages over sampling periods of one hour. The figure also shows the relationship between V_a and u_* calculated from Deacon's model for conditions of neutral atmospheric stability, Eq. (5), (9) and (10). (Variations in atmospheric stability have only minor effects on Deacon's V_a which is dominated by V_v). For the most part, the model overestimates V_a by ~25%. Our observations extend up to wind speeds of 8 m s^{-1} at 1.1 m. For the wind speeds usually encountered over inland water bodies, say 1–5 m s^{-1}, Deacon's model may be too inaccurate for predicting evaporation rates, but may be accurate enough for predicting the exchanges of dissolved trace gases where the gas phase is only part of the overall pathway.

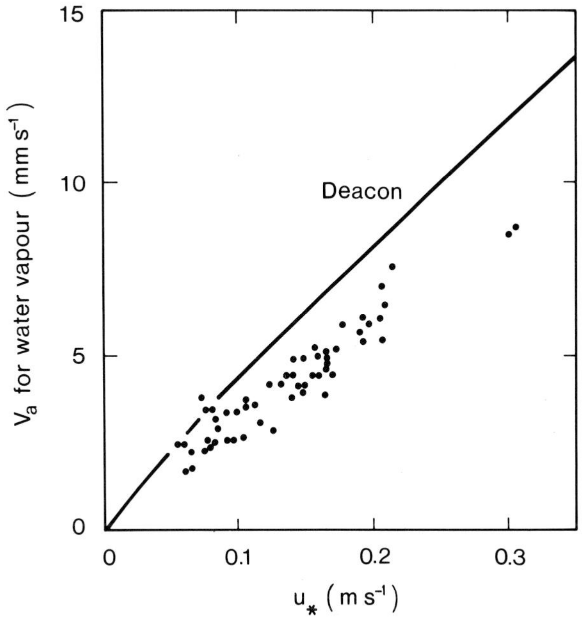

Fig. 4. Measured transfer coefficients for water vapour between the surface and a height of 1.1 m V_a (·) and those predicted by model of Deacon (1977) (–) plotted against friction velocity in the air u_*.

Ammonia

Average daytime values of c_w, the equilibrium gas density in the bulk water, and c_m, the atmospheric gas density at 1.1 m, are shown in Fig. 5. After an initial period of urea hydrolysis and pH increase due to algal growth, c_w increased rapidly to be ~30 times c_m, at

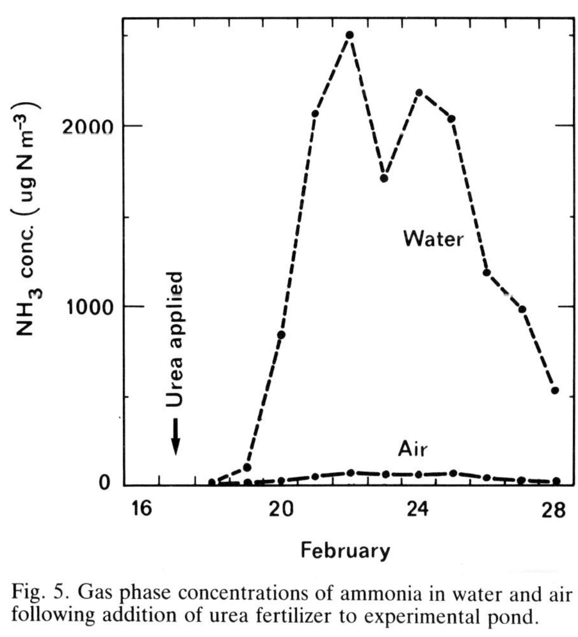

Fig. 5. Gas phase concentrations of ammonia in water and air following addition of urea fertilizer to experimental pond.

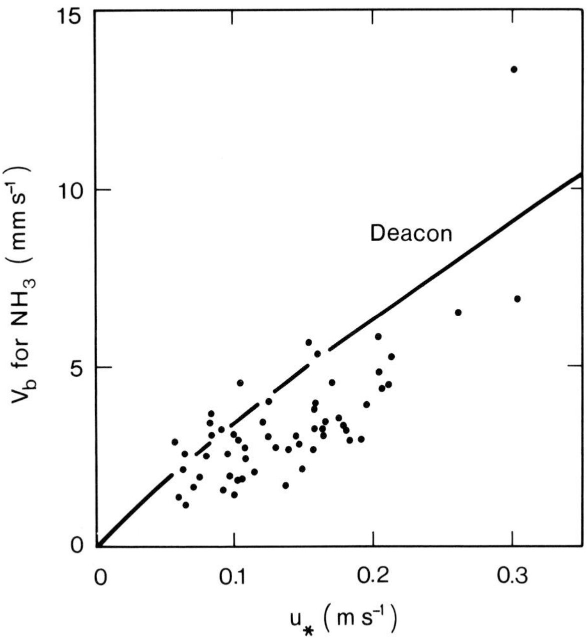

Fig. 6. Measured bulk transfer coefficients for ammonia V_b (·) and those predicted by model of Deacon (1977) (–) plotted against friction velocity in air u_*.

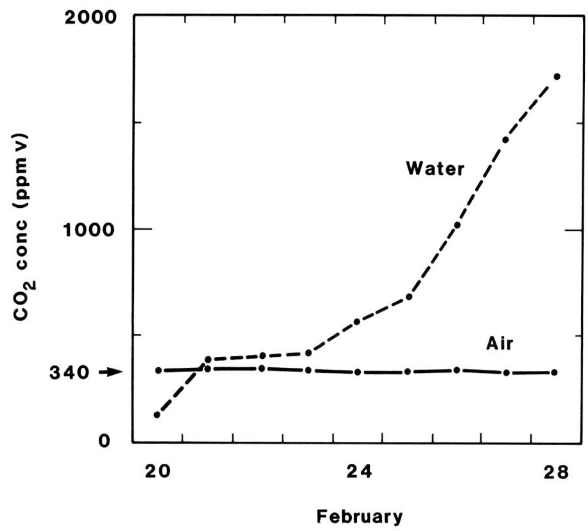

Fig. 7. Gas phase concentrations of CO_2 in water and air during experiment.

which time the NH_3 fluxes were large and able to be measured with high precision.

Observed and modelled bulk transfer coefficients for NH_3 are shown in Fig. 6. Again, Deacon's model overestimates the observed values at low wind speeds but there is just a hint that the opposite may be true in high winds when wave action and bubble emission probably play important parts in gas exchange. Further calculations indicate that the discrepancies between Deacon's model and the observations lie more in the atmospheric transfer coefficient V_a than in the liquid phase coefficient V_L, which is encouraging for use of the model for other gases where liquid phase transport dominates the overall exchange rate.

Carbon dioxide

Figure 7 shows the deficit in the dissolved CO_2 content of the water, relative to air, resulting from algal growth in the first days of the experiment (mentioned earlier), and the very large excess, arising from microbial activity in the soil, in the last part. The daily average c_w first dropped to ~100 ppmv and then increased to >1900 ppmv. Concentrations in the air remained steady at ~340 ppmv.

As shown in Denmead (1989), the fluxes of CO_2 were very small. For a concentration difference of 100 ppmv between water and air, the expected upper limit for the oceans, the typical flux was only ~5 μg m^{-2} s^{-1}. This value is very much in line with the probable CO_2 flux predicted for such a concentration difference by Broecker et al. (1986) from radioisotope observations, and is close to the value of 10 μg m^{-2} s^{-1} suggested by those authors as the lower detection limit for micrometerological observations. The precision of the flux measurements in the present study depended very much on the ability of the IRGA to resolve small concentration differences. This is largely a matter of technique, but with care, a resolution of 0.01 to 0.02 ppmv is

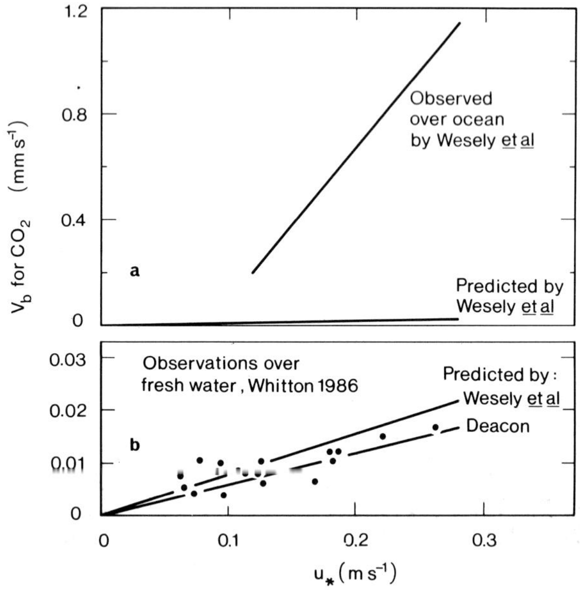

Fig. 8. a) Bulk transfer coefficients for CO_2 over the ocean as measured and predicted by Wesely et al. (1982). b) Measured bulk transfer coefficients for CO_2 (·) and those predicted by models of Deacon (1977) and Wesely et al. (1982) (–) plotted against friction velocity in air u_*.

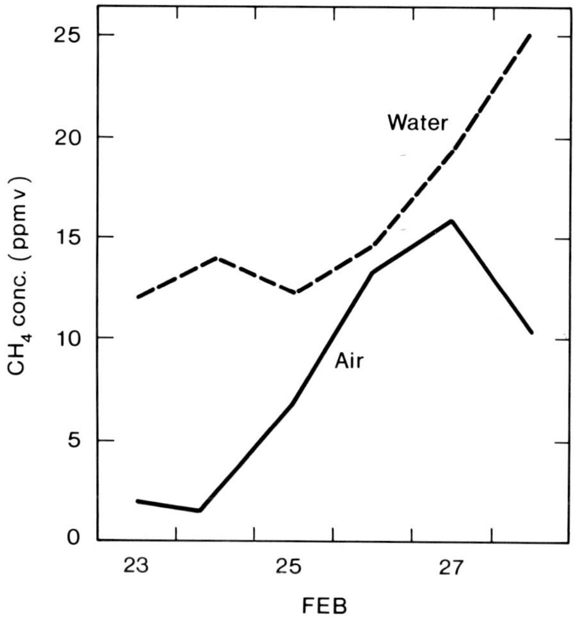

Fig. 9. Gas phase concentrations of CH_4 in water and air during experiment.

possible. Data presented in Denmead (1989) show that for the light winds typical of this study, a difference in atmospheric CO_2 concentration of 0.02 ppmv between 0.3 and 1.1 m would correspond to a CO_2 flux of ~3 µg m^{-2} s^{-1}.

Nineteen one-hour sampling periods out of the 45 considered by Denmead (1989) have been selected for detailed analysis, particularly for the dependence of the transfer coefficient on wind speed. To remove as much equivocation as possible from the results, runs were chosen in which corrections to the fluxes for stability effects were small, and the dissolved CO_2 concentrations exceeded the ambient by 600 ppmv or more. The smallest difference in atmospheric CO_2 concentration between 0.3 and 1.1 m was 0.08 ppmv. The largest was 0.49 ppmv, so that uncertainties in the calculated transfer coefficients arising from the IRGA measurements were between 4 and 25%. The bulk transfer coefficients determined from Eq. (6) for these periods are plotted against friction velocity in Fig. 8b, along with predictions of transfer coefficients from the theories of Wesely et al. (1982) and Deacon (1977), but with the latter doubled, as Deacon suggests, for hydration effects. There is no doubt that both theories provide adequate prescriptions for calculating transfer coefficients in the experiment described here. Our observations did not encompass the whole range of wind speeds encountered by Wesely et al. (1982), but they did span the lower part of that range. Fig. 8a shows that even at a friction velocity of 0.2 m s^{-1}, our transfer coefficients are 1/50 those measured by Wesely et al. (1982). Wherein lies the difference is obviously an urgent topic for research.

Methane

As seen in Fig. 9, high concentrations of CH_4 were developed in the water. Over the period in which CH_4 emissions were measured, c_w increased from 12 to 25 ppmv (from 7 to 15 times normal ambient). It is interesting that the emissions resulted in substantial modification of the methane concentration of the surface boundary layer. Over the measurement period, c_m increased from the normal background concentration of 1.6 ppmv to as much as 15 ppmv. This may have resulted from very rapid transport of CH_4 from the soil source to the atmosphere as discussed below. An examination of Table 1 will indicate that as well, the very high ambient concentrations of CH_4 on Feb 26 and 27 were associated with high CH_4 fluxes and low wind speeds.

We believe this is the first time micrometeorological methods have been used to measure CH_4 fluxes over a water surface. As a guide to others contemplating a similar exercise, we have listed in Table 1 the wind speeds, fluxes and differences in atmospheric CH_4 concentration between our measuring heights (0.3 and 1.1 m) during 17 one-hour sampling periods with long fetches and small corrections for atmospheric stability. The range of flux densities is similar to those observed in other studies of CH_4 emission from water bodies, e.g. Bartlett et al. 1988, Crill et al. 1988, Devol et al. 1988. Given that and the small gradients of atmospheric CH_4 concentration which were developed, it is evident that flux measurement in this way will require very good technique. We note, for instance, that the smallest concentration difference listed in Table 1, 0.02 ppmv, is right on the limits of detection of the IRGA system we employed.

Bulk transfer coefficients were calculated for 11 of

Table 1. Methane fluxes and atmospheric concentration differences for one-hour periods commencing at the indicated time.

Date	Time	$u_{1.1}$ m s^{-1}	F µg m^{-2} s^{-1}	$s_{0.3} - s_{1.1}$ ppmv
23 Feb	1700	4.8	18.6	0.25
	1800	4.4	8.5	0.13
24 Feb	1200	1.6	2.9	0.11
25 Feb	1130	1.9	2.9	0.09
	1530	2.8	2.2	0.08
	1630	3.8	1.1	0.03
	1730	3.4	1.1	0.03
26 Feb	1100	2.6	4.1	0.19
	1900	2.3	8.9	0.33
	2100	1.3	3.3	0.12
	2300	1.1	2.5	0.08
27 Feb	1600	1.4	4.4	0.17
28 Feb	1000	4.8	5.4	0.10
	1100	4.0	7.2	0.21
	1200	3.3	1.4	0.06
	1300	3.4	0.4	0.02
	1500	3.5	1.5	0.06

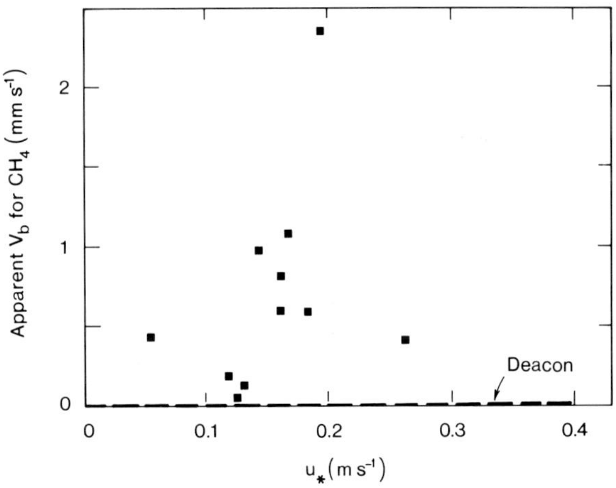

Fig. 10. Measured bulk transfer coefficients for CH_4 (·) and those predicted by model of Deacon (1977) (−) plotted against friction velocity in air u_*.

the periods in Table 1, selected on a basis similar to that used for CO_2: large fetch, small stability effects, and $c_w \gg c_m$. The results from these runs are compared with the predictions of Deacon's model for CH_4 in Fig. 10. There we see that the apparent V_b are orders of magnitude larger than Deacon would predict. In fact, <1% of the observed flux can be accounted for by Deacon's model.

The explanation appears to be that in this situation, most of the CH_4 produced in the soil escaped in bubbles directly to the atmosphere without entering solution. Some measurements on captured bubbles showed that they were more than one-third CH_4. The predominance of ebullition over diffusion in CH_4 transport in water bodies which are mostly devoid of vegetation has been observed by others in flooded rice fields, rivers, lakes and wetlands, as noted in the introduction. Obviously, Deacon's model is inappropriate in these circumstances.

Conclusions

In our experimental situation (shallow water, high insolation, little temperature stratification in the water, low winds), Deacon's model provided a reasonably good description of gas transport in both liquid and gaseous phases. It generally overestimated the exchanges of water vapour and ammonia between the pond and the air by ~25%, but gave an excellent description of the exchange of carbon dioxide. We note that for water vapour, all of the resistance to transfer is in the gaseous phase; for ammonia, most is in the gaseous phase, and for carbon dioxide, most is in the liquid phase. The model was inappropriate for methane where most of the water–air exchange occurred by ebullition, but overall, Deacon's model should provide a very useful means of predicting the exchange of dissolved trace gases between freshwater bodies and the atmosphere.

References

Bartlett, K. B., Crill, P. M., Sebacher, D. I., Harriss, R. C., Wilson, J. O. and Melack, J. M. 1988. Methane flux from the central Amazonian floodplain. – J. Geophys. Res. 93: 1571–1582.

Bouwmeester, R. J. B. and Vlek, P. L. G. 1981. Rate control of ammonia volatilization from rice paddies. – Atmos. Environ. 15: 131–140.

Broecker, W. S., Ledwell, J. R., Takahashi, T., Weiss, R., Merlivat, L., Memery, M., Peng, T.-H., Jahne, B. and Munnich, K. O. 1986. Isotopic versus micrometeorologic ocean CO_2 fluxes: a serious conflict. – J. Geophys. Res. 91: 10517–10527.

Crill, P. M., Bartlett, K. B., Wilson, J. O., Sebacher, D. I., Harriss, R. C., Melack, J. M., MacIntyre S., Lesack, L. and Smith-Morrill, L. 1988. Tropospheric methane from an Amazonian floodplain lake. – J. Geophys. Res. 93: 1564–1570.

Deacon, E. L. 1977. Gas transfer to and across an air-water interface. – Tellus 29: 363–374.

Denmead, O. T. 1989. Transfer of CO_2 across the air-water interface. – In: Proc. fourth Australasian conf. on heat and mass transfer, Christchurch, N.Z. Secretariat, pp. 449–458.

– , Freney, J. R. and Simpson, J. R. 1976. A closed ammonia cycle within a plant canopy. – Soil Biol. Biochem. 8: 161–164.

– , Trevitt, A. C. F., Muirhead, W. A., Freney, J. R. and Simpson, J. R. 1987. Modelling ammonia loss from flooded soil systems. – In: Bacon, P. E., Evans, J., Storrier, R. R. and Taylor, A. C. (eds), Nitrogen cycling in temperate agricultural systems, Vol. II. Australian Soc. of Soil Sci., Riverina Branch, Wagga Wagga, N.S.W. pp. 353–359.

Devol, A. H., Richey, J. E., Clark, W. A., King, S. L. and Martinelli, L. A. 1988. Methane emissions to the troposphere from the Amazonian floodplain. – J. Geophys. Res. 93: 1583–1592.

Emerson, K., Russo, R. C., Lund, R. E. and Thurston, R. V. 1975. Aqueous ammonia equilibrium calculations: effect of pH and temperature. – J. Fish. Res. Bd Can. 32: 2379–2383.

Fuchs, M. and Tanner, C. B. 1966. Infrared thermometry of vegetation. – Agron. J. 58: 597–601.

Hales, J. M. and Drewes, D. R. 1979. Solubility of ammonia in water at low concentrations. – Atmos. Environ. 13: 1133–1147.

Holzapfel-Pschorn, A., Conrad, R. and Seiler, W. 1986. Effects of vegetation on the emission of methane from submerged paddy soil. – Plant Soil 92: 223–233.

Jones, E. P. and Smith, S. D. 1977. A first measurement of sea-air CO_2 flux by eddy correlation. – J. Geophys. Res. 85: 5990–5992.

Ledwell, J. J. 1984. The variation of the gas transfer coefficient with molecular diffusivity. – In: Brutsaert, W. and Jirka, G. H. (eds), Gas transfer at water surfaces. D. Reidel Publ. Co., Dordrecht, pp. 293–302.

Leuning, R., Denmead, O. T., Simpson, J. R. and Freney J. R. 1984. Processes of ammonia loss from shallow floodwater. – Atmos. Environ. 18: 1583–1592.

Liss, P. S. and Slater, P. G. 1974. Flux of gases across the air-sea interface. – Nature (Lond.) 247: 181–184.

List, R. J. 1957. Smithsonian meteorological tables, 6th revised edition. – Smithsonian Inst., Washington, DC.

Miyake, M., Donelan, M., McBean, G., Paulson, C., Badgely, F. and Leavitt, E. 1970. Comparison of turbulent fluxes

over water determined by profile and eddy correlation technique. – Q.J.R. Meteorol. Soc. 96: 132–139.
Moeller, M. B. and Vlek, P. L. G. 1982. The chemical dynamics of ammonia volatilization from aqueous solution. – Atmos. Environ. 16: 709–717.
O'Brien, J. J. 1986. An important scientific controversy: oceanic CO_2 fluxes. – J. Geophys. Res. 91: 10515.
Sebacher, D. I., Harriss, R. C. and Bartlett, K. B. 1983. Methane flux across the air-water interface: air velocity effects. – Tellus 35B: 103–109.
Smith, S. D. and Jones, E. P. 1986. Isotopic and meteorological ocean CO_2 fluxes: different time and space scales. – J. Geophys. Res. 91: 10529–10532.
Takahashi, T. 1961. Carbon dioxide in the atmosphere and the Atlantic Ocean water. – J. Geophys. Res. 66: 477–494.
Thom, A. S. 1975. Momentum, mass and heat exchange of plant communities. – In: Monteith, J. L. (ed.), Vegetation and the atmosphere, Vol. 1. Academic Press, London, pp. 57–109.
Trevitt, A. C. F., Freney, J. R., Denmead, O. T., Zhu Zhao-Liang, C. G.-X. and Simpson, J. R. 1988. Water-air transfer resistance for ammonia from flooded rice. – J. Atmos. Chem. 6: 133–147.
van Wijk, W. R. and de Vries, D. A. 1963. Periodic temperature variations in a homogeneous soil. – In: van Wijk, W. R. (ed.), Physics of plant environment. North-Holland, Amsterdam, pp. 102–143.
Webb, E. K., Pearman, G. I. and Leuning, R. 1980. Correction of flux measurements for density effects due to heat and water vapour transfer. – Q.J.R. Meteorol. Soc. 106: 85–100.
Wesely, M. L. 1986. Response to "Isotopic versus micrometeorological ocean CO_2 fluxes: a serious conflict" by W. Broecker et al. – J. Geophys. Res. 91: 10533–10535.
– , Cook, D. R., Hart, R. L. and Williams, R. M. 1982. Air-sea exchange of CO_2 and evidence for enhanced upward fluxes. – J. Geophys. Res. 87: 8827–8832.

Ecological Bulletins 42: 42–52. Copenhagen 1992

Parameterization of trace gas dry deposition for a regional mesoscale diffusion model

Monika Baer

Baer, M. 1992. Parameterization of trace gas dry deposition for a regional mesoscale diffusion model. – Ecol. Bull. (Copenhagen) 42: 42–52.

Dry deposition is parameterized for the three dimensional regional diffusion simulation model DRAIS. The modeling domain is subdivided into 53×49 horizontal grid areas of 5×5 km each, and 25 vertical layers, 35 chemical species are considered. From these species 24 are stable enough to undergo atmospheric transport.
Dry deposition in DRAIS is parameterized as follows; a big leaf multiple resistance model is applied to calculate the uptake of trace gases by different land use types. The canopy resistance is then combined with calculated aerodynamic and surface boundary-layer resistances to determine the dry deposition velocity which is computed at each grid point. The canopy resistance dependencies on physical, chemical and biological surface properties as leaf area index (LAI), temperature, humidity, photosynthetically active radiation (PAR), solubility and reactivity of the trace gas are taken into account.
The dry deposition model DEPOMOD has given reasonable diurnal cycles and spatial distributions of dry deposition velocities and canopy resistances for chemical species as SO_2, O_3, NO, HNO_3, H_2O_2, PAN and organic acids. DEPOMOD seems to be more adequate for use in regional mesoscale models than commonly used look up tables for the canopy resistance and dry deposition velocity.

M. Baer, Inst. für Meteorologie und Klimaforschung, Kernforschungszentrum Karlsruhe/Univ. Karlsruhe, P.O. Box 3640, D-7500 Karlsruhe, Germany. Present address; BASF AG, Emissionsüberwachung Luft und Lärm, D-6700 Ludwigshafen, Germany.

Introduction

The deposition of gaseous pollutants to vegetated surfaces is perceived to be one of the major environmental problems in Germany, Europe and eastern North America. Gaseous deposition was identified to be a major factor damaging forests. Assessments of area-wide deposition budgets are frequently prevented by the inability to model or compute trace gas exchange with specific kinds of land use types and vegetation.

The development of area estimates of trace gas dry deposition fluxes requires linkage of adequate canopy dry deposition models with georeferenced data sets containing variables relevant to these models (e.g. NDVI, Normalized Difference Vegetation Index), canopy physical characteristics, land use type classification (LANDSAT TM). An important task was therefore to develop better methods to quantify dry deposition to landscapes for inclusion in mesoscale numerical models.

For many chemical species of interest, mean concentrations can be measured using available technology. The dry deposition velocity, on the other hand, is difficult to determine, since it depends on the chemical species and is a function of meteorological, biological and surface variables.

The main objective of this paper is to derive parameterizations for estimating surface resistances. The substances of interest include H_2O_2, HCHO and other aldehydes, organic peroxide, peroxyacetic acid, organic acids, peroxyacetyl nitrate (PAN), HNO_3 besides SO_2, O_3, NO and NO_2.

Since the experimental data base on surface resistances of these chemical species is very sparse, parameterizations for computing surface resistances to uptake of gases have been developed for use in a regional mesoscale diffusion model. They are primarily based on information available on fluxes of water vapor, heat and CO_2 in and above agricultural and forest stands and on

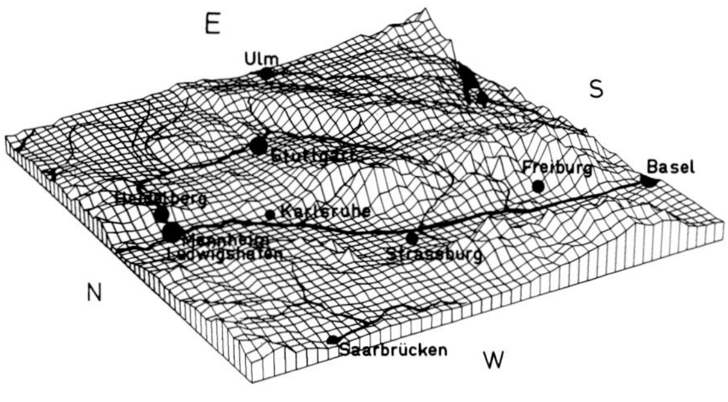

Fig. 1. Modeling area for KAMM and DRAIS (53×49 grid points, 5×5 km each).

resistances to SO_2 and O_3 uptake. Dry deposition, a consequence of the same atmospheric exchange mechanisms responsible for the surface fluxes of heat, moisture and momentum, is strongly influenced by physical, chemical and biological surface properties.

The surface or canopy resistance dependencies on specific surface properties have been parameterized for the conditions during the mesoscale field experiment TULLA (Transport und Umwandlung von Luftschadstoffen im Lande Baden-Württemberg und aus Anrainer Staaten = transport and conversion of airborne pollutants in the state of Baden-Württemberg and from neighbouring states) performed in March 1985 in the state of Baden-Württemberg (Fiedler 1987). The modeling area includes Baden-Württemberg. It is an irregular terrain with altitudes up to 1400 m (Fig. 1).

The model DRAIS

The conservation equation for gaseous or aerosol material in the atmosphere which emerges as a result of the averaging operations and closure approximations and which is the basis of most airshed models can be expressed in Cartesian coordinates by the equation:

$$\frac{\delta \bar{c}_i}{\delta t} + \nabla(\bar{\vec{u}}\bar{c}_i) = \nabla(\hat{k}\nabla\bar{c}_i + f_i(\bar{c}_1 \dots \bar{c}_p) \qquad (1)$$

$\bar{c}_i$ mean concentration of chemical species i
$\bar{\vec{u}}$ mean advective velocity field $\bar{\vec{u}} = (\bar{u}, \bar{v}, \bar{w})$
$\hat{k}$ second order eddy diffusivity tensor
f_i includes chemical formation or loss rate of species i, sources and sinks of species i

In a Cartesian coordinate system the corresponding lower boundary condition can be formulated as follows:

$$(\hat{\kappa} \nabla \bar{c}_i)\vec{n}_h = Q_i(x,y,h(x,y),t = v_{D,i}\, \bar{c}_{r,i} \qquad (2)$$

where $\vec{n}_h$ is the unit vector normal to the terrain directed into the atmosphere, Q_i is the mass flux of species i to the surface, and h(x,y) is the terrain elevation above sea level at x, y (Reynolds et al. 1973), $v_{D,i}$ is the dry deposition velocity of chemical species i and $\bar{c}_{r,i}$ is the mean concentration at a reference height, where $v_{D,i}$ is determined conveniently.

At every grid point the mean advective velocity field $\bar{\vec{u}}$ and the Eddy diffusivity are computed by KAMM (Karlsruhe Atmospheric Mesoscale Model) (Dorwarth 1986). The KAMM model is three-dimensional, non-hydrostatic and instationary. A terrain following coordinate system is used. It calculates the flow, temperature and diffusion coefficient fields over an irregular terrain by solving the filtered Navier-Stokes equation of motion, the continuity equation, and the thermodynamic equation.

The chemical production and loss terms are determined by the RADM gas phase chemical reaction mechanism (Stockwell 1986, Chang et al. 1987). According to the chemical mechanism used, 24 chemical species are stable enough to undergo atmospheric transport (Table 1). Most of them are expected to be deposited. Dry deposition of these species is taken into account as a lower boundary condition in DRAIS.

Emission rates of SO_2, NO_x and VOC (volatile organic compounds) available as hourly averages on a 1×1 km grid for the whole period of the TULLA experiment in the state of Baden-Württemberg (Table 1) are considered as source terms. DRAIS has in its actual form 53×49 horizontal grid points with a horizontal grid size of 5×5 km and 25 vertical levels, 24 trace gases are transported.

Table 1. Transported chemical species in DRAIS.

Species name	Species symbol	Deposited	Emitted
Sulfur dioxide	SO_2	x	x
Sulfuric acid	H_2SO_4	x	
Nitric oxide	NO	x	x
Nitrogen dioxide	NO_2	x	x
Nitrate radical	NO_3		
Dinitrogen pentoxide	N_2O_5		
Nitric acid	HNO_3	x	
Ozone	O_3	x	
Hydrogen peroxide	H_2O_2	x	
Carbon monoxide	CO		x
Formaldehyde	HCHO	x	x
Aldehydes	ALD	x	x
Organic acids	ORA	x	x
Organic peroxide	OP	x	
Peroxyacetic acid	PAA	x	
Peroxyacetyl nitrate	PAN	x	
Other alkanes	HC		x
Ethane	ETH		x
Ethene	OL2		x
Propene	OL3		x
Butene and other reactive olefins	OL4		x
Toluene	TOL		x
Xylene	XYL		x

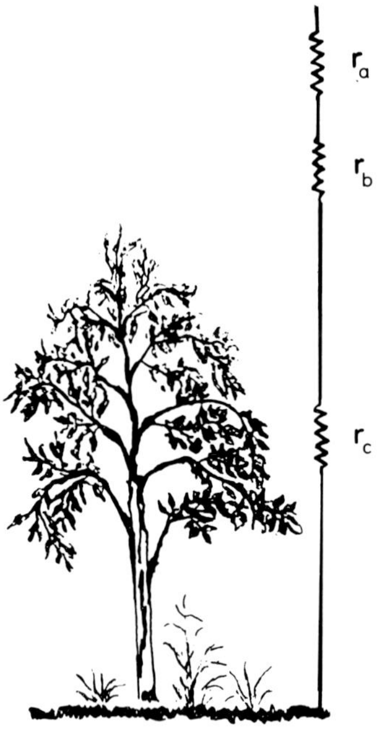

Fig. 2. "Big-leaf" multiple resistance model: r_a: aerodynamic resistance, r_b: surface layer resistance, r_c: canopy resistance.

Additionally the data of the TULLA experiment are used to verify the diffusion model DRAIS (Dreidimensionales Regionales Ausbreitungs- und Immissions-Simulationsmodell = threedimensional regional diffusion simulation model) (Baer and Nester 1987).

Comparisons of measured concentrations of SO_2 with values calculated by the DRAIS model proved that this model is able to simulate the diffusion of SO_2 over irregular terrain in a regional area with satisfying accuracy. Furthermore the simulations showed, that mass balances of SO_2 can be derived from the TULLA data and that dry deposition is a very important removal process.

Dry deposition was therefore modeled carefully in order to get a good estimate of total dry deposition in an area of c. 250×250 km. The DRAIS dry deposition model DEPOMOD is described in detail in the following sections.

Parameterization of the dry deposition velocity

In most models the deposition rate is described by a single quantity, the pollutant deposition velocity v_D. The flux of material, F, directed towards the lower boundary surface, is defined by:

$$F = v_D \, c(z_r) \tag{3}$$

where $c(z_r)$ is the concentration of the material at a reference height z_r. The dry deposition velocity involves complex linkage between turbulent diffusion in the surface boundary layer molecular scale motion at the air-ground interface and interaction of the material with the surface.

In most efforts to model and parameterize v_D for the dry deposition of gases to vegetated surfaces a 'big-leaf' multiple resistance model is used (Fig. 2) (Baldocchi et al. 1987). The most important individual resistances to pollutant transfer are the aerodynamic resistance (r_a) associated with atmospheric turbulence, a quasi-laminar boundary resistance (r_b) which takes into account the resistance to mass transfer through the quasi-laminar layer of air in contact with the surface elements and is influenced by the diffusivity of the material being transferred, and a net canopy resistance (r_c), which is dominated by biological surface factors and includes the stomatal, mesophyll and cuticular uptake resistances; however, the stomatal resistance tends to dominate for gases. The aerodynamic, quasi-laminar boundary layer and canopy resistances are in series. Thus, v_D is given by:

$$v_D = \frac{1}{r_a + r_b + r_c} \tag{4}$$

Determination of the aerodynamic resistance

The aerodynamic resistance is a function of wind speed, surface properties and atmospheric stability. The parameterization used is given by:

for stable conditions: $\frac{z_r}{L_*} > 0$

$$r_a(z_r) = \frac{1}{\varkappa u_*}\left(0.74 \ln\left(\frac{z_r}{z_0}\right) + \frac{4.7}{L_*}(z_r - z_0)\right) \tag{5}$$

for neutral conditions: $\frac{z_r}{L_*} = 0$

$$r_a(z_r) = \frac{1}{\varkappa u_*}\left(0.74 \ln\left(\frac{z_r}{z_0}\right)\right)$$

for unstable conditions: $\frac{z_r}{L_*} < 0$

$$r_a(z_r) = \frac{1}{\varkappa u_*} 0.74 \left[\ln \left(\frac{\sqrt{1 - 9\frac{z_r}{L_*}} - 1}{\sqrt{1 - 9\frac{z_r}{L_*}} + 1} \right) - \ln \left(\frac{\sqrt{1 - 9\frac{z_0}{L_*}} - 1}{\sqrt{1 - 9\frac{z_0}{L_*}} + 1} \right) \right]$$

where u_* is the friction velocity, $\varkappa$ the von Karman constant ($\varkappa = 0.4$), z_r the reference height, z_0 the roughness length, and L_* the Monin-Obukhov stability length. The displacement height is taken into account in the topographic height.

Values for z_0 were parameterized by an interpolation of the z_0-values for the different known ground covers within one grid area of 5×5 km size. The method has been described by van Dop (1983). Only the weighting factors are not taken into account. The mean z_0 for the whole modeling area was c. 1 m. This arises because in Baden-Württemberg many areas are urban or forested.

At every grid point values for u_* and L_* were calculated by KAMM. In KAMM the fluxes of momentum and heat are determined by the scaling parameters u_*, Θ_* and L_* of the Monin-Obukhov theory of the Prandtl layer. These parameters are determined by fitting the analytical profile functions for the wind velocity and the potential temperature to the model solution (Adrian and Fiedler 1991).

Determination of the quasi-laminar surface boundary layer resistance

This near-surface resistance is associated with transfer through the quasi-laminar layer in contact with the surface. Modeling studies and wind-tunnel investigations showed that r_b is strongly influenced by the diffusivity of the material being transferred. Specialized surface transfer models (Brutsaert 1975, 1979) predict a functional dependence of r_b on the Schmidt-number, Sc, such that in general.

$$r_b \propto (Sc)^p \qquad (6)$$

The exponent p is often taken to be 2/3. The uncertainty associated with the exponent is not critical for most trace gases. Only for very slowly diffusing species such as aerosol particles the uncertainties can become large. Other models have been developed to describe the variation of r_b with surface roughness (Garratt and Hicks 1973).

For the present purposes the quasi-laminar boundary layer resistance r_b in DRAIS is assumed to be given by:

$$r_b = \frac{2}{\varkappa u_*} (Sc)^{\frac{2}{3}} = \frac{1}{\varkappa u_*} 2 \left(\frac{\nu}{D_i}\right)^{\frac{2}{3}} \qquad (7)$$

ν: kinematic viscosity of the transport medium (ν_{air} 0.15 cm^2/s (20°C, p = 1 atm))

D_i: molecular diffusion coefficient of chemical species i

Experimentally determined molecular diffusion coefficients in air for various substances can be found in Landolt-Börnstein (1969).

Parameterization of the canopy resistance

Surface exchange processes occurring between the atmosphere and plant canopies are varying with the physiological state of the plants, the canopy architecture, the atmospheric turbulence, the plant species composition and the chemical characteristics of the trace gases. The canopy resistance dependencies on physical, chemical and biological surface properties as leaf area index, temperature, humidity, photosynthetically active radiation (PAR), solubility and reactivity of the chemical species, have therefore to be taken into account.

Special caution must be exercised with respect to nomenclature and terminology. Many plant physiologists quantify resistances associated with biological pathways via stomata or leaf cuticle in terms of a biological resistance to transfer expressed as the effective resistances per unit area of foliar surface. In the presented big leaf model we are mainly interested in the integrated form of these resistances. The link between the two will be the LAI of the stand. Measurements of canopy resistances to gaseous deposition are available in the literature (Wesely and Hicks 1977, Fowler 1978, Wesely et al. 1978, Fowler and Unsworth 1979, Hicks et al. 1982, Fowler and Cape 1982). All studies have shown, that the stomatal resistance is the most dynamic and most influential resistance to transfer of O_3, SO_2, NO_x if the canopy is dry.

Stomatal resistance

Transfer through stomata is by diffusion. The stomatal resistance is thus inversely dependent on the molecular diffusivity of the trace gas (Winner and Mooney 1980 a, b). The effects of individual factors such as radiation,

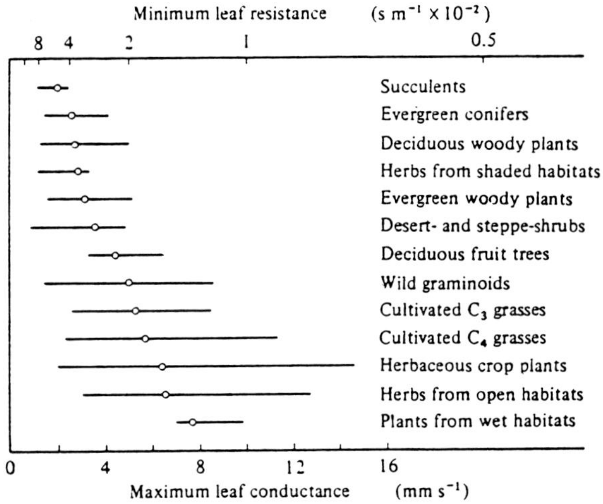

Fig. 3. Maximum leaf conductance in different groups of plants. The lines cover c. 90% of individual values reported. The open circles represent group average conductances (from Körner et al. 1979).

temperature, humidity or leaf water status on stomatal resistance can be studied best in controlled environments or leaf chambers where each factor may be varied independently.

Perhaps the most consistent and well documented stomatal response is the opening that occurs in most plant species as irradiance increases. In the DRAIS dry deposition model we assumed the following dependence on incident PAR (I_{PAR}).

$$r_{st} = r_{st,min}\left(1 + \frac{b}{I_{PAR}}\right)\frac{1}{f_e f_T f_w f_D} \tag{8}$$

$r_{st,min}$ is some minimum value, which varies with the plant species. In Körner et al. (1979), a large number of reported measurements of maximum leaf conductance (minimum resistance) is summarized for different groups of plants (Fig. 3). The correction factors f_e, f_T and f_w were introduced by Jarvis (1976) in order to correct for effects of humidity, temperature and water stress. These factors vary between 0.0 and 1.0. The plant specific constant b is equal to the PAR flux density at twice the minimum stomatal resistance. The correction factor f_D accounts for the differences in molecular diffusivity between the chemical species in question and water vapor.

The analyses summarized by Jarvis (1976), suggest a linear reduction of stomatal conductance with increasing atmospheric water vapor pressure deficit:

$$f_e = 1 - b_e(e_s(T) - e) \tag{9}$$

$e_s(T)$ saturated water vapor pressure at air temperature T
e actual atmospheric vapor pressure
b_e plant specific constant

Stomatal resistance is relatively independent of the leaf water potential W until the latter drops below a threshold value. During the TULLA experiment the stands were well watered. Influences of plant water stress could therefore be neglected ($f_w = 1$).

To account for temperature effects on r_{st}, the following correction factor was used

$$f_T = \frac{T - T_{min}}{T_0 - T_{min}}\left(\frac{T_{max} - T}{T_{max} - T_0}\right)^{b_T} \tag{10}$$

where

$$b_T = \frac{T_{max} - T_0}{T_0 - T_{min}} \tag{11}$$

T_{max} and T_{min} are plant species dependent higher and lower temperature extremes at which stomata no longer open (typically 40°C and 5°C respectively) and T_0 is the temperature at which photosynthesis and stomatal exchange is optimized (typically 25°C). Experimentally determined values for T_0, T_{min}, T_{max} and b were summarized from literature by Baldocchi et al. (1987). The factor f_D in Eq. (8) is equal to

$$f_D = \frac{D_i}{D_0} \tag{12}$$

Table 2. Plant specific constants for the determination of the stomatal resistance (Eq. 8) in the TULLA region.

Ground cover / land use	$r_{st,min}$ (s m^{-1})	b (W m^{-2})	T_{min} (°C)	T_{max} (°C)	T_0 (°C)
High density residential	–	–	–	–	–
Low density residential	350	25	5	45	25
Forest	350	25	0	40	15
Scrub	350	25	0	40	15
Agricultural land dry	150	40	5	45	25
Agricultural land wet	120	40	5	45	25
Water surface	–	–	–	–	–

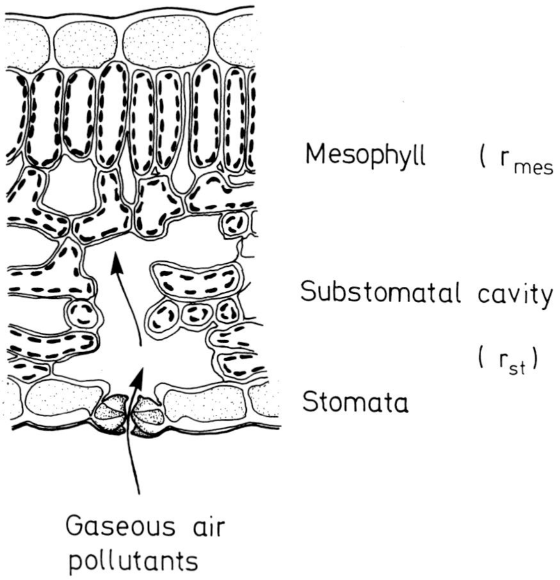

Fig. 4. Resistances along the diffusive pathway into the leaf.

where D_0 is the molecular diffusivity of the chemical specie for which $r_{st,min}$ in Eq. (8) was determined (usually H_2O or CO_2) and D_i is the molecular diffusiyity of the trace gas in question.

The topographic data base for the TULLA modeling region contains seven different land use types. The values listed in Table 2 have been chosen for the plant specific constants in Eq. (8) and (10). In order to get any information on the physiological state of vegetation we use NDVI satellite scenes of Blümel et al. (1983–1985). The NDVI satellite scenes have a resolution of 1×1 km. Different growth and canopy characteristics of vegetation at different seasons of the year can therefore be taken into account.

Once the pollutants enter substomatal cavities, they must either be destroyed by chemical reaction or transferred to plant cells. The consequence is, that a mesophyllic resistance has to be combined with the diffusive stomatal resistance given in Eq. (8) (Fig. 4).

Because r_{st} and r_{mes} are additive the overall resistance to transfer through stomata is thus

$$r_s = r_{st} + r_{mes} \tag{13}$$

Since the site of deposition for a gas molecule within the leaf interior is a hydrated surface of a mesophyll cell, an air-to-water interface has to be crossed.

In order to describe the transfer of chemical species to these eventual sinks (Liss and Slater 1974, O'Dell et al. 1977, Taylor et al. 1983) the effective Henry's Law constants are used to estimate uptake by hydrated plant cell surfaces and extracellular water in the substomatal cavities of plant leaves.

The more water soluble a compound, the more effective the uptake will be (Taylor et al. 1983). The ability of a gas to go into such solutions is described by the effective Henry's Law constant, which strongly varies with chemical species, temperature and pH value. Since in plants we typically find a pH near-neutral (Canny 1984), the effective Henry's Law constants were adjusted accordingly. To account for chemical reactivity as a further sink, the chemical reactivity of the transported species in DRAIS (Table 3) was scaled with respect to O_3. Wesely (1988) estimated the mesophyll resistance from considerations of CO_2 uptake by vegetation (Monteith 1973). Following Wesely (1988) the dependencies of the mesophyll resistance on effective Henry's Law

Table 3. Effective Henry's Law constants and reactivity factors.

Species name	Species symbol	Reactivity factor f_i	Henry's law constant in M atm^{-1} (water)	Reference
Sulfur dioxide	SO_2	0	$1.6\ (1 + 2.1\ 10^{-2} / H+)$	Chameides (1987)
Nitric oxide	NO	0	$1.9\ 10^{-3}$	Loomis (1928)
Nitrogen dioxide	NO_2	0.1	10^{-2}	"
Nitric acid	HNO_3	0	$5.8\ 10^{6} / H+$	Chameides (1987)
Ozone	O_3	1.0	$1.5\ 10^{-2}$	"
Hydrogen peroxide	H_2O_2	1.0	$1.8\ 10^{5}$	"
Formaldehyde	HCHO	0	$3.26\ 10^{-4}$	Lebury and Blair (1925)
Aldehydes	ALD	0	76	Snider and Dawson (1985)
Organic acids	ORA	0	$1.45\ 10^{-4}$	Weast (1978), Jacob (1986)
Organic peroxide	OP	0.25	665	Lind and Kok (1986)
Peroxyacetic acid	PAA	0.5	1635	"
Peroxyacetyl nitrat	PAN	0.1	3.6	Lee at al. (1983)
Other alkanes	HC	0	$1\ 10^{-3}$	Macckay and Shiu (1981)
Ethane	ETH	0	$1.9\ 10^{-3}$	"
Ethene	OL2	0	$4.9\ 10^{-3}$	Loomis (1928)
Propene	OL3	0	$4.7\ 10^{-3}$	Macckay and Shiu (1981)
Butene and other reactive olefins	OL4	0	$1.3\ 10^{-3}$	"
Toluene	TOL	0	0.15	"
Xylene	XYL	0	0.1	"

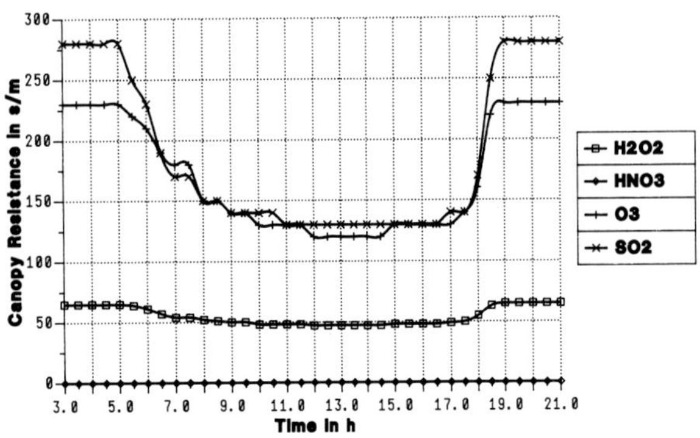

Fig. 5. Modeled diurnal cycles of canopy resistances of trace gases, shown are mean values in the modelling region (see Fig. 1).

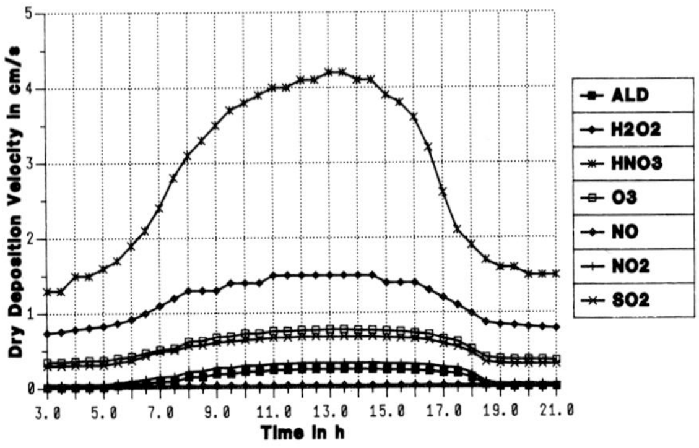

Fig. 7. Modeled diurnal cycles of dry deposition velocities of trace gases, shown are mean values in the modelling region (see Fig. 1).

constant of chemical species i, H^i_{eff}, and chemical reactivity f_i are assumed as follows:

$$r^i_{mes} = \frac{1}{\dfrac{H^i_{eff}}{3000\ \mathrm{M/atm}} + 100\ f_i}\ \mathrm{s/m} \qquad (14)$$

In Table 3 values for H_{eff} and f_i are summarized for near neutral pH.

The formulation allows two parallel pathways to the sink, one analogous to a water soluble species, like SO_2, the other one analogous to a chemically reactive species, like O_3. The pathway based on f_i will produce small values of r_{mes} if f_i is near 1 or 0.1. The other pathway will result in low values of r_{mes} if H^i_{eff} is sufficiently large, so that transfer from the gaseous to the aqueous phase represents a large sink of the trace gas.

The time dependence of r_{st} shows clearly the opening and closing behaviour of stomata shortly after sunrise and before sunset.

Cuticular resistance

Deposition directly to leaf cuticle is likely to be important for several trace gases. But up to now little is known about how r_{cut} varies with the chemical reactivity of the material being transferred, the adsorption and possible chemical Red/Ox reactions with the structural materials of the leaf cuticle. In the DRAIS deposition model the cuticular resistance of chemical species i is computed as:

$$r^i_{cut} = \frac{3500\ \mathrm{s/m}}{\left(\dfrac{H^i_{eff}}{H^{SO_2}_{eff}} + f_i\right)} \qquad (15)$$

The value 3500 s m^{-1} stands for an upper value of cuticular resistances, estimated experimentally for SO_2 (Spedding 1969).

Baldocchi (1988) showed that the dry deposition velocity is quite insensitive to variations in cuticular resistance as long as r_{cut} is >2000 s m^{-1}. In Taylor et al. (1987) reported ratios of leaf surface to internal dry deposition

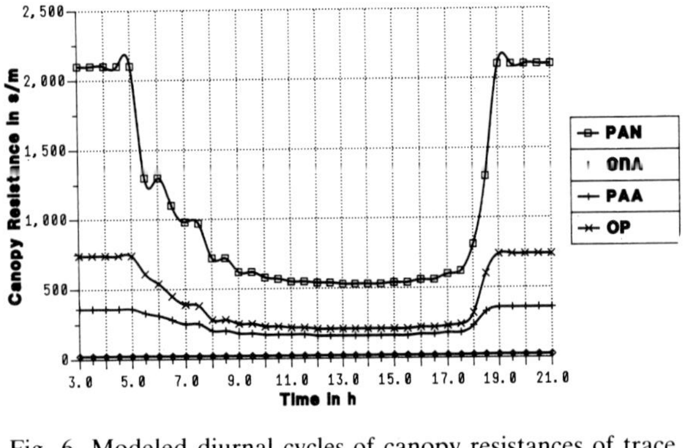

Fig. 6. Modeled diurnal cycles of canopy resistances of trace gases, shown are mean values in the modelling region (see Fig. 1).

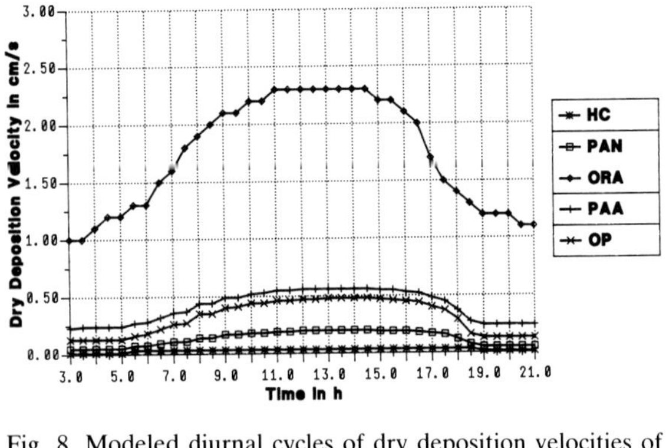

Fig. 8. Modeled diurnal cycles of dry deposition velocities of trace gases, shown are mean values in the modelling region (see Fig. 1).

Table 4. Mean values r_c and v_D in the TULLA modelling region.

Species	mean r_c (s^{-1})	mean v_D (cm s^{-1})	Measured		Reference
SO_2	130	0.68	Deciduous	0.2 – 0.8 cm s^{-1}	Baldocchi 1988
O_3	124	0.77	Coniferous forest		Wesely et al. 1978
NO	3100	0.03	Grass	0.02 cm s^{-1}	Hoefken et al. 1986
NO_2	436	0.33	Grass	0.2 cm s^{-1}	"
HNO_3	0	4.1	Grass	1.3 cm s^{-1}	"
H_2O_2	47	1.5		1.6–2 cm s^{-1}	Heikes et al. 1986 Walcek 1987
HCHO	123	0.74	–	–	–
ALD	599	0.25	–	–	–
OP	211	0.47	–	–	–
PAA	163	0.56	–	–	–
PAN	536	0.20	Grass	v_D: 0.14–0.25 cm s^{-1} r_c: 400–700 s m^{-1}	Garland and Penkett 1976
ORA	21	2.3	–	–	–

ranged from 0.1 (NO_2) to greater 0.8 (HNO_3) for gases of varying physicochemical properties.

Soil surface resistance

Since during the period of the TULLA experiment the ground surfaces were well water-saturated, the water solubility of the trace gases should be a determining quantity for the parameterization of soil surface resistances. As a secondary quantity the reactivity of the trace gas in contact with the surface should play an important part.

For some few trace gases, mainly SO_2 and O_3 experimentally determined soil surface resistances are available in the literature (Turner et al. 1973, Payrissat and Beilke 1975, Judeikis and Wren 1978, Fowler 1978, Fowler and Unsworth 1979, Wesely et al. 1981).

Taking into account the water solubility and reactivity of the trace gas relative to SO_2 and O_3 the soil surface resistance is modeled.

$$r^i_{soil} + \frac{1}{\left(\frac{H^i_{eff}}{H^{SO_2}_{eff}} \frac{1}{r^{SO_2}_{soil}} + \frac{f_i}{r^{O_3}_{soil}}\right)} \quad (16)$$

In a similar way the resistances for water surfaces and surfaces wetted by rain or dew are estimated (Chameides 1987).

Combining Eq. (8), (14), (15) and (16) the total canopy resistance is given by:

$$r_c = \frac{1}{\left(\frac{1}{r_{st} + r_{mes}} + \frac{1}{r_{cut}}\right) LAI + \frac{1}{r_{soil}}} \quad (17)$$

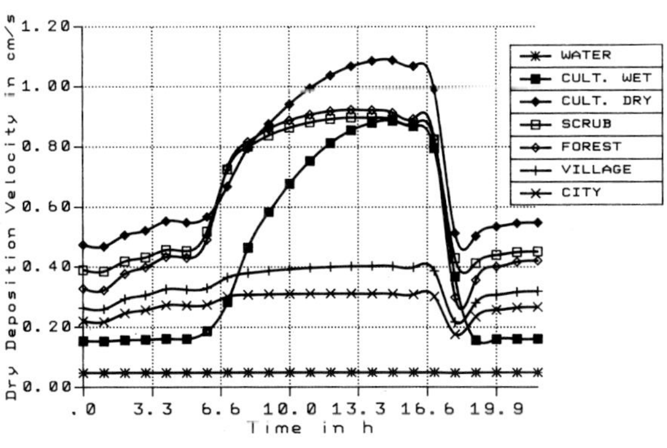

Fig. 9. Modeled diurnal cycles of dry deposition velocities of ozone for seven different land-use types.

Results

Typical diurnal cycles of the friction velocity, Monin Obukhov stability length and air temperature, calculated for clear sky conditions in early spring, were used to determine the corresponding dry deposition velocities and canopy resistances (Figs 5–8). If the pollutant transfer is controlled by stomata, the canopy resistances decrease significantly after sunrise and increase after sunset. Substances with an extremely high Henry's Law constant, like HNO_3, are in contrast characterized by very low values of r_{cut} and r_{soil}. The corresponding dry deposition velocity in this case is mainly dependent on the aerodynamic resistances r_a and r_b.

The day-time mean values of v_D and r_c (Figs 5–8) generally conform to expectations, especially if the uptake through stomata is the controlling resistance in r_c and if the Henry's Law constant of the trace gas is

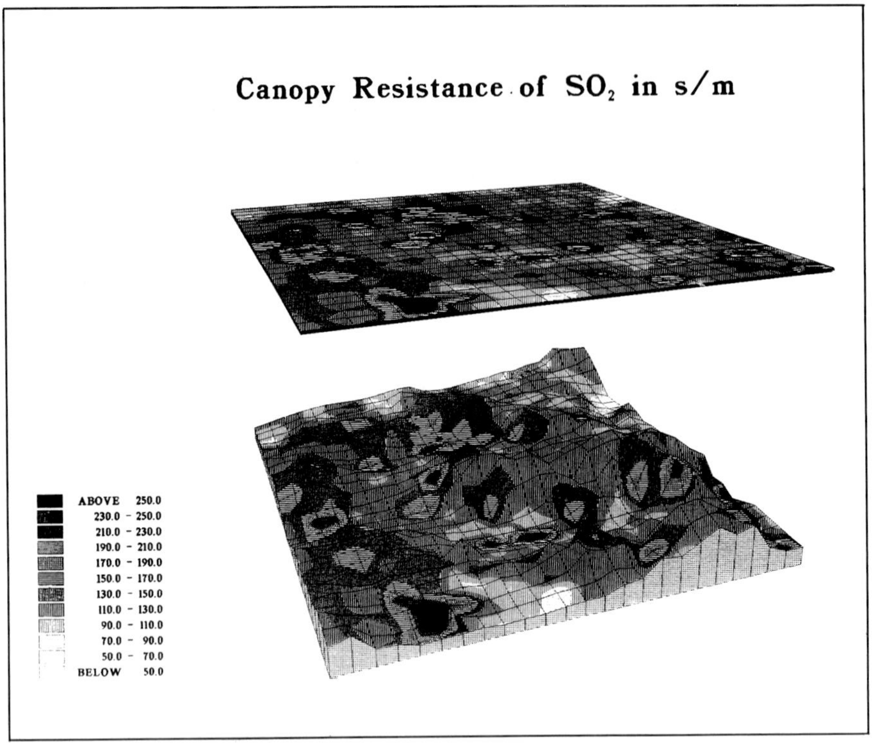

Fig. 10. Spatial variability of canopy resistance of SO_2 in the TULLA modeling region (clear day in March 1985 at 12.00 UTC of SO_2).

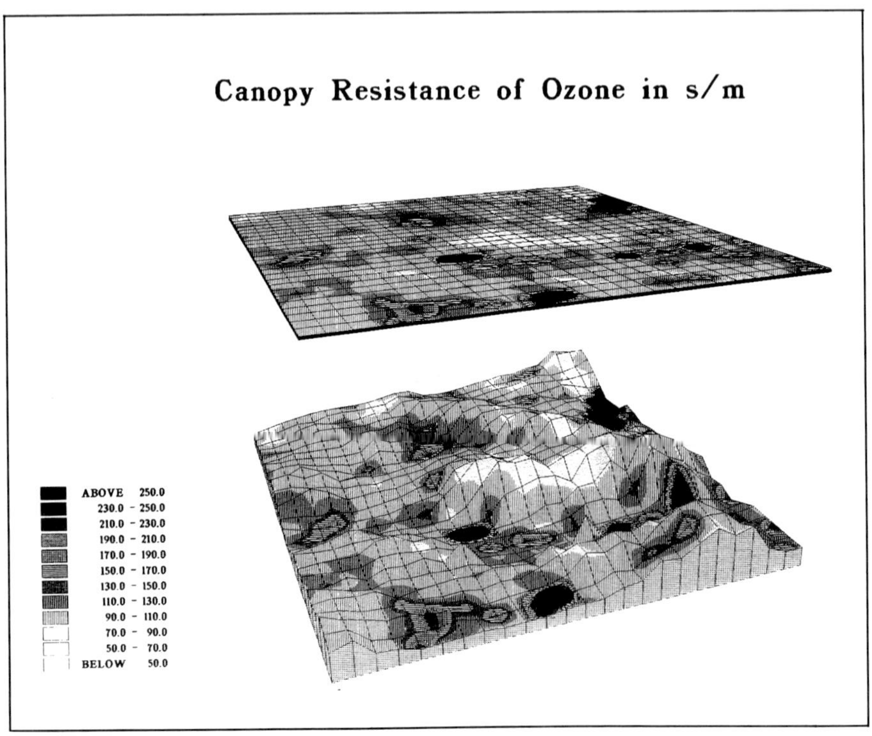

Fig. 11. Spatial variability of canopy resistance of O3 in the TULLA modeling region (clear sky day in March 1985 at 12.00 UTC of O_3).

comparable to $H_{eff}^{SO_2}$ or if the species could at least be regarded as being slightly reactive. Measured and modeled values of r_c and v_D are listed in Table 4 for comparison.

Diurnal cycles of dry deposition velocities have been calculated for seven different land use types that are available in the TULLA topographical data base. As an example the diurnal cycles of ozone dry deposition velocities are shown in Fig. 9. The canopy resistance was calculated at each grid-point by knowledge of the corresponding land-use type contribution within one grid-area of 5×5 km size according to:

$$r_c(x,y) = \sum_{i=1}^{7} p_i(x,y) r_{c,i}(x,y) \tag{18}$$

$p_i(x,y)$ contribution of land-use type i at grid point (x,y)

$r_{c,i}(x,y)$ canopy-resistance calculated for land-use type i at grid-point (x,y)

The spatial variability of canopy resistances calculated in that way are shown in Figs 10 and 11.

Lowest r_c values are usually to be found in forest and country regions; highest values at high residential sites. In Fig. 10 the locations of highest r_c can be related to the cities Mannheim, Karlsruhe, Stuttgart (see also Fig. 11). Trace gases with an extremely low water solubility will have maximum r_c values in the region of Lake Constance and other river and lake sites (Fig. 11).

Conclusion

The model presented has given diurnal cycles and spatial distributions of dry deposition velocities in accordance to modeling studies and experimentally determined dry deposition velocities of Fowler and Unsworth (1979), Hicks et al. (1985, 1987), Baldocchi et al. (1987), Baldocchi (1988), Hicks et al. (1988), McMillen (1990). The dry deposition Model DEPOMOD seems therefore to be more adequate for use in regional mesoscale models than commonly used look up tables for r_c or v_D. Current dry deposition models compute deposition fluxes using parameterizations based on a relatively sparse database of SO_2, SO_4^{2-}, HNO_3, NO, NO_2, NO_x and O_3 dry deposition measurements. For gases such as H_2O_2, PAN and other organic species, measurements that allow to study the general deposition behaviour have not yet been performed. For those species further studies of dry deposition are necessary to improve and evaluate the existing parameterizations of trace gas dry deposition in regional mesoscale models.

References

Adrian, G. and Fiedler, F. 1991. Simulation of unstationary wind and temperature fields over complex terrain and comparison with observations. – Beitr. Phys. Atmosph. 64: 27–48.

Baer, M. and Nester, K. 1987. Simulation of concentration, dry deposition and fluxes of SO_2 during TULLA. – Ann. Geophysicae 5B: 497–502.

Baldocchi, D. D. 1988. A multi-layer model for estimating sulfur dioxide deposition to a deciduous oak forest canopy. – Atmos. Environ. 22: 869–884.

– , Hicks, B. B. and Camara, P. 1987. A canopy stomatal resistance model for gaseous deposition to vegetated surfaces. – Atmos. Environ. 21: 91–101.

Blümel, K., Bolle, H.-J., Eckardt, M., Lesch, L. and Tonn, W. 1983–1985. Der Vegetationsindex für Mitteleuropa – Inst. für Meteorologie der Freien Universität Berlin.

Brutsaert, W. P. 1975. The roughness length for water vapor, sensible heat, and other scalars. – J. Atmos. Sci. 32: 2028–2031.

– 1979. Heat and mass transfer to and from surfaces with dense vegetation or similar permeable 'roughness'. – Boundary-Layer Meteorol. 16: 365–388.

Canny, M. C. 1984. Translocation of nutrients and hormones. – In: Wilkens, M. B. (ed.), Advanced plant physiology. Pitman Publishing Inc., Marsfield, MA, pp. 227–296.

Chameides, W. L. C. 1987. Acid dew and the role of chemistry in the dry deposition of reactive gases to wetted surfaces. – J. Geophys. Res. 92: 11895–11908.

Chang, J. S., Brost, R. A,. Isaksen, I. S. A., Madronich, S., Middleton, P., Stockwell, W. R. and Walcek, C. J. 1987. A three-dimensional Eulerian acid deposition model: physical concepts and formulation. – J. of Geophys. Res. 92: 14681–14700.

van Dop, H. 1983. Terrain classification and derived meteorological parameters for interregional transport models. – Atmos. Environ. 17: 1099–1105.

Dorwarth, G. 1986. Numerische Berechnung des Druckwiderstandes typischer Geländeformen. – Wiss. Ber. Inst. für Meteorol. Klimaforschung, Univ. Karlsruhe Nr. 6.

Fiedler, F. 1987. Atmospheric transport of air pollutants in the mesoscale over hilly terrain: a review of the TULLA-experiment. – In: Sandroni, S. (ed.), Regional and long range transport of air pollution. Elsevier Sci. Publ., Amsterdam.

Fowler, D. 1978. Dry deposition of SO_2 on agricultural crop. – Atmos. Environ. 12: 369–373.

– and Unsworth, M. H. 1979. Turbulent transfer of sulfur dioxide to a wheat crop. – Quart. J. Roy. Met. Soc. 105: 767–783.

– and Cape, J. N. 1982. Air pollutants in agriculture and horticulture. – In: Unsworth, M. H. and Ormrod, D. P. (eds), Effects of gaseous air pollution in agriculture and horticulture. Butterworth Sci., London, pp. 3–26.

Garratt, J. R. and Hicks, B. B. 1973. Momentum, heat and water vapor transfer to and from natural and artificial surfaces. – Quart. J. Royal Meteorol. Soc. 99: 680–687.

Garland, J. A. and Penkett, J. S. 1976. Absorption of peroxyacetylnitrate and ozone by natural surfaces. – Atmos. Environ. 10: 1127–1131.

Heikes, B. G., Delany, A., Lazrus, A. L. and Penkett, S. A. 1986. Measured surface flux of H_2O_2 vapor to wheat. – EOS Trans. Am. Geophys. Union 67: 885.

Hicks, B. B., Wesely, M. L., Coulter, R. L., Hart, R. L., Dunham, J. L., Speer, R. E. and Stedman, D. H. 1982. An experimental study of sulfur deposition to grassland. – In: Pruppacher, H. R., Semonin, R. G. and Slinn, W. G. N. (eds), Precipitation scavenging dry deposition and resuspension. Elsevier, New York, Vol. 2, pp. 933–942.

– , Baldocchi, D. D., Hosker, R. P., Hutchison, B. A., Matt, D. R., McMillen, R. T. and Satterfield, L. C. 1985. On the

use of monitored air concentrations to infer dry deposition. – NOAA Tech. Memo. ERL ARL-141, p. 65.
– , Baldocchi, D. D., Meyers, T. P., Hosker, R. P. and Matt, D. R. 1987. Multiple resistance routine for deriving dry deposition velocities from measurable quantities. – Water, air and soil pollution, 36, pp. 311–330.
– and Meyers, T. P. 1988. Measuring and modelling dry deposition in mountainous areas. – In: Unsworth, M. H. and Fowler, D. (eds), Acid deposition at high elevation sites. Kluwer Academic Publ., pp. 541–552.
Hoefken, K. D., Meixner, F., Mueller, K. P. and Ehhalt, D. H. 1986. Investigation on dry deposition and emission of atmospheric NO, NO_2 and HNO_3 and from natural surfaces (Thesis). – JUEL-2054.
Jacob, D. J. 1986. Chemistry of OH in remote clouds and its role in the production of formic acid and peroxymonosulfate. – J. Geophys. Res. 91: 9807–9826.
Jarvis, P. G. 1976. The interpretation of the variations in leaf water potential and stomatal conductance found in canopies in the field. – Phil. Trans. Royal Soc., London B 273: 593–610.
Judeikis, H. S. and Wren, A. G. 1978. Laboratory measurements of SO_2 and NO_2 depositions onto soil and cement surfaces. – Atmos. Environ. 12: 2315–2319.
Körner, C., Scheel, J. A. and Bauer, H. 1979. Maximum leaf diffusive conductance in vascular plants. – Photosynthetica 13: 45–82.
Landolt-Börnstein. 1969. Zahlenwerte und Funktionen aus Physik – Chemie – Astronomie – Geophysik und Technik. II. Band: Eigenschaften der Materie in ihren Aggregatszuständen. 5. Teil: Transportphänomene – Kinetik – Homogene Gasgleichgewichte. – Springer, Heidelberg, 6. Auflage.
Lebury, W. and Blair, E. W. 1925. The partial formaldehyde vapour pressure of aqueous solutions of formaldehyde. II. – J. Chem. Soc.: 2832–2839.
Lee, Y.-N., Senum, G. I. and Gaffney, J. S. 1983. Peroxyacetyl nitrate (PAN) stability, solubility, and reactivity. – Implications for tropospheric nitrogen cycles and precipitation chemistry. – Fifth international conference of the commission on atmospheric chemistry and global pollution; symposium on tropospheric chemistry, Oxford, England, Aug 28 – Sept 2, 1983.
Lind, J. A. and Kok, L. G. 1986. Henry's law determinations for aqueous solutions of hydrogen peroxide, methylhydroperoxide and peroxyacetic acid. – J. Geophys. Res. 91: 7889–7895.
Liss, P. S. and Slater, P. G. 1974. Flux of gases across the air-sea interface. – Nature (Lond.) 247: 181–184.
Loomis, A. G. 1928. Solubilities of gases in water. – In: International critical tables. McGraw-Hill, New York, Vol. III, pp. 255–261.
Macckay, D. and Shiu, W. Y. 1981. A critical review of Henry's law constants for chemicals of environmental interest. – J. Phys. Chem. Ref. Data, pp. 1175–1199.
McMillen, R. T. 1990. Estimating the spatial variability of trace gas dry deposition velocities. – NOAA Techn. Memorandum ERL ARL-181.
Monteith, J. L. C. 1973. Principles of environmental physics. – Am. Elsevier Publ. Co., Inc., New York, pp. 200–212.
O'Dell, R. A., Taheni, M. and Kabul, R. L. 1977. A model for uptake of pollutants by vegetation. – J. Air Pollut. Control Assoc. 27: 1104–1109.
Payrissat, M. and Beilke, S. 1975. Laboratory measurements of the uptake of sulphur dioxide by different european soils.– Atmos. Environ. 9: 211–217.
Reynolds, S. D., Roth, P. M. and Seinfeld, J. H. 1973. Mathematical modeling of photochemical air pollution. I. Formulation of the model. – Atmos. Environ. 7: 1033–1061.
Snider, J. R. and Dawson, G. A. 1985. Tropospheric light alcohols, carbonyls and acetonitrile: concentrations in the southwestern United States and Henry's law data. – J. Geophys. Res.: 3797–3805.
Spedding, D. J. 1969. Uptake of sulphur dioxide by barley leaves at low sulphur dioxide concentrations. – Nature, 224: 1229–1230.
Stockwell, W. R. 1986. A homogeneous gas phase mechanism for use in a regional acid deposition model. – Atmos. Environ. 20: 1615–1632.
Taylor, G. E., Mc Laughlin, S. B., Shriner, D. S. and Selvidge, W. J. 1983. The flux of sulfur containing gases to vegetation. – Atmos. Environ. 17: 789–796.
– , Hanson, P. J. and Baldocchi, D. D. 1987. Pollutant deposition to individual leaves and plant canopies: Sites of regulation and relationship to injury. – In: Heck, W.W., Taylor, O. C. and Tingly, D. T. (eds), Assessment of crop loss from air pollutants. Elsevier, London and New York.
Turner, N. C., Rich, S. and Waggoner, P. E. 1973. Removal of ozone by soil. – J. Environ. Qual. 2: 259–264.
Walcek, C. J. 1987. A theoretical estimate of O_3 and H_2O_2 dry deposition over the northeast united states. – Atmos. Environ. 21: 2649–2659.
Weast, R. C. (ed.). 1978. Handbook of chemistry and physics, 59th edition – Chem. Rubber Co., Cleveland OH.
Wesely, M. L. 1988. Improved parameterizations for surface resistance to gaseous dry deposition in regional-scale, numerical models. – EPA/600/3-88/025, U.S. Environ. Protection Agency (available as PB88-225099 from NTIS).
– and Hicks, B. B. 1977. Some factors that effect the deposition rates of sulfur dioxide and similar gases on vegetation. – J. Air Pollut. Control Ass. 27: 1110–1116.
– , Eastman, J. A., Cook, D. R. and Hicks, B. B. 1978. Daytime variations of ozone eddy fluxes to maize. – Bound. Layer Meteorol. 15: 361–373.
– , Cook, D. R. and Williams, R. M. 1981. Field measurement of small ozone fluxes to snow, bare wet soil and lake water. – Boundary-Layer Meteorol. 20: 459–471.
Winner, W. E. and Mooney, H. A. 1980a. Ecology of SO_2 resistances: I. Effects of fumigations on gas exchange of deciduous and evergreen shrubs. – Oecologia 44: 290–295.
– and Mooney, H. A. 1980b. Ecology of SO_2 resistances: II. Photosynthetic change of shrubs in relation to SO_2 absorption and stomatal behavior. – Oecologia 44: 296–302.

Ecological Bulletins 42: 53–61. Copenhagen 1992

The dynamics of gas exchange between soil and atmosphere in relation to plant–microbe interactions: fluxes measuring and modelling

N. S. Panikov and A. J. Gorbenko

Panikov, N. S. and Gorbenko, A. J. 1992. The dynamics of gas exchange between soil and atmosphere in relation to plant–microbe interactions: fluxes measuring and modelling. – Ecol. Bull. (Copenhagen) 42: 53–61.

Dynamic observations were carried out on arable grey forest soil under barley. Fifteen parameters were determined continuously for 44 days: gas composition of soil air with membrane probes, plant photosynthetic activity and dark respiration separately for soil and plants by the chamber method, microbial biomass by kinetic method, number of protozoa by direct microscopy, standing crop of the above- and belowground phytomasses, content of soluble organic matter in soil, moisture and temperature of soil, insolation and precipitation. All dynamic variables, which are related to gas exchange and microbial activity, were found to oscillate with the period of 2–5 days. This feature of dynamic pattern was reproduced in laboratory soil-plant microcosm. Statistical analysis of time-series and deterministic mathematical model were used to explain the observed dynamics. Both approaches allowed to hypothesize that dynamic pattern of gas exchange was controlled by some components of sun radiation via plant photosynthetic activity.

N. S. Panikov, Inst. of Microbiol., Russia Acad. of Sci., Pr. 60-Letija Octjabrja 7, Moscow 117811, Russia. – A. J. Gorbenko, Inst. for the Biol. of Inland Waters, Acad. of Sci., Borok, Jaroslavl region, Russia.

Introduction

The aim of the present investigation was to explain the dynamic pattern of gas exchange between the soil and atmosphere as a result of trophic interactions between the members of ecosystem community. The solving of this problem seems to be very important for the entire IGBP, as it gives clues for understanding, describing and predicting the biogeochemical processes that regulate the total Earth system. The main attention in the present work was paid to studying the CO_2 fluxes as a test case for integrated modelling (Schimel et al. 1992). In spite of the tremendous number of investigations in this field, the quantitative aspects of the CO_2 exchange between the soil and atmosphere remain to be poorly understood. Usually the CO_2 evolution from the soil is used as an index of soil microbiological activity. There have been fewer attempts to integrate the CO_2 fluxes into the context of overall ecosystem nutrient cycling models.

In order to understand the driving forces of gas exchange between the soil and atmosphere we have attempted to measure all the main components of the C balance in one particular terrestrial ecosystem. The measurements of key processes include the rates of plant photosynthesis and dark respiration, root deposition (exudation and root sloughing), microbial growth, and grazing of microbes by Protozoa, which are the most active predators. The dynamics of these variables are poorly understood; for instance the observed short-term fluctuations of microbial (mainly bacterial) biomass were attributed to the action of autoinhibitors such as periodin (Khudyakov 1972), ethylene (Smith 1976), or ethylene oxide (Nikitin and Nikitina 1978). Other authors suggested prey-predator interactions (Kutuzova 1982) or environmental factors such as precipitation and nutrient supply (Clarholm and Rosswall 1980) as the main controlling mechanism. Moreover some authors doubt that there are any significant fluctuations of soil microbial biomass, since observed generation rates could not be supported by available organic matter supply to soil (Hisset and Gray 1976). To resolve the contradictions we relied on two methodological principles: 1) the use of unambiguous and preferably instrumental methods for measurements of dynamic variables (the rates of CO_2 exchange, C content in soil, biomass by kinetic methods etc.) and 2) the application of a system approach, which is based on the combination of laboratory experiments, field observation and mathematical modelling.

Methods

The study was undertaken on the field experimental station of the Inst. of Soil Science and Photosynthesis USSR Academy of Sciences, Pushchino, Moscow region (55°32′N, 37°39′E). The field experiment was designed to test different soil management systems (various mechanical treatments or fertilizers), but we made measurements only on the 20×20 m^2 control plot (without fertilization). The soil was tilled in early spring by a rotary cultivator to the depth of 10 cm, seeded with barley (cultivar Trumph) on 27 April 1985 and harvested on 5 Aug 1985. The soil was arable grey forest silty clay loam, C_{org} 0.92%, N_{org} 0.11%, pH_{KCl} 5.5.

The CO_2 fluxes were determined using static chambers, which were based on a design by Conrad et al. (1983). The chamber consisted of two parts: a permanent square 40×40 cm stainless steel base, which was inserted 10 cm into the soil just after spring seeding, and a removable Plexiglas cover of 30 to 50 cm height. The chamber top was connected with the base via water seal only during measuring (usually for 20–30 min), the rest of the time the chambers were open to air. Plant growth was not significantly affected by this manipulation and was practically the same inside and outside of the chamber.

The CO_2 fluxes were measured under alternately light and dark conditions. Immediately after the mounting of the Plexiglas section it was covered by nontransparent screen (aluminum foil), and the samples of the inside air were taken by syringes 3 times every 5–10 min. The dark respiration of soil-plant community (DRSPC) was calculated from increase of CO_2 concentration (dp/dt) with the correction for accumulating gas escape. The correction was made by fitting the experimental points to the following equation.

$$dp/dt = DRSPC/h - (\Psi/h)\,(p-p_o)/z \qquad (1)$$

or after integration

$$p = p_o + DRSPC[1-\exp(-\Psi' t)]/(h'\Psi'), \qquad (2)$$

where t is the time, h is the height of the above-ground part of the chamber, z is the depth of below-ground part of the chamber, p and p_o are the CO_2 concentration in air within and outside of the chamber respectively, is CO_2 diffusion coefficient in soil, Ψ' is the specific CO_2 permeability of the inserted chamber ($\Psi' = \Psi/hz$).

The temperature inside the chamber was usually increased during measurement. However, due to light reflection from aluminium foil this increase did not exceed 1–2°C and was neglected in future considerations.

After removal of aluminum foil (i.e. under light conditions) we measured the plant photosynthtic activity (PPA). Air inside of the chamber was mixed by a small fan (power supply from 4.5 V electrical cell); air samples were taken every 1 min, the actual values of PPA were calculated from observed decrease of CO_2 concentration (–dp/dt), the correction being made for elevated substrate concentration according to Michaelis-Menten equation:

$$PPA = (-dp/dt)p_o/(K_p + p_o), \qquad (3)$$

where K_p is the saturation constant and p_o is an ambient CO_2 concentration.

In spite of air mixing there was a significant temperature increase inside the chamber depending on exposure time and insolation intensity (5–8°C during 10 min incubation). The other problem was water condensation on the chamber inner wall which decreased the actual illumination. To minimize these chamber artifacts we confined the measurement time by 3–5 min and extrapolated PPA-rates to zero incubation time. The described static chamber method was compared on several occasions with dynamic leaf cuvettes method, adapted for field conditions (Karpushkin 1971); the PPA was measured from sunrise till sunset and expressed per unit of leaves surface.

Soil respiration (SR) was measured by small stainless steel chambers (8 cm ∅ and 11 cm in height), which were inserted into the soil between plant rows. SR was assumed to be the difference between DRSPC and dark respiration of shoots, however, this assumption can be questioned when inter-row and intra-row soil properties differ significantly.

The CO_2 concentration in soil air as well as other soil gases were determined by means of membrane probe technique (Kasparov et al. 1985). The membrane probe is a hollow perforated stainless steel tube covered by polythene membrane, which is placed horizontally in the soil layer. PVC tubing make a connection between the membrane probe and a sampling port above the soil surface. The membrane is permeable to most soil gases and excludes only water vapors and heavy organic molecules. The equilibrium in the system is generally established within one day.

The concentrations of soil gases and CO_2 in the air were determined by GC method using LHM–8MD (USSR) and Chrom–5 (Czechoslovakia) models of gas chromatographs, equipped by thermoconductivity (for CO_2 and O_2) and flame ionization detectors (for CH_4

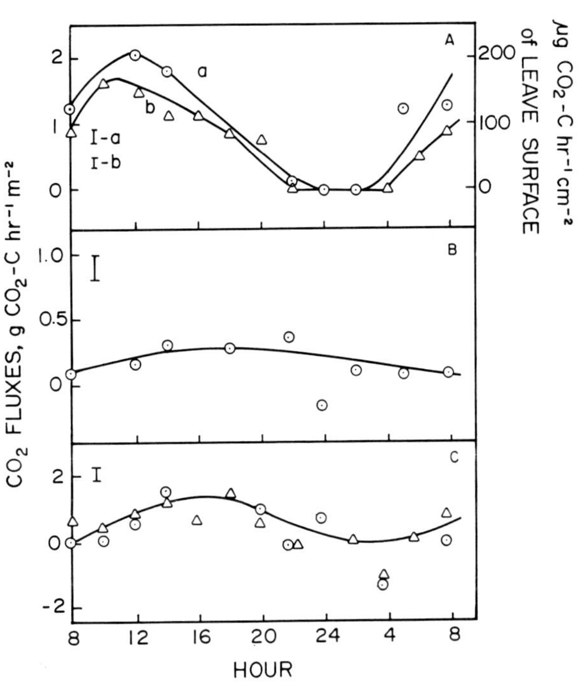

Fig. 1. Diurnal cycles of CO_2 exchange between the soil and atmosphere: A) plant photosynthetic activity estimated by static chamber (a) and leaf-cuvette (b) methods; B) soil respiration; C) dark respiration of soil-plant community. The bars represent 95% confidence limits.

and other volatile organic substances). There were very low concentrations of organic compounds in the air of the aerobic soil. Gas samples were delivered to the laboratory in glass bottles under salt water locks and analysed within 2–3 h.

Shoots and roots phytomasses were determined gravimetrically after drying at 70°C. The roots were washed from soil particles on 0.25 mm screen. Before drying, the dark respiration of shoots and roots (DRS and DRR respectively) were estimated by GC as CO_2 evolution rate during plant material incubation at 25°C in 500 cc glass bottles. The obtained rates were then corrected for temperature difference between in situ and in vitro conditions according to Arrhenius equation.

The soil microbial biomass were estimated by a kinetic method (Mirchink and Panikov 1985), which is a refined version of a known physiological method (Anderson and Domsch 1978). Twenty g of fresh soil were placed on the filter paper bed inside of a serum bottle. The soil moisture content was adjusted to 24 wt.% or 60% of WHC using a 1% glucose solution. The substrate induced soil respiration (SIR) was measured by GC as CO_2 evolution rate during 1 h soil incubation at 25°C. Microbial biomass x was calculated as

$$x = SIR/k, \qquad (4)$$

where conversion factor k is the average specific respiration rate of soil microorganisms. It is known (Pirt 1975), that k is related to specific growth rate μ: $k = \mu/Y_{x/p}$, where $Y_{x/p}$ is stoichiometric coefficient equal to the ratio of biomass carbon produced to substrate carbon respired. The direct determination of k-value was as follows: soil was incubated with glucose longer than for usual SIR estimation (8–10 h) to provoke the beginning of microbial growth in situ, μ was found from curvature of SIR dynamics (μ=d(1n SIR)/dt), and $Y_{x/p}$ – from carbon balance of microbial growth ($Y_{x/p}$= (glucose uptake – CO_2 evolution)/CO_2 evolution). On average, k-value for studied grey forest soil was 0.055 μg CO_2-C h^{-1}(μg biomass C$)^{-1}$.

Soluble organic matter in the soil (SOM) were determined by wet combustion in 0.06 M K_2SO_4 extract (1:2 soil to solution ratio). Dichromate oxidation was used as described by Harris and Adams (1979) with the modification of the final step: chemical consumption of $Cr_2O_7^{2-}$ was estimated by absorbency at 350 nm (Panikov et al. 1987).

Populations of protozoans were enumerated by direct count in stained soil smears (Dogel et al. 1927). Soil moisture was determined gravimetrically. Soil temperature was measured at 2 cm depth with thermistors. Data for solar insolation and precipitation were supplied by the meteorological station of ISSP AN USSR for the experimental field.

All determinations, apart from gas exchange rates were made on the composite samples of fresh soil. Five random cores (5 cm ∅) were collected from depth 0–10 cm and mixed before analysis.

Microcosm experiments were carried out in growth chambers with the same soil and plants as in the field studies. 300 g of soil was placed in a beaker and packed to a density of 1.2 g cm^{-3}. Barley seeds were sown and soil was incubated at 25°C under constant illumination 10,000 lux (light period 14 h). Moisture content was adjusted to 50% of WHC and was controlled by daily weighing. All the dynamic variables were determined by the methods described above. CO_2 exchange rates were measured not only in static chambers, but also in continuous air-flow systems with CO_2 determination by infrared gas analyser (GIAM-5, USSR, ventilation rate 0.5–5.0 1 min^{-1}).

Results and discussion

Variations in CO_2 fluxes were measured over a 24 h period on several occasions (Fig. 1). The two used methods (static chamber and dynamic leaf cuvettes) gave similar dynamic patterns of the diurnal cycle of PPA with maximum at 10–12 am. Maximal respiration (DRSPC and SR) were shifted as compared with PPA to the later part of the light period (16–20 h). During the night the evolution of CO_2 from soil sometimes became negative probably because of CO_2 absorption in dew

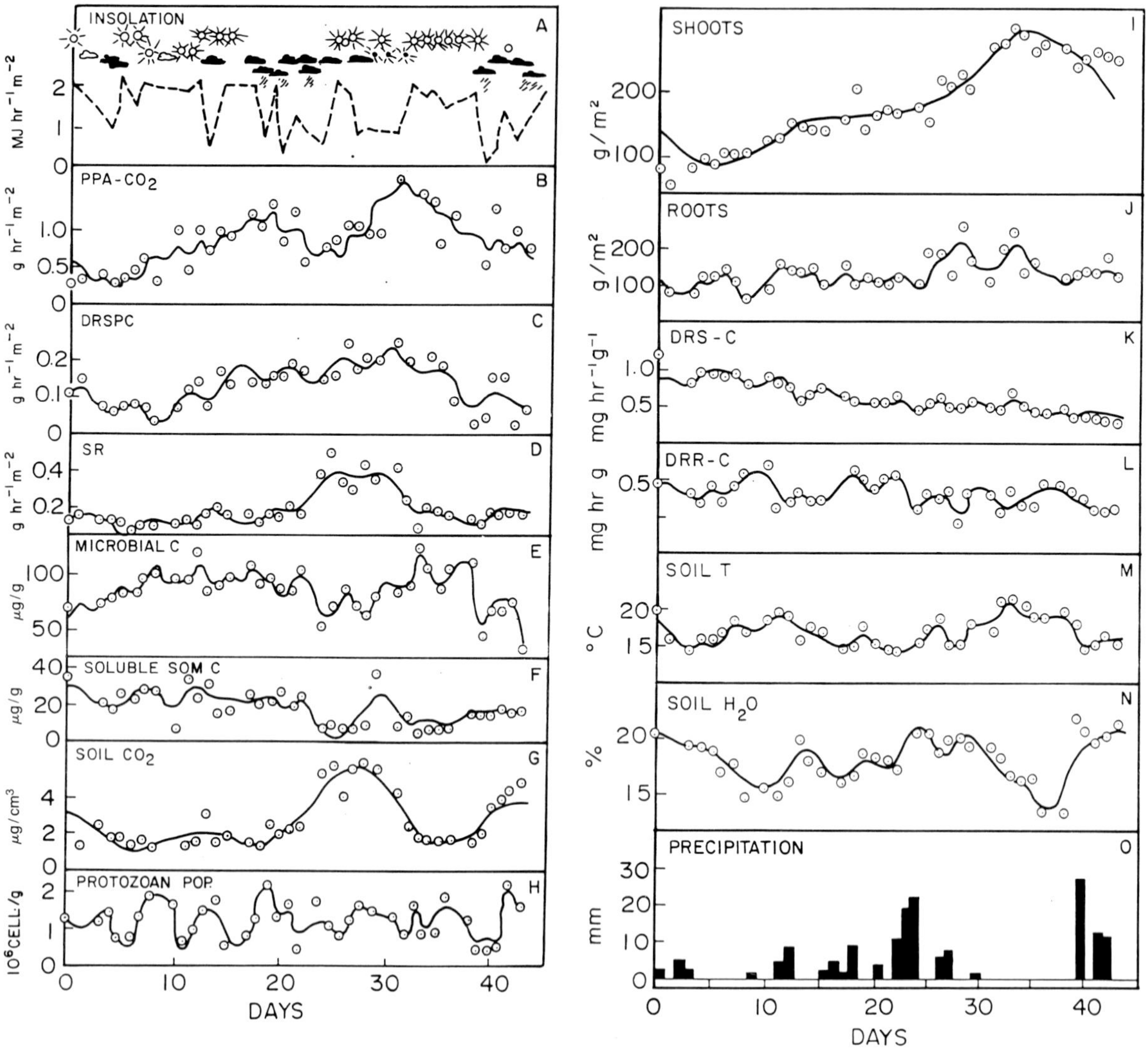

Fig. 2. Seasonal dynamics of CO_2 exchange rates and related variables of arable grey forest soil under barley: A. – insolation (MJ h^{-1} m^{-2}); B. – plant photosynthetic activity (PPA, g CO_2-C h^{-1} m^{-2}); C. – dark respiration of soil-plant community (DRSPC, g CO_2-C h^{-1} m^{-2}); D. – soil respiration (SR, g CO_2-C h^{-1} m^{-2}); E. – microbial biomass in soil (µg C (g of soil)$^{-1}$); F. – soluble organic matter (SOM, µg C (g of soil)$^{-1}$); G. – CO_2 in soil air at depth of 10 cm (µg C cm^{-3}); H. – protozoan population (10^6 cell (g of soil)$^{-1}$); I. – shoots phytomass (g dr.wt. m^{-2}); J. – roots phytomass(g dr.wt. m^{-2}); K. – dark respiration of shoots (DRS, mg CO_2-C h^{-1}(g of phytomass)$^{-1}$); L. – dark respiration of roots (DRR, mg CO_2-C h^{-1}(g of phytomass)$^{-1}$); M. – soil temperature at depth of 2 cm (°C); N. – moisture (%); O. – precipitation (mm). The curves are calculated according to Eq. (1) taking into account only statistically valid (on the level 0.05) periodic components.

water. Other variables, e.g. microbial biomass, protozoan population, and phytomasses were characterized by nonsignificant fluctuations during the one day span.

Bearing in mind the existence of diurnal cycles, all seasonal measurements were made at a fixed time of the day, 9.00 to 9.45 am. Results are shown in Fig. 2 as a time-series for 15 variables. All variables, except phytomass, exhibited frequent fluctuations with rises and falls every 3–6 d. At the same time the general trends were observed for PPA, DRSPC and SR to increase parallel to plant growth.

In order to statistically analyse this complicated pattern, we used time-series analysis (Box and Jenkins 1970, Gorbenko and Kryshev 1985). The contributions of periodic components, each from 15 time-series were approximated by the equation

$$z(t) = \bar{z} + \sum_{k=1}^{n} R_k \cos 2\pi f_k (t - \varphi_k) + l_t \qquad (5)$$

Table 1. Correlation and cluster analysis of field data.

Group $r_{xy}/_{\tau}$*	Parameter	1	2	3	4
1	1. Soil respiration	x			
	2. CO_2 in soil air	0.83/0	x		
	3. Dark respiration of soil-plant community	0.65/–3	0.58/–3	x	
	4. Plant photosynthetic activity	0.55/–2	0.49/–3	0.60/0	x
2	1. Shoots phytomass	x			
	2. Dark respiration of shoots	–8.83/0	x		
	3. Soluble organic matter	–0.62/0	0.54/4/	x	
	4. Roots phytomass	0.52/1/	–0.42/–1	–0.55/–2	x
3	1. Microbial biomass	x			
	2. Moisture	–0.83/0x			
	3. Temperature	0.51/0	–0.54/0	x	

* Nominator is the cross-correlation coefficent r_{xy}, denominator is the time delay.

where $\bar{z}$ is the mean, 1_t is so called white-noise, and R_k, φ_k and f_k are respectively amplitude, phase and frequency of a harmonic with index k. Fitting of data to Eq. (5) was made by the nonlinear least-square procedure, which was realized on computer BESM–6.

Calculated curves are plotted in Fig. 2. Statistically unreliable (on the level 0.05) variations were discarded according to the Fisher criterion. Equation (5) explained 70–85% of the total variance, the rest being attributed to noise. All variables could be divided in respect to harmonic composition into two groups. One group is characterized by predominance of low-frequency component (periods are equal to 22 and 44 d), variables of the second group are dominated by high-frequency components (with periods from 2 to 14.7 d). To the first group belong all variables, which are related to plant growth and CO_2-exchange: PPA, DRSPC, SR, DRS and DRR, CO_2 in soil air and phytomasses. The second group is compiled by microbial biomass, protozoan population and SOM.

In the next step of statistical analysis we calculated the cross correlation functions r_{xy} (τ) for all possible paired combinations of 15 time-series with different time delays between them:

$$r_{xy}(\tau) = \frac{c_{xy}(\tau)}{S_x S_y} \tag{6}$$

$$c_{xy}(\tau) = (1/N) \sum_{t=1}^{N-\tau} [(x(t) - \bar{x})\ (y(t+\tau) - \bar{y})]$$

where x and y are means, x(t) and y(t+τ) are the values of parameters x at moments t and y at moment t+τ respectively. The delay or lag time τ is used as a measure of inertia, which is very characteristic for natural systems.

The summary of the cross-correlation and following cluster analysis of field data is presented in Table 1. The first cluster was formed by parameters PPA, DRSPC, SR and CO_2 in soil air. PPA and DRSPC were correlated with each other withoud time delay. However both these variables forestalled by 2–3 d SR and CO_2 concentration in soil air (the negative values of τ for cross-correlations between PPA and SR, PPA and CO_2, as well as between DRSPC and SR, DRSPC and CO_2, see Table 1). So we can propose the following mechanistic explanation of this observation: an increase of PPA is accompanied by concomitant increase of root exudation and (or) root deposition, microbial growth on the excreted organic matter gives rise to elevated CO_2 production from soil. Time delay in this case corresponds to residence time of CO_2 in the plant-soil system. It is worthwhile to notice, that radiolabel $^{14}CO_2$ gives similar estimate of turnover time, c. 40 h (Warembourg 1975).

The second cluster was formed by parameters which characterized the physiological state and biomass of the plant. The negative correlation of dark respiration of plant with phytomass is the result of plant aging with concomitant decrease of metabolically active components. The reverse relation between phytomass and SOM gives the evidence for decrease of root exudation in late stages of plant development. The analysis indicates the linkage between SOM and plant-derived parameters, i.e. SOM depends strongly on physiological state of plants.

The third cluster was formed by microbial biomass, temperature and moisture of soil. The negative correlation between temperature and moisture is expected due to increased evaporation under warm conditions. The reverse relation between microbial biomass and soil moisture content could be tentatively explained by higher grazing activity of protozoa under moist conditions.

It is interesting that time-series analysis did not show

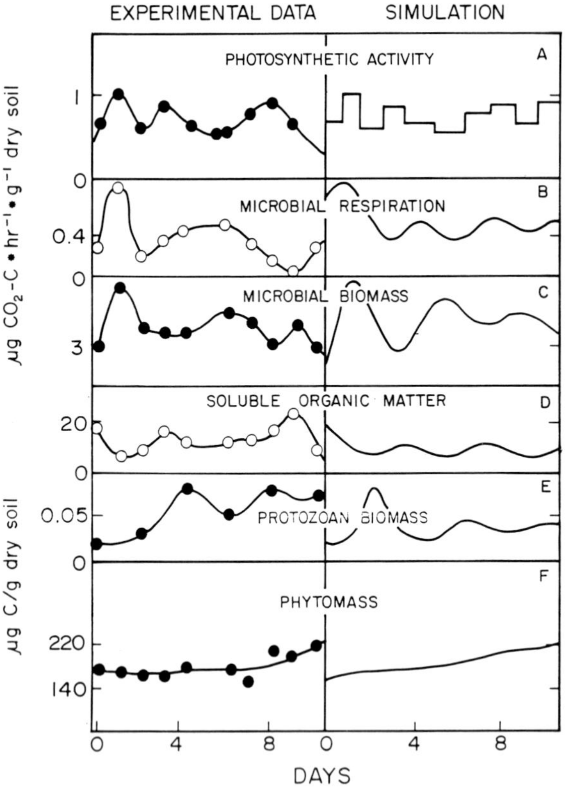

Fig. 3. Experimental modelling and simulation of CO_2 exchange rates and related variables in laboratory soil-plant microcosms.

any significant cross-correlation between microbial biomass and SR (in spite of the fact that c. 80% of the total SR was accounted for by microbial respiration). Contrary to dynamic data, single-point determinations made from numerous soil samples collections indicate a high degree of correlation between microbial biomass and SR (e.g. Ross et al. 1980).

This apparent contradiction can be partially resolved by focusing on the specific metabolic activity of the microorganisms, which is a function of environment factors (first of all of the supply of available organic substrates). In the case of single-point determinations of CO_2 evolution from soils with wide range of biological properties the correlation between biomass and SR is derived from the interconnection between microbial population and the content of soil organic matter, the latter being the primary factor controlling SR.

In this study, a narrow range of microbial biomass variation is exhibited (30–130 μg C g^{-1} soil), the main factor of variability of SR being the specific metabolic activity of soil microorganisms. The substrate-activity relationship is nonlinear (see below Eq. 7), so the relationship between biomass and SR could not be revealed by correlation analysis. This result can be deduced from observations that this soil has a limited "rhizosphere effect" on microbial biomass; biomass was estimated on one occasion to be 106 and 89 μg C g^{-1} of bare and rooted soil respectively. At the same time SR was significantly larger in soil under a plant – 144 mg C-CO_2 m^{-2} h^{-1} as compared to 13 mg C-CO_2 m^{-2} h^{-1} of bare soil.

Now we can outline the main results of time-series analysis of the field data. All indexes of CO_2 exchange and microbial activity showed statistically significant high frequency oscillations. Cross-correlation and cluster analysis revealed a strong positive and delayed relationship between PPA, DRSPC and SR. It was tentatively interpreted as a functional connection between soil microorganisms and plant via root deposition, delay being attributed to the time of photoassimilates movement from plant shoots to microbial colonies in the rhizosphere zone.

In order to confirm the proposed mechanism we have attempted to simulate the observed field dynamics in laboratory experiments, run under controlled conditions. In preliminary studies with different microcosms we have found that sustained oscillations of CO_2 exchange could be simulated only in soil with growing plants. In other types of microcosms (e.g. soil amended with glucose, cellulose or plant litter, soil column percolated by nutrient solution, plant litter decomposing by zoo-microbial complexes) we observed only monotonous dynamics of SR or damped oscillations of CO_2 evolution from soil (Panikov 1989).

In soil-plant microcosms even under constant environmental conditions (temperature, moisture, illumination) almost all measured parameters (PPA, DRSPC, SR, microbial biomass, protozoan population and SOM) exhibited fluctuations, which resemble those in the field. Again the only exception were the phytomass and dark respiration of plants (Fig. 3). After cutting the shoots, the soil respiration sharply decreased, damping oscillations being sustained for 2–3 weeks.

To describe the experimental points we used one of the simplest mechanistic mathematical models:

$$\text{Microbial biomass x} \quad \frac{dx}{dt} = \mu x - \eta xy/Y_a, \ \mu = \mu_m s/(K_s+s)$$

$$\text{Protozoan biomass y} \quad \frac{dy}{dt} = \eta xy - \lambda y \qquad (7)$$

$$\text{SOM concentration s} \quad \frac{ds}{dt} = kF - \mu x/Y$$

where Y and Y_a are the yield factors for growth of microorganisms and protozoa respectively, μ and η are the specific growth rates of microbes and their predators, λ is the specific death rate of protozoa and kF is SOM exudation rate (F is gross PPA, and k is a proportionality coefficient).

Verbal translation of the mechanism described by Eq. (7) includes the following points: 1) microbial growth is supported by the consumption of root exudation prod-

Table 2. The carbon budget of the arable grey forest soil under barley.

C contribution	C-flux $g\ C\ m^{-2}$ $(44\ d)^{-1}$
CO_2 photoassimilation	+ 374
Plant growth	– 110
Plant respiration	– 40
Root deposition	– 224
Soil respiration	– 166
incl. microbial respiration	– 158
animal's respiration	– 8
C incorporation into soil organic matter (by difference)	– 58

ucts, any other organic substances in the rhizosphere, which are available for microbes (e.g. root sloughings, exopolimers) are assumed to be equivalent to SOM, the contribution of soil humus and other indigenous soil organic matter to microbial growth is assumed to be insignificant; 2) root exudation is proportional to gross PPA; 3) elimination of growing microbial cells is accomplished via their grazing by protozoa, the rate of predation is proportional to the probability of contact between prey and predators, predators themselves are subjected to spontaneous accidental death; and 4) microbial growth rate depends on SOM concentration according to Monod equation, maintenance requirements are proportional to the rate of substrate consumption, so the yield factor could be regarded as a constant.

Qualitative stability analysis of Eq. (7) showed that under constant kF autonomous sustained oscillations of x, s and y were permissible only under unrealistic conditions when $kF > \mu_m\lambda/Y\eta$, i.e. when microbial growth is not limited by substrate. Otherwise (when $kF <= \mu_m\lambda/Y\eta$) oscillations are damped and all variables are moving to stable steady-state values $\bar{x}=\lambda/\eta$, $\bar{y}=\bar{Y}_aYkF/\lambda$ and $\bar{s}=Y\eta K_skF/\mu_m\lambda\text{-}Y\eta kF)$. Therefore sustained oscillations in Eq. (7) could be only induced by external excitation sources. The behaviour of Eq. (7) as dependent on input excitation signals (amplitude and frequency of F periodic changes) was studied in numerical experiments. Integration of Eq. (7) was made by 4th order Runge-Kutta method. The main results are summarized as follows: 1) SR and PPA are described by the curves of the same shape, SR-curves being always delayed; 2) time-delays in response to PPA changes increase in the sequence SR = SOM < microbial biomass < protozoan number; 3) amplitude of induced oscillations depends on the amplitude of kF periodic changes; 4) periods of free damped oscillations for SR, x, s and y are reversely related to absolute value of kF; 5) steadystate χ values depend on grazing activity of protozoa and does not depend on kF values. All listed predictions of Eq. (7) are in good qualitative agreement with experimental data. To demonstrate quantitative agreement we made the numerical integration of Eq. (7) with F-values, set from day-average experimental points of PPA. For simplicity diurnal cycles of PPA and SR were neglected. Fig. 3 gives the comparison between experimental curves and simulations for the following set of model coefficients: $\mu_m=0.09$ and $\lambda=0.135\ h^{-1}$; $K_s=13\ \mu g\ C\ g^{-1}$ of soil; $\eta=0.035\ g\ (\mu g\ C)^{-1}\ h^{-1}$; $k=0.5$, $Y=0.2$ and $Y_a=0.047$. (It was assumed that 10^6 protozoa cells equalled 1 µg C). The dynamics of SR served for verification of the model by means of comparison of the observed and calculated values (the experimental SR-curves were not used for parameter identification). Calculations of SR were made according to material balance equation:

$$SR = kF - \frac{dx}{dt} - \frac{ds}{dt} - \frac{dy}{dt} \tag{8}$$

As can be seen from Fig. 3 there is adequate reproduction by the model of amplitudes and frequencies of oscillations, as well as time-delays for all studied variables. Disparity between model and observed results are due to simplifications made that may not adequately take into account input of SOM from hydrolysis of soil humus, elimination of microbes by microfauna and phages, and heterogeneity of state variables. Nevertheless, Eq. (7) appeared to be useful as a first step approximation of C-fluxes in the plant-microbe community; mass balance was maintained and natural patterns were adequately simulated. A particularly interesting result was the relatively high microbial growth rate (generation time g = (ln2)/µ was equal to 20–40 h), in contrast to most previous estimates (Hisset and Gray, 1976). Turnover time for C-CO_2 calculated as sY/µx was found to be in average 36 h. This is close to both Warembourg (1975) and our statistical time-series estimates.

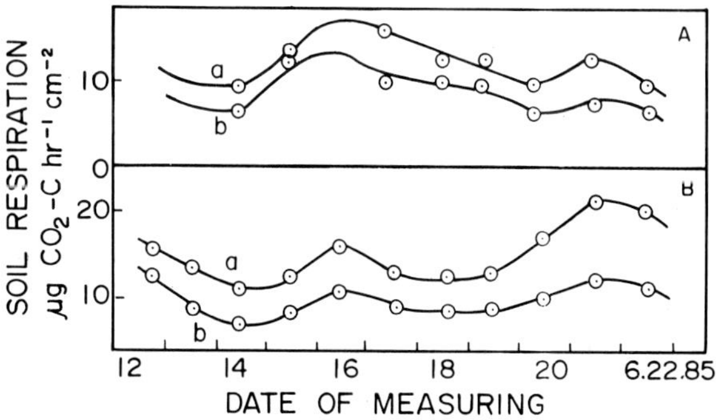

Fig. 4. Evidence for temporal synchronization of CO_2 evolution from soils of different geographical locations. The measurements of SR were made in Moscow, field experimental station of Moscow State University (A) and in Pushchino, field experimental station of ISSP USSR AS 100 km from Moscow (B). Small letters a and b refer to results of SR measurements with two individual chamber 3 m apart.

Identified coefficents Eq. (7) were used for the quantitative interpretation of field data: Y, Y_a and k values were applied to calculate the C-budget of the studied ecosystem for observation time of 44 days (Table 2). There appeared to be a surplus of C photoassimilated, which formed the pool of "soil humus". Therefore during summer arable soil acts as a net CO_2 sink. However, in the winter there should be no CO_2 photoassimilation and soil respiration should remain positive. As average winter SR accounts for $\leq 5\%$ of summer values (unpubl. data), the annual contribution from these soils should be considered to be a net source of CO_2.

The factors which induce PPA fluctuations are not well understood. Firstly, we noted that oscillations of SR (and therefore PPA) take place synchronously in distinct locations (Fig. 4). So oscillations of PPA are probably not free but are induced in different places by the common "master" oscillator. Secondly as was indicated by time-series analysis, no one of the known environmental parameters (moisture, temperature, illumination) has a significant effect on the CO_2-fluxes periodicity (not to be confused with the absolute values of the fluxes). It may be that the regulating nature of the environment factors has not been shown because of some shortcomings of used techniques – "chamber effects", incomplete data on temperature and moisture history, short length of time-series etc. For the moment there is no obvious explanation for this phenomenon.

Conclusions

In conclusion we will make some comments related to the problem of trace gas exchange in a global perspective. One of the most important tasks in this respect was stated to be the mechanistic process studies which interpret temporal and spatial variability in the trace gas fluxes. Our attention have been concentrated on temporal aspects. The mechanistic model was provided which explained main features of observed CO_2 exchange dynamics as a result of C-flows between plants, soil microorganisms and microbovore predators. In addition to well known environment parameters (temperature, humidity, PAR) we have postulated the participation of some cosmic factors (components of sun radiation) in control of the periodicity of PPA and SR. Obviously proposed models and considered situations (short-term dynamics of one particular type of agroecosystem) were too simple to be useful immediately for global scale implementations. For this purposes it is necessary to continue time-series observations several years, to use more sophisticated techniques (micrometeorological instead of chamber methods) and to increase the number of studied core sites. Nevertheless, we hope that even at its present state our study will be useful in general terms, for instance to answer the question, how frequently the trace gas fluxes have to be measured. Dynamic patterns of gas exchange were shown to be extremely complicated, they included the variations that occur on 1) daily 2) weekly (every 5–10 d), and 3) seasonal cycles. Neglecting of any one of these cycles would definitely lead to severe miscalculations in global trace gas budgets. The compromise between the reliability of calculations and the cost of measurements could be met by employing a stratified sampling scheme. For example, 1–2 times per season, field measurements can be carried out during 1–2 weeks every day at a fixed time (say, about the noon); in addition, once during this period, a diurnal cycle of the fluxes can also be monitored. In this case one has the possibility to estimate flux variations at all 3 levels and hence, can make a more reliable interpolation of the measured values across the whole year.

References

Anderson, J. P. E. and Domsch, K. H. 1978. A physiological method for the quantitative measurement of microbial biomass in soils. – Soil Biol. Biochem. 10: 215–221.

Box, G. E. P. and Jenkins, G. M. 1970. Time series analysis: forecasting and control. – Holden-Day, San Francisco.

Clarholm, M. and Rosswall, T. 1980. Biomass and turnover of bacteria in a forest soil and a peat. – Soil Biol. Biochem. 12: 49–57.

Conrad, R., Seiler, W. and Bunse, G. 1983. Factors influencing the loss of fertilizer into atmosphere as N_2O. – J. Geophys. Res. 88: 6709–6718.

Gorbenko, J. A. and Kryshev, I. I. 1985. Statistical analysis of marine microbial ecosystem dynamics. – Naukova dumka, Kiev. (in Russian).

Dogel, V. A., Rammelmaer, E. S. and Strelkov, A. A. 1927. On the method of soil Protozoa observation. – Trudy Otdela celhosmicrobiologii 11: 155–167.

Harris, R. F. and Adams, S. S. 1979. Determination of carbon-bound electron composition of microbial cells and metabolites by dichromate oxidation. – Appl. Env. Microbiol. 37: 237–243.

Hisset, R. and Gray, T. R. G. 1976. Microsites and time changes in soil microbial ecology. – In: The role of terrestrial and aquatic organisms in decomposition processes. Blackwell, Oxford, pp. 23–47.

Karpushkin, L. T. 1971. Application of infrared analyzer for study of CO_2 exchange in plants. In: Biophysical methods in plant physiology. Nauka, Moscow, pp. 44–71, (in Russian).

Kasparov, S. V., Panikov, N. S. and Minko, O. I. 1985. Membrane sampling device for the soil air gas composition analysis. – Pochvovedenije 11: 145–150, (in Russian).

Khudyakov, J. P. 1972. Periodicity in microbial processes in soil and its explanation. – In: Problems of abundance, biomass and productivity of microorganisms in soil. Nauka, Leningrad, pp. 20–36, (in Russian).

Kutuzova, R. S. 1982. On the nature of short-term periodic changes of the quantity of soil bacteria. – Microbiologia 51: 138–142, (in Russian).

Mirchink, T. G. and Panikov, N. S. 1985. Advances in measuring of the biomass and productivity of bacteria and fungi in soil. – Uspechi microbiologii 20: 198–226, (in Russian).

Nikitin, D. I. and Nikitina, E. S. 1978. Processes of the environment self purification and parasitic bacteria. – Nauka, Moscow, (in Russian).

Panikov, N. S. 1989. The kinetics of microbial growth: general principles and ecological implementation. – D.Sc. Thesis, Moscow Univ., Moscow, (in Russian).

– , Gorbenko, A. J., and Svetlov, S. V. 1987. The method of measuring of the soluble organic matter content in soils. – USSR patent N 1318909, G 01 N 33/24, (in Russian).

Pirt, S. J. 1975. Principles of microbe and cell cultivation. – Blackwell Sci. Publ., Oxford.

Ross, D. J., Tate, K. R., Cairns, A. and Pansier, E. A. 1980. Microbial biomass estimated in soils from tussock grassland by three biochemical procedures. – Soil Biol. Biochem. 12: 375–383.

Schimel, J., Robertson, G. P., Baldocchi, D., Bogner, J., Davidson, E., Duyzer, J., Ehhalt, D., Fowler, D., Groffman, P., Haider, K., Isadorov, V., Klemedtsson, L., Melillo, J., Smith, K., Su, W.-H. and Wieprechts, W. 1992. Impacts of trace gas fluxes in mid-latitude ecosystems. – Ecol. Bull. (Copenhagen) 42: 124–132.

Smith, A. M. 1976. Ethylene in soil biology. – Ann. Rev. Phytopathol. 14: 53-73.

Warembourg, F. R. 1975. Application de techniques radioisotopiques á l'etude de l'activité biologique dans la rhizosphere des plantes. – Rev. ecol. et biol. du sol 12: 261–272.

Ecological Bulletins 42: 62–70. Copenhagen 1992

High-latitude ecosystems as CH_4 sources

William S. Reeburgh and Stephen C. Whalen

Reeburgh, W. S. and Whalen, S. C. 1992. High-latitude ecosystems as CH_4 sources. – Ecol. Bull. (Copenhagen) 42: 62–70.

Because of their high stored soil carbon contents, high-latitude soils are expected to play an important role as a source of the radiatively active gases CH_4 and CO_2 under the warmer dryer conditions projected by climate models. Estimates of the present day CH_4 source strength of tundra are uncertain due to a lack of data and coverage, but several independent estimates seem to be converging on a source strength of c. 35 Tg yr^{-1}. Microbial oxidation of CH_4 is an important modulator of tundra fluxes. The process is restricted to the region adjacent the water table in waterlogged soils, but occurs at high rates in moist soils, which equilibrate rapidly with the atmosphere. Methane oxidation is expected to become more important under warmer, dryer conditions, and preliminary work suggests that the process could not only control tundra CH_4 emissions, but also serve as a sink for atmospheric CH_4. The emission ratio of CO_2 to CH_4 will likely increase under these conditions, so the overall greenhouse impact of high latitudes is unknown.

W. S. Reeburgh and S. C. Whalen, Inst. of Marine Sci., Univ. of Alaska, Fairbanks, AK 99775–1080, USA.

Introduction

The recently observed increase in the concentration of atmospheric CH_4 (~1% yr^{-1}) is widely accepted and is supported by evidence from three independent sources: time series measurements at clean air sampling stations from a range of latitudes (Rasmussen and Khalil 1984, Steele et al. 1987, Blake and Rowland 1988), investigations of historical solar infra-red spectra (Rinsland et al. 1985), and polar ice core records (Craig and Chou 1982, Raynaud et at. 1988, Stauffer et al. 1988). This CH_4 increase is of concern because CH_4 is more effective as a greenhouse gas, molecule for molecule, than CO_2 (20–30-fold; Blake and Rowland 1988, 3.7-fold; Lashof and Ahuja 1990), and both have the potential to change global climate through atmospheric warming (Ramanathan et al. 1985). Major terms in the global CH_4 budget remain uncertain (Cicerone and Oremland 1988) and the cause of the increase is not well-understood. High latitude wetlands have occupied a prominent spot in global CH_4 budgets, but have always been an uncertain term (Ehhalt and Schmidt 1978, Cicerone and Oremland 1988, Khalil and Rasmussen 1990). The uncertainty results from three problem areas: a lack of basic flux data; lack of understanding of the processes controlling fluxes; and uncertainties in scaling-up from sites to regional and global scales.

The tundra and taiga or boreal forest biomes cover large areas of the high latitude landscape. Tundra is the treeless arctic and sub-arctic plain populated with dwarf shrubs, graminoids, and cryptogams and is often underlain by permafrost. Boreal forest is a general term describing the mixed deciduous-coniferous forest biome located between the tundra and temperate forest. Taiga is an equivalent term used by Soviet and Alaskan workers. Tundra and taiga environments have several features that suggest their importance may increase with changing global climate. Tundra and peatland environments presently play an important role in the global CH_4 budget, accounting for 10% of the total budget. The interhemispheric gradient and high seasonal amplitudes in atmospheric CH_4 concentration (Rasmussen and Khalil 1981, Steele et al. 1987) show that high latitude natural environments are a major contributor to the seasonal signal in the northern hemisphere. The soil carbon contents of tundra and taiga soils are among the highest on earth (Post et al. 1982), and combined, represent an inventory of 455 Gt (Gorham 1991). Much of this carbon is stored as peat or in permafrost, and is subject to minimal biogeochemical degradation. Model

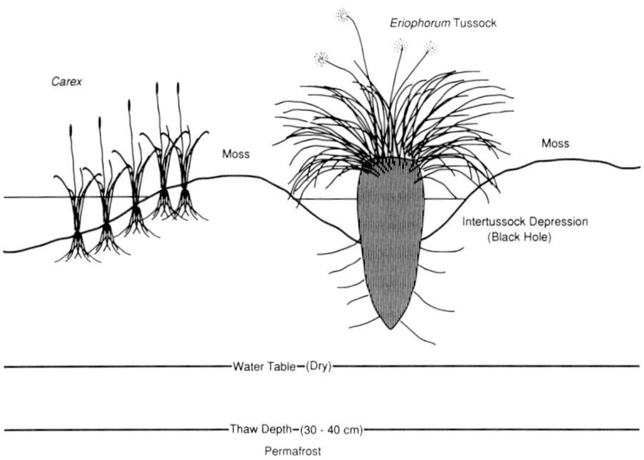

Fig. 1. Schematic diagram of tundra CH_4 flux sites.

projections (Manabe and Bryan 1985, Manabe and Wetherald 1987, Wilson and Mitchell 1987, Schlesenger and Zhao 1989) suggest that high latitude environments could become warmer and dryer. Large-scale shifts in the area of tundra and taiga are possible (Emanuel et al. 1985). These altered conditions are expected to cause changes in permafrost thaw depth and water tables. These changes in turn are expected to mobilize the stored soil carbon, making it available to biogeochemical attack and conversion into radiatively active gases like CO_2 and CH_4. The fate of the stored boreal soil carbon reservoir is a major uncertainty.

High interannual variability in climate is expected for high latitudes (Hansen et al. 1988, Mitchell 1989), so detection of a greenhouse-driven warming trend will be difficult. The best evidence to data for a high latitude temperature increase is temperature distributions in permafrost (Lachenbruch and Marshall 1986). Permafrost serves as an integrator, and recently-measured temperature distributions are consistent with a decadal scale rise in temperature.

High latitude environments are also expected to play an important role in feedbacks. Several models (Hameed and Cess 1983, Guthrie 1986, Khalil and Rasmussen 1989, Lashof 1989) suggest that increasing temperatures will lead to increased CH_4 production and a positive climate feedback. Recent process studies indicate that soil microbial communities may be capable of not only reducing CH_4 fluxes by oxidation in soils, but that high permeability soils equilibrate rapidly with the atmosphere and may be capable of consuming atmospheric CH_4 (Whalen and Reeburgh 1990a).

Although this paper's focus is on CH_4, it should be recognised that CO_2 and CH_4 will both play important roles and should be considered together in future work (Gorham 1991). Carbon dioxide flux measurements are not as straightforward as CH_4 flux measurements, and are complicated by photosynthesis within chambers; photosynthetic CO_2 uptake in chambers must be accounted for by light: dark measurements and estimates of photosynthetic rates to obtain net fluxes. The best information on CO_2 storage and release by tundra comes from budget studies (Chapin et al. 1980, Billings et al. 1982, 1983, Billings 1987), and laboratory (Peterson et al. 1984) and field (Silvola 1986, Gruelke et al. 1990, Oechel and Billings 1992) manipulations. Laboratory studies on peat-filled soil columns by Moore and Knowles (1989) show the relationships between CO_2 and CH_4 emission under varying moisture conditions and water table levels.

This paper reviews our recent measurements of CH_4 fluxes from tundra and taiga environments in Alaska and extends them to global estimates. Processes important in modulating CH_4 fluxes, and their role as potential feedbacks are also considered.

Table 1. Tundra and taiga sites.

Tundra sites
Eriophorum tussocks (1, 2)
Carex (1, 2, 3)
Moss (1, 2)
Intertussock depression (Black Hole) (1, 2)
Hummocks (2)

(UA Arboretum[1], Toolik Lake Long Term Ecological Research site/Kuparuk River[2], Prudhoe Bay[3])

Taiga/ boreal forest sites
(Bonanza Creek Long Term Ecological Research site)

Floodplain sites	*Upland sites*
Sandbar, open willow/poplar	White spruce (S exposure)
Alder/poplar	Aspen (S exposure)
Poplar w/spruce understory	Birch (S exposure)
mature white spruce	Birch/aspen (N exposure)
climax black spruce	Black spruce (N exposure)
	Burn site

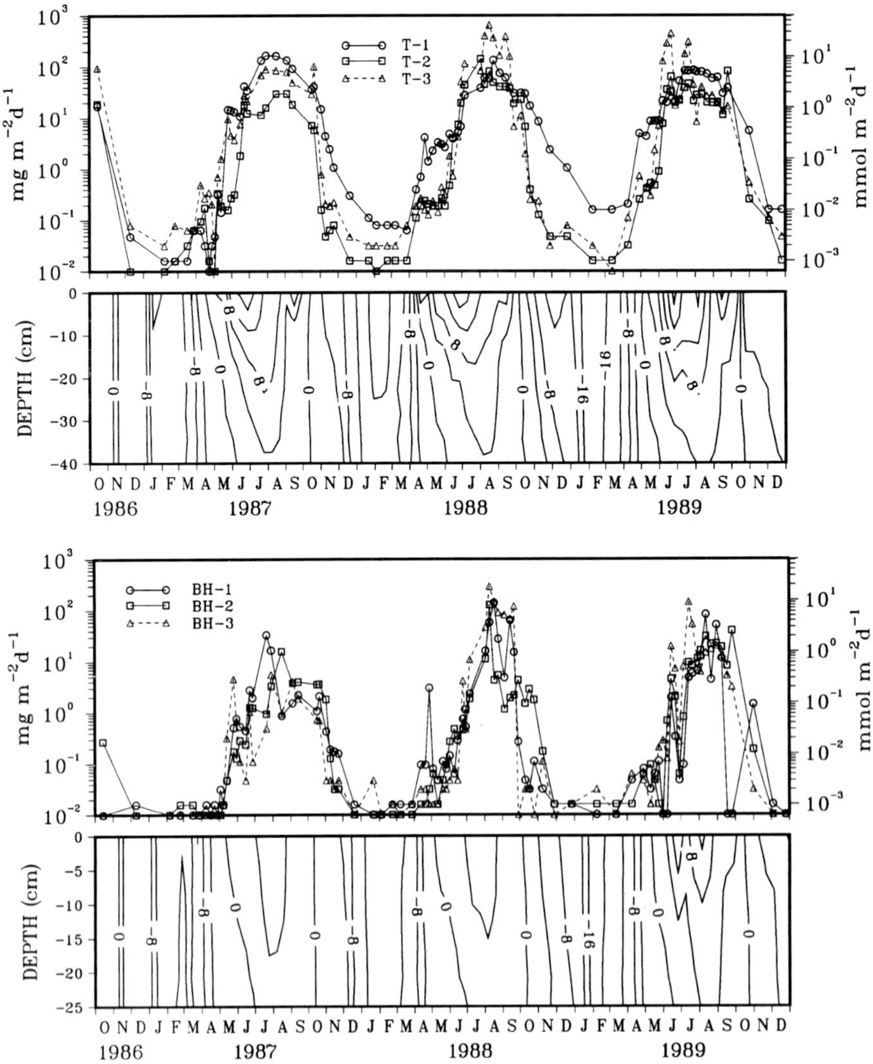

Fig. 2. Methane fluxes from permanent Tussock (T) and Black Hole (BH) sites in Univ. of Alaska Arboretum, 1986–1989. Note that the flux axis is logarithmic. Integrated annual fluxes are given in Table 2. Bottom panels show contours of temperature (°C) as a function of soil depth and date.

Methods and sites

Net CH_4 fluxes are determined using static chambers (Hutchinson and Mosier 1981) incorporating modifications suggested by Conrad et al. (1983). The chambers consist of three parts: a permanent base made from sheet aluminum, a topless plexiglas vertical section that can be stacked to vary volumes, and a plexiglas cover fitted with a septum for sampling. A channel that can be filled with water is located at the top of each base and vertical section, providing a seal between chamber parts. Syringe samples are withdrawn from the chambers at intervals and are analyzed by gas chromatography. Details on the chamber dimensions and operations are covered in Whalen and Reeburgh (1988). We have made a number of consumption rate measurements on soil cores placed in Mason jars. The jar lids are fitted with a septum to permit syringe injections for modification of the jar headspace and for subsequent sampling. The soil cores are retained in plastic coring tubes and the core bottoms are capped; the jar experiments are thus physically equivalent to a chamber flux measurement. The chamber and jar area:volume ratios agree to within 5%, so results are comparable. Relaxation experiments were used to establish equilibration times for added headspace gases (Sparks 1989). Methane oxidizers were inhibited with acetylene, and the

Table 2. CH_4 emission by tundra site types (g m^{-2} yr^{-1}).

	1987	1988	1989
Eriophorum	8.05±2.5	11.38±2.88	8.11±1.80
Carex	4.88±0.73	0.81±1.09	4.27±0.67
Moss	0.47±0.16	4.38±1.35	4.78±1.56
Black Hole	0.62±0.28	3.90±1.09	4.27±0.67

Tab. 3. Global tundra methane emission estimate.

Site	% cover	Annual emission, $Tg\ yr^{-1}$ 1987	1988	1989
Tussock and low shrub tundra ($6.46 \times 10^{12}\ m^2$)				
Eriophorum	24–45	12.4 –23.3	17.7 –33.1	12.6 –23.6
Carex	3–7	1.0 – 2.3	0.2 – 0.4	0.8 – 1.9
Moss	37–63	1.2 – 2.0	10.6 –18.0	1.4 –19.5
Black Hole	30	1.2	7.6	4.0
Total tussock		15.8 –28.8	36.1 –59.1	18.8 –49.0
Wet meadow tundra ($0.884 \times 10^{12}\ m^2$)				
Carex	80–90	3.5 – 3.9	0.57– 0.64	2.9 – 3.3
Moss	10–20	0.03– 0.09	0.4 – 0.8	0.04– 0.1
Total wet meadow		3.5 – 4.0	0.97– 1.44	3.3 – 4.1
Total tundra		19.3 –32.8	37.0 –60.5	22.1 –53.1

Total global methane sources:
400–600 $Tg\ yr^{-1}$ (Cicerone and Oremland 1988)
450 $Tg\ yr^{-1}$ (Fung et al. 1991)

rate of equilibration of added CH_4 was determined by sampling the jar headspace until concentration changes could not be detected. Equilibrium in 10–12 cm core sections was reached within 6 min in most cases. The equlibration occurred as a two-rate process: a fast process presumably involving porous surface material or large voids ($k=-1.54\ min^{-1}$), followed by a slower process ($k=-0.75\ min^{-1}$) involving more compact material or smaller, tributary voids. These equilbration rates are faster than biological consumption rates, so equilibration with the headspace does not limit CH_4 oxidation (Whalen and Reeburgh 1990a).

Tundra sites

Our work has involved measuring a CH_4 flux time series at sites representing important components of Arctic tundra. We have approached the scaling-up problem by collecting two types of flux data: time series measurements at permanent sites near Fairbanks, and also along a north-south transect in Arctic Alaska.

The permanent site studies have considered several general site types (Table 1) at a range of latitudes in Alaska; the tundra sites are described in detail in Whalen and Reeburgh (1988). Fig. 1 is a schematic diagram showing seasonal water table and thaw depth relationships and vegetation types in the tundra sites. The Univ. of Alaska Arboretum sites have been sampled at approximately weekly intervals through the thaw season and at monthly intervals during winter for four years, and the Toolik Lake and Prudhoe Bay sites have been visited about twice a year. We established that winter fluxes do occur, and that they are quantitatively unimportant, so we eliminated winter sampling in 1989–90. Methane flux measurements in the Univ. of Alaska Arboretum continue.

We measured net CH_4 fluxes on a detailed transect along the Trans Alaska Pipeline Haul Road (Whalen and Reeburgh 1990b). This work was undertaken during a three week period in 1987 and involved measuring CH_4 fluxes at sites 10 km apart along a 500 km transect.

Taiga sites

We established CH_4 flux sites within the Bonanza Creek Long Term Ecological Research (L.T.E.R.) site in 1989. The scientific focus within the Bonanza Creek L.T.E.R. is on community succession in floodplain and fire-affected environments (Van Cleve et al. 1986). We established permanent CH_4 flux sites within the L.T.E.R. floral control sites, which represent major stages in community succession and are also listed in Table 1. A great deal of ancillary data (meterological and PAR data, soil temperatures, water table) is collected at each control site and is available to us. The floodplain succession begins with river bars with a sparse willow-balsam poplar cover. This vegetation traps suspended material during high river stages, resulting in better-drained terraces, which support willow and poplar. These species are replaced by balsam poplar, which is in turn replaced by white spruce. The thermal regime changes when the evergreen spruce become dominant; the ground is shaded and an insulating moss community results, so permafrost becomes established. The climax floodplain community is black spruce underlain by permafrost. Erosion by the meandering river is the major disturbance. The upland taiga is a mosaic, and community composition is dependent on drainage and aspect. White spruce is the climax community, and the major disturbance is wildfire, which occurs on a 100 year cycle. The undisturbed floodplain and upland successions require ~250 years (Van Cleve et al. 1991).

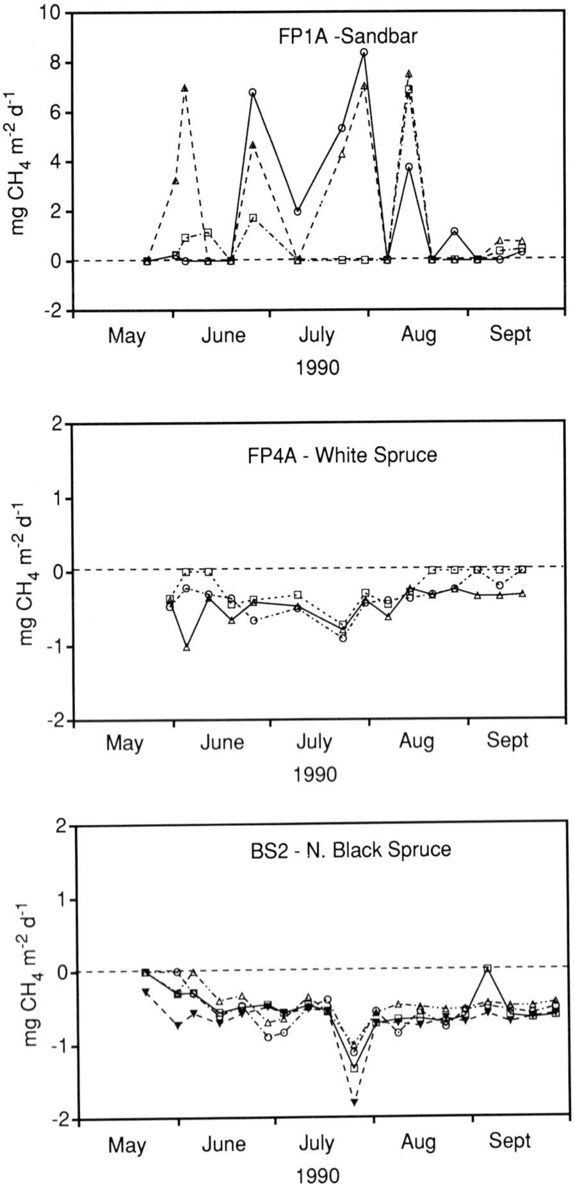

Fig. 3. Methane fluxes from permanent sites in the Bonanza Creek Long Term Ecological Research site, near Fairbanks, Alaska. The Sandbar sites (FP1A) and Alder-Poplar sites (not shown) have small, intermittent CH_4 fluxes and are the only sites with positive fluxes. Other floodplain (FP4A, white spruce) and upland (BS2, N. black spruce) sites have either zero or negative CH_4 fluxes.

Results and discussion

Estimates of present methane fluxes

Tundra

Whalen and Reeburgh (1988) estimated the tundra contribution to the global CH_4 budget based on one year of time-series flux measurements at permanent sites located within the Univ. of Alaska Arboretum. This estimate established the mean annual flux for each site type by integrating the flux measurements over time. Typical CH_4 flux and soil temperature records for the Tussock and Black Hole sites are shown in Fig. 2. Table 2 gives the mean annual CH_4 emission for each of the site types for three years. The flux data from each of three sites was integrated and used to calculate a mean and standard deviation. The low annual fluxes for the 1987 Moss and Black Hole sites, and the 1988 *Carex* site all result from dry conditions and low water tables. Annual precipitation was about half the long-term mean in 1987; the *Carex* sites were located near a lake whose level dropped in 1988. These results show that the vascular components of the tundra community, *Eriophorum* and *Carex*, are the largest contributors to the overall CH_4 emission. These site fluxes were weighted areally using areal coverage estimates (Walker et al. 1987) for two broad types of tundra: tussock-shrub tundra, which is typical of upland areas and has 2–3 *Eriophorum* tussocks m^{-2}, and wet meadow tundra, which is restricted to the coastal plain (c. 10% of tundra area) and is dominated by *Carex* to produce the estimate of global tundra methane emissions given in Table 3.

Taiga

We have completed 1 year of CH_4 flux measurements at the Bonanza Creek sites (Whalen et al. 1991). All of these sites, with the exception of the sandbar and alder-poplar sites, have negative fluxes and consume atmospheric CH_4 at rates of c. 0.2 mg m^{-2} d^{-1}. Typical results from sites in the Bonanza Creek L.T.E.R. are shown in Fig. 3. Jar experiments on core segments show that the CH_4 oxidation rate maximum is located at the mineral soil interface, some 10–25 cm below the surface. We have also found that the fluxes are zero after rainfalls, and become negative when the soils dry. We have made no attempt to extend the Bonanza Creek data set to a global CH_4 emission estimate, but it is clear that large areas of the boreal forest are a CH_4 sink and that point-source bogs and fens are the only important CH_4-emitting sites. Taiga sites in our transect study (Whalen and Reeburgh 1990b), which also included forested bogs, gave a taiga CH_4 emission estimate of 15 (9–25) Tg yr^{-1}.

Global estimates

Matthews and Fung (1987) estimated the importance of wetlands to the global methane budget by applying literature estimates of methane emission (flux×emission period) from a variety of wetland types to a global wetland data base. Only two methane flux data sets for high latitudes were available to Matthews and Fung, results from a seasonal study at the Stordalen (Sweden) IBP site (Svensson and Rosswall 1984) and results from a transect along the Trans-Alaska Pipeline Haul Road (Sebacher et al. 1986). Matthews and Fung estimate the

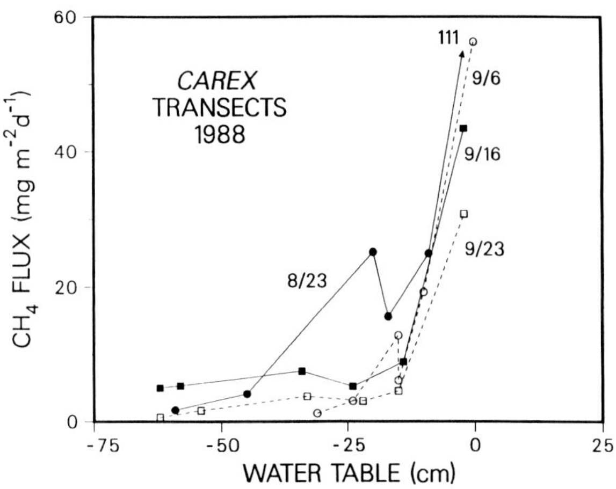

Fig. 4. Methane flux from a *Carex* site transect as a function of water table depth. The transect was perpendicular to a small pond margin and the sites were one meter apart.

methane flux from areas between 50° to 70°N to be 62 Tg yr^{-1}. Sebacher et al. (1986) estimated that methane emission from all boreal and arctic wetlands was 45 to 106 Tg yr^{-1}. Crill et al. (1988) estimated methane emission from undrained peatlands north of 40° to be 72 Tg yr^{-1}.

Results from Table 2 were extended to a global estimate using the same approach as Whalen and Reeburgh (1988). The results are shown in Table 3. The mean annual fluxes were used in this estimate; the ranges in the year columns result from areal coverage ranges. A total tundra mean calculated using the annual ranges is ~35 Tg yr^{-1}.

Our Alaska Pipeline Haul Road transect date were used to derive an independent estimate of global tundra emissions (Whalen and Reeburgh 1990b). Areal weighting was accomplished by classifying the sites into six broad categories and testing their areal coverage using vegetation maps. The transect approach minimizes variations due to areal coverage, but relies on a "snapshot" flux data set. The Whalen and Reeburgh (1990b) tundra methane emission estimate is 38 (24–60) Tg yr^{-1}.

Fung et al. (1991) used the GISS global 3-D tracer model, compilation of geographic and seasonal methane sources and sinks, and methane carbon isotope ratios to provide constraints on global methane budget terms. Several budget scenarios were tested for their ability to reproduce the meridional gradient and seasonal variations in atmospheric methane observed at the GMCC sites. Their preferred budget calls for methane emissions of 35 Tg yr^{-1} from wetlands and tundra north of 50°.

High latitude environments have been the focus of a great deal of study over the past five years, and tundra, peatlands, and to a much smaller extent, taiga environments, have been studied using ground-based chamber techniques. Two major NASA programs involving chamber studies, tower-based eddy correlation measurements, and aircraft-based boundary layer measurements have been conducted in high latitude environments recently. One Atmospheric Boundary Layer Experiment (ABLE 3A) was conducted in the Yukon Delta in 1988, and ABLE 3B was conducted in the Hudson Bay lowlands in 1990. Although a great deal of work remains to be done, there is very good agreement between independent methods for measuring fluxes and the global estimate for high latitude CH_4 emissions seems to be converging on a value of c. 35 Tg yr^{-1}.

Processes controlling methane fluxes

Temperature

Temperature is clearly a dominant control on tundra methane emissions, as the temperature record follows methane fluxes. It is difficult to assign a single temperature to tundra soils for model calculations. The depth interval involved in producing methane also lies on a steep temperature gradient, ranging from ambient air temperatures at the surface to the freezing point at the thaw depth some 40 cm below the surface.

Vascular transport

Vascular transport has been identified as an important methane transport process in rice (Cicerone and Shetter 1981, Seiler et al. 1984) and other aquatic plants (Sebacher et al. 1985), where methane from waterlogged anoxic soils is transported through the vascular system to the atmosphere. Vascular transport appears to be the cause of the higher fluxes from *Eriophorum* and *Carex* sites (Table 2). Whether vascular transport in tundra plants is active or passive is unclear. An important feature of vascular transport is the fact that it bypasses microbial oxidation.

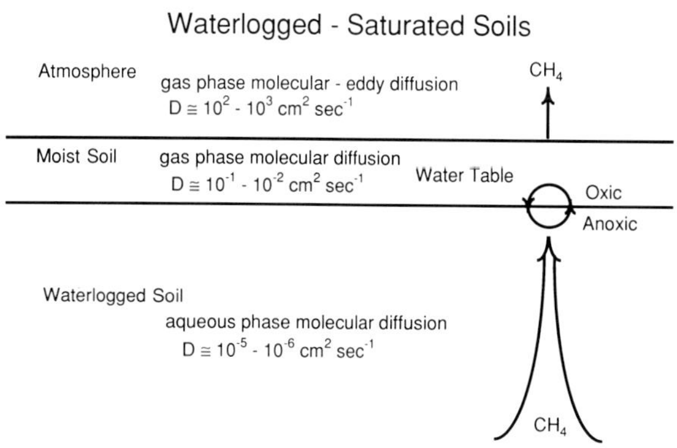

Fig. 5. Schematic diagram showing controls on CH_4 oxidation in waterlogged and saturated soils. Oxidation occurs in the vicinity of the water table, which is the oxic:anoxic boundary. Transport of methane is controlled by aqueous phase molecular diffusion. Methane fluxes to the atmosphere are positive.

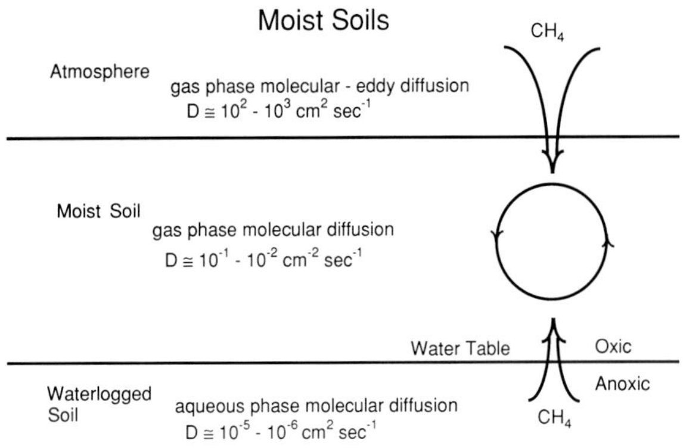

Fig. 6. Schematic diagram showing controls on CH_4 oxidation in moist soils. Oxidation occurs throughout the moist soil layer. Transport is controlled by gas phase molecular diffusion. The moist soil interval equilibrates with the atmosphere at rates that are fast relative to biological consumption rates. Moist soils are capable of consuming atmospheric methane.

Microbial oxidation

Our studies on microbial oxidation in Moss and Black Hole sites indicate that microbial oxidation occurs at the water table, which is the oxic:anoxic interface in waterlogged tundra systems. Fig. 4 shows results from seasonal measurements along a transect of *Carex* sites located 1 m apart near a pond margin. These results show that water table level is also strongly related to methane flux. Similar observations have been reported for the Great Dismal Swamp by Harriss et al. (1982) and for northern fens by Roulet et al. (1991). We have made methane oxidation rate measurements at several other sites, namely in a heath tundra in the Aleutian Islands (Whalen and Reeburgh 1990a), and in the cover soil of a retired municipal landfill at Berkeley, CA (Whalen et al. 1990). Our studies in these two environments indicate that moist soils equilibrate rapidly with the atmosphere and that methane oxidizing bacteria present in these environments are able to consume atmospheric methane. The rates are sufficient to consume methane to a threshold concentration of 0.1 ppm in a matter of 1 to 8 h. Methane-oxidizing bacteria isolated from the landfill environment have kinetic properties that lie within the envelope of other methanotrophs, indicating that they are not unusual organisms. The zero and negative methane fluxes from our dry taiga sites and the moist tundra sites (Reeburgh and Whalen 1990a) appear to result from low methane production and microbial oxidation. The methane oxidizing capacity and the ability of these oxidizers to function at low concentrations suggest that tundra environments may have the ability to function as net consumers of atmospheric methane under projected warmer, dryer conditions. Fig. 5 and 6 show these relationships schematically.

Future work

Several lines of evidence suggest that methane fluxes from high northern latitudes are converging on a value of c. 35 Tg yr^{-1}. The data used in these global estimates derives almost solely from North American peatlands and tundra sites. Major effort should be placed on establishing cooperative transects and time-series studies in tundra regions of Russia to verify the estimates.

Another major problem is our lack of knowledge of the sensitivity of high latitude systems to changes that may accompany climate change. Important aspects are the magnitude and direction soil carbon storage (Prentice and Fung 1990) and possible enhanced release of radiatively active trace gases. Long- and short-term field manipulations are needed to understand how changes in temperature, soil moisture and other parameters affect these terms. Field and laboratory manipulations are also essential in finding and quantifying potential feedbacks.

Future work should focus on radiatively and chemically active gases in addition to methane. Our understanding of how CO_2 and methane respond to changes is incomplete because simultaneous measurements are rare. One technique that avoids problems with competing reactions in chambers and has the potential for widespread application has been described by Dörr and Münnich (1987). This technique involves measurements of ^{222}Rn profiles and fluxes to estimate fluxes of other gases from measured soil profiles. Radon-222 is inert and has a well-understood source function in soils, so the only processes affecting its profiles are radioactive decay ($t_{1/2}$=3.8 d) and transport. Thus, the technique can be used to estimate fluxes of other gases collected at the same time. These authors (1990) and Born et al. (1990) have applied the technique to CO_2 and methane.

Acknowledgements – This work was supported by NASA grant IDP-88-032. Contribution number 894 from the Inst. of Marine Science, Univ. of Alaska Fairbanks.

References

Billings, W. D. 1987. Carbon balance of Alaskan tundra and taiga ecosystems: past present and future. – Quat. Sci. Rev. 6: 165–177.

– , Luken, J. O., Mortensen, D. A. and K. M. Peterson, K. M. 1982. Arctic tundra: a source or sink for carbon dioxide in a changing environment? – Oecologia (Berl.) 53: 7–11.

– , Luken, J. O., Mortensen, D. A. and Peterson, K. M. 1983. Increasing atmospheric carbon dioxide: possible effects on arctic tundra. – Oecologia (Berl.) 58: 286–289.

Blake, D. R. and Rowland, F. S. 1988. Continuing worldwide increase in tropospheric methane, 1978–1987. – Science 239: 1129–1131.

Born, M., Dörr, H. and Levin, I. 1990. Methane concentration in aerated soils in West-Germany. – Tellus 42B: 2–8.

Chapin, F. S. III, Miller, P. C., Billings, W. D. and Coyne, P. 1980. Carbon and nutrient budgets and their control in coastal tundra. – In: Brown, J., Miller, P. C., Tieszen, L. L.

and Bunnell, F. L., (eds), An arctic ecosystem: the coastal tundra at Barrow, Alaska. Dowden, Hutchinson and Ross, Stroudsburg, PA, pp. 452–482.
Cicerone, R. J. and Shetter, J. D. 1981. Sources of atmospheric methane: measurements in rice paddies and a discussion. – J. Geophys. Res. 86: 7203–7209.
– and Oremland, R. S. 1988. Biogeochemical aspects of atmospheric methane. – Global Biogeochem. Cycl. 2: 299–327.
Conrad, R., Seiler, W. and Bunse, G. 1983. Factors influencing the loss of fertilizer into the atmosphere as N_2O. – J. Geophys. Res. 88: 6709–6718.
Craig, H. and Chou, C. C. 1982. Methane: the record in polar ice cores. – Geophys. Res. Lett. 9: 477–480.
Crill, P. M., Bartlett, K. B., Harriss, R. C., Gorham, E., Verry, E. S., Sebacher, D. I., Mazdar, L. and Sanner, W. 1988. Methane flux from Minnesota peatlands. – Global Biogeochem. Cycles 2: 371–384.
Dörr, H. and Münnich, K. O. 1987. Annual variation in soil respiration in selected areas of the temperate zone. – Tellus 39B: 114–121.
– and Münnich, K. O. 1990. ^{222}Rn flux and soil air concentration profiles in West-Germany. Soil ^{222}Rn as tracer for gas transport in the unsaturated soil zone. – Tellus 42B: 20–28.
Ehhalt, D. and Schmidt, U. 1978. Sources and sinks of atmospheric methane. – Pure Appl. Geophys. 116: 452–464.
Emanuel, W. R., Schugart, H. H. and Stevenson, M. P. 1985. Climatic change and the broad-scale distribution of terrestrial ecosystem complexes. – Climatic Change 7: 457–460.
Fung, I., John, J., Lerner, J., Matthews, E., Prather, M., Steele, L. P. and Fraser, P. J. 1991. Three-dimensional model synthesis of the global methane cycle. – J. Geophys Res. 96: 13033–13065.
Gorham, E. 1991. The role of subarctic and boreal peatlands in the global carbon cycle, and their probable responses to "greenhouse" climate warming. – Ecol. Appl. 1: 182–195.
Gruelke, N. E., Reichers, G. H., Oechel, W. C., Hjelm, U. and Jaeger, C. 1990. Carbon balance in tussock tundra under ambient and elevated CO_2. – Oecologia (Heidelb.) 83: 485–494.
Guthrie, P. D. 1986. Biological methanogenesis and the CO_2 greenhouse effect. – J. Geophys. Res. 91: 10847–10851.
Hameed, S. and Cess, R. D. 1983. Impact of global warming on biospheric sources of methane and its climatic consequences. – Tellus 35B: 1–7.
Hansen, J., Fung, I., Lacis, A., Rind, D., Lebedeff, S., Ruedy, R. and Russell, G. 1988. Global climate changes as forecast by Goddard Institute for space studies three-dimensional model. – J. Geophys. Res. 93: 9341–9364.
Harriss, R. C., Sebacher, D. I. and Day, F. P. 1982. Methane flux in the Great Dismal Swamp. – Nature (Lond.) 297: 673–674.
Hutchinson, G. L. and Mosier, A. R. 1981. Improved soil cover method for field measurement of nitrous oxide fluxes. – Soil Sci. Soc. Am. J. 45: 311–316.
Khalil, M. A. K. and Rasmussen, R. A. 1989. Climate-induced feedbacks for the global cycles of methane and nitrous oxide. – Tellus 41B: 554–559.
– and Rasmussen, R. A. 1990. Constraints on the global sources of methane and analysis of recent budgets. – Tellus 42B: 229–236.
Lachenbruch, A. H. and Marshall, B. V. 1986. Climate change: geothermal evidence from permafrost in the Alaskan arctic. – Science 34: 689–696.
Lashof, D. A. 1989. The dynamic greenhouse: feedback processes that may influence future concentrations of atmospheric trace gases and climate change. – Clim. Change 14: 213–242.
– and Ahuja, D. R. 1990. Relative contributions of greenhouse gas emissions to global warming. – Nature (Lond.) 344: 529–531.
Manabe, S. and Bryan, J. Jr. 1985. CO_2-induced change in a coupled atmosphere model and its paleoclimatic implications. – J. Geophys Res. 90: 11689–11708.
– and Wetherald, R. T. 1987. Large-scale changes of soil wetness induced by an increase in atmospheric carbon dioxide. – J. Atmos. Sci. 44: 1211–1235.
Matthews, E. and Fung, I. 1987. Methane emission from natural wetlands: global distribution, area and environmental characteristics of sources. – Global Biogeochem. Cycl. 1: 61–86.
Mitchell, J. F. B. 1989. The "greenhouse" effect and climate change. – Rev. Geophysics 27: 115–139.
Moore, T. R. and Knowles, R. 1989. The influence of water table levels on methane and carbon dioxide emissions from peatland soils. – Can. J. Soil Sci. 69: 33–38.
Oechel, W. C. and Billings, W. D. 1992. Anticipated effects of global change on carbon balance of arctic plants and ecosystems. – In: Chapin, F. S. III, Jefferies, R. L., Reynolds, J., Shaver, G. R. and Svoboda, J., (eds), Arctic ecosystems in a changing climate: an ecophysiological perspective. Academic Press, New York, pp. 139–168.
Peterson, K. M. , Billings, W. D. and Reynolds, D. N. 1984. Influence of water table and atmospheric carbon dioxide concentration on the carbon balance of arctic tundra. – Arctic Alp. Res. 16: 331–335.
Post, W. M., Emanuel, W. R., Zinke, P. J. and Stangenberger, A. J. 1982. Soil carbon pools and world life zones. – Nature (Lond.) 298: 156–159.
Prentice, K. C. and Fung, I. 1990. The sensitivity of terrestrial carbon storage to climate change. – Nature (Lond.) 346: 48–51.
Ramanathan, V., Cicerone, R. J., Singh, H. B. and Kiehl, J. T. 1985. Trace gas trends and their potential role in climate change. – J. Geophys. Res. 90: 5547–5566.
Rasmussen, R. A. and Khalil, M. A. K. 1981. Atmospheric methane (CH_4): trends and seasonal cycles. – J. Geophys. Res. 86: 9826–9832.
– and Khalil, M. A. K. 1984. Atmospheric methane in the recent and ancient atmospheres: concentrations, trends and the interhemispheric gradient. – J. Geophys. Res. 89: 11599–11605.
Raynaud, D., Chappelez, J., Barnola, J. M., Korotkevich, Y. S. and Lorius, C. 1988. Climatic and CH_4 cycle implications of glacial-interglacial change in the Vostok ice core. – Nature (Lond.) 333: 655–657.
Rinsland, C. P., Levine, J. S. and Miles, T. 1985. Concentration of methane in the atmosphere deduced from 1951 infrared solar spectra. – Nature (Lond.) 318: 245–249.
Roulet, N., Moore, T., Bubier, J. and LaFleur, P. 1991. Northern fens: CH_4 flux and climate change. – Tellus 44B: 100–105.
Schlesinger, M. E. and Zhao, Z.-C. 1989. Seasonal climate change induced by doubling CO_2 as simulated by the OSU atmospheric GCM/mixed-layer ocean model. – J. Climate 2: 459–495.
Sebacher, D. I., Harriss, R. C. and Bartlett, K. B. 1985. Methane emissions to the atmosphere through aquatic plants. – J. Environ. Qual. 14: 40–46.
– , Harriss, R. C., Bartlett, K. B., Sebacher, S. M. and Grice, S. S. 1986. Atmospheric methane sources: Alaskan tundra bogs, an alpine fen and a subarctic boreal marsh. – Tellus 38B: 1–10.
Seiler, W., Holzapfel-Pschorn, A., Conrad, R. and Scharffe, D. 1984. Methane emission from rice paddies. – J. Atmos. Chem. 1: 241–268.
Silvola, J. 1986. Carbon dioxide in mires reclaimed for forestry in eastern Finland. – Ann. Bot. Fenn. 23: 59–67.
Sparks, D. L. 1989. Kinetics of soil chemical processes. – Academic Press, San Diego.
Stauffer, B., Lochbronner, E., Oeschger, H. and Schwander, J. 1988. Methane concentration in the glacial atmosphere

was only half that of the preindustrial Holocene. – Nature (Lond.) 332: 812–814.
Steele, L. P., Fraser, P. J., Rasmussen, R. A., Khalil, M. A. K., Conway, T. J., Crawford, A. J., Gammon, R. H., Masarie, K. A. and Thoning, K. W. 1987. The global distribution of methane in the troposphere. – J. Atmos. Chem. 5: 125–171.
Svensson, B. H. and Rosswall, T. 1984. In situ methane production from acid peat in plant communities with different moisture regimes in a subarctic mire. – Oikos 43: 341–350.
Van Cleve, K., Chapin, F. S. III, Flanagan, P. W., Viereck, L. A. and Dyrness, C. T. 1986. Forest ecosystems in the Alaskan taiga: a synthesis of structure and function. – Ecol. Studies 57, Springer, New York.
– , Chapin, F. S. III, Dyrness, C. T. and Viereck, L. A. 1991. State factor control of element cycling in taiga forests. – BioScience 41: 78–88.
Walker, D. A., Lederer, N. D. and Walker, M. D. 1987. Permanent vegetation plots, data report. – Dept. of Energy R4D Project, INSTAAR, Univ. of Colorado, March 31 1987.
Whalen, S. C. and Reeburgh, W. S. 1988. A methane flux time-series for tundra environments. – Global Biogeochem. Cycl. 2: 399–409.
– and Reeburgh, W. S. 1990a. Consumption of atmospheric methane by tundra soils. – Nature (Lond.) 346: 160–162.
– and Reeburgh, W. S. 1990b. A methane flux transect along the trans-Alaska pipeline haul road. – Tellus 42B: 237–249.
– , Reeburgh, W. S. and Sandbeck, K. A. 1990. Rapid methane oxidation in a landfill cover soil. – Appl. Environ. Microbiol. 56: 3405–3411.
– , Reeburgh, W. S. and Kizer, K. S. 1991. Methane consumption and emission from taiga sites. – Global Biogeochem. Cycles 5: 261–273.
Wilson, C. A. and Mitchell, F. B. 1987. A doubled CO_2 climate sensitivity experiment with a GCM including a simple ocean. – J. Geophys. Res. 92: 13315–13343.

Ecological Bulletins 42: 71–76. Copenhagen 1992

Non-methane hydrocarbons in the atmosphere of boreal forests: composition, emission rates, estimation of regional emission and photocatalytic transformation

V. A. Isidorov

Isodrov, V. A. 1992. Non-methane hydrocarbons in the atmosphere of boreal forests: composition, emission rates, estimation of regional emission and photocatalytic transformation. – Ecol. Bull. (Copenhagen) 42: 71–76.

Studies of the emission of volatile organic compounds from 17 species of plants, characteristic of the boreal forest are presented. Isoprene and terpene emissions were measured from both coniferous and deciduous plants. The estimated regional emissions of terpenes were calculated by using the measured emission rates and correcting for temperature dependence of emissions. The terpene emissions from coniferous forests of the northern and middle taiga zones of the European portion of the USSR were estimated to be 1.0 to 3.5 Tg · yr^{-1}. The results show that the photocatalytic oxidation of terpenes and isoprene on the surface of natural aerosols may provide a significant atmospheric sink.

V. A. Isidorov, Chem. Dept, St. Petersburg Univ., 198904, St. Petersburg, Russia.

Introduction

The study of the emission of volatile organic compounds (VOC) from living plants into the atmosphere started over 60 years ago (Nilov 1928). From the very first days these investigations were drawn to the solution of different ecological problems, in particular, inter- and intraspecific interactions of organisms (Kholodnii 1944). More recently, geophycists and atmospheric chemists have significant interest in shown phytogeneous organic compounds of the atmosphere. This interest is due to the suggestion that phytogeneous VOC play a considerable role in some physico-chemical processes occuring in the atmosphere (Graedel 1979).

The vegetation of the continents is one of the largest sources of organic carbon (C_{ogr}) to the Earth's atmosphere. However, the available estimates of global emission of phytogeneous C_{org} are highly uncertain. Rasmussen and Went (1965) have estimated emission of C_{org} to be 437 Tg · yr^{-1} based on measurements of concentrations atmospheric terpene. Zimmerman et al. (1978) estimated total emission of isoprene and terpenes to be 941 Tg · yr^{-1}. Ioffe and Isidorov (1982) have estimated a lower limit of organic compound emission from terrestrial vegetation to be 1550 Tg · yr^{-1}.

All of these estimates were obtained by extrapolating results of measurements carried out for a few species of plants and plant associations to the globe. As a result, there is considerable error associated with the calculations. In order to quantify the global emissions of phytogeneous C_{org}, it is evidently necessary to accumulate and summarize as complete data as possible for various regions.

Given the tremendous rates of emission, it is important to study possible paths of chemical transformations of phytogeneous substances in the atmosphere. Numerous investigations have shown that such phytogeneous components as isoprene and terpenes actively react with HO· and NO_3 radicals and with ozone. However, chemical transformations on the surface of natural aerosols have not been examined, although they may to be one of the main atmospheric sinks for these substances.

Composition and emission rates of phytogeneous VOC into the atmosphere

In this paper, I have focused on VOC emitted to the atmosphere by boreal forest vegetation which occupies

Table 1. Organic compounds in volatile emission of arboreous plants, shrubs, deciduous moss and fern.

Compounds	Plant species*	Compounds	Plant species*
Pentane	8	Chloroform	16
2-Methylbutane	1–8, 12–16	Dimethyl sulfide	15
Propylene	1, 8, 9	3-Hexene-1-ol acetate	1, 4–8, 13
Butenes	1–12	Diethyl ether	16
Pentene	1, 2, 5–9, 13–16	3-Octanone	15, 16
Nonene	1, 6, 11	Santene	8, 11
Isoprene	1–17	Cyclofenchene	8–10
2, 3-Dimethyl butadiene	8, 17	Bornilene	8–13
p-Cymene	7–12	Tricyclene	7–10, 16
Ethanol	3–5, 10, 12–16	α-Thujene	9, 10, 13
3-Hexen-1-ol	1, 2, 5, 7, 8, 13, 14	α-Pinene	3–13, 16–17
Acetaldehyde	1, 3	β-Pinene	3–11, 16, 17
Propanal	6, 15	δ-Fenchene	8, 10, 12
Butanal	15	ε-Fenchene	9–12
α-Methylacroleine	1–4, 7, 9, 15, 16	α-Fenchene	7–13
Acetone	1–17	Camphene	3–13, 16, 17
2-Butanone	8, 15	Sabinene	12
3-Butene-2-on	2, 4	Myrcene	7–12, 16, 17
3-Methyl-−3-butene-2-on	15	3-Carene	7–12, 16, 17
2-Pentanone	6, 8, 13–16	α-Phellandrene	8–11, 13
3-Pentanone	5, 7–10, 16	β-Phellandrene	8–11
Ethyl acetate	9, 13, 14, 16	α-Terpinene	8–10
1-Octene-3-on	15, 16	β-Terpinene	3, 8–10
Dimethyl cyclopentenon	15, 16	γ-Terpinene	9
Furan	6	Limonene	3, 7–12, 16, 17
2-Methyl furan	1–5, 12, 17	Terpinolene	9–12, 17
3-Methyl furan	1–5, 12, 14	Alloocymene	17
Ethyl furan	1, 2, 4, 8, 13–15, 17	1, 8-Cineol	3, 8
Vinyl furan	1, 2, 4, 8, 17	Camphor	8
Hexyl furan	17	Anethole	17

*1. willow, 2. aspen, 3. balsam poplar, 4. European oak, 5. European birch, 6. mountain ash (*Sorbus sibirica*), 7. European larch, 8. European fir, 9. Scotch pine, 10. Siberian pine, 11. silver fir, 12. common juniper, 13. red bilberry shrub, 14. bilberry shrub, 15. fern, 16. decidiuous moss, 17. marsh tea.

Table 2. Isoprene and monoterpenes emission rates from various plant species.

Plant species	Air temperature C°	Emission rate $\mu g \cdot g^{-1} \cdot h^{-1}$
Sibirian pine (*Pinus sibirica*)	15–32	1.8–26.5
Scotch pine (*P. sylvestris*)	14–28	1.6–18.8
Scotch pine (*P. sylvestris*)*	23	2.6
Fir (*Picea exelsa*)	16	0.8
Fir (*Picea exelsa*)*	2	1.6
Fir (*Picea exelsa*)*	27	5.8
Fir (*Picea exelsa*)*	21	1.1
Larch (*Larix sibirica*)	22–27	2.9–6.4
Silver fir (*Abies sibirica*)	20–25	2.1–9.9
Oak (*Quercus robur*)	15–26	1.3–36.2**
Oak (*Quercus robur*)*	18–32	2.4–50.4**
Aspen (*Populus tremula*)	20–26	1.8–32.5**
Aspen (*Populus tremula*)*	18–40	2.4–51.7**
Willow (*Salix fragilis*)	2–26	2.2–8.6**
Birch (*Betula vericosa*)*	18–32	0.1–3.95
Pine (*Pinus caribea*)	23–26	23.5–37.4
Pine (*P. tropicalis*)	26	5.1
Lemon	26–27	15.8–18.3
Tangerine	26–27	21.4–23.5
Orange	26–27	1.3–4.9

* Armalis and Nika (1984).
** Emission rate of isoprene (depending on illumination level).

a territory of c. 11×10^6 km^2. It is unlikely that the composition of volatile emissions of all plant species of this zone will ever be completely investigated. Although the boreal forests of Eurasia and America are very diverse, these forests consists of a few dominant species of coniferous trees (larch, pine, spruce, silver fir and tsuga) with a relatively small component of small-leaved species (birch, alder, aspen, willow and some other). In the understory, bushes (black and red bilberry bushes, marsh tea), mosses and ferns predominate. Hence, a general idea of the composition of phytogeneous VOC of boreal forests may be obtained from the study of emissions of a relatively small number of plants.

Table 1 gives the composition of VOC emitted by these plants. VOC determinations were made by gas chromatography-mass spectrometry (GC–MS). VOC were concentrated by passing through sorption tubes with polymer adsorbent Tenax GC of 0.5–1.0 l of air from polyethylene packs put on branch ends of living plants. Analysis was carried out with an LKB–2091 GC–MS instrument (Sweden) after the thermal desorption of concentrated compounds into a capillary column. Identification was done by comparing the recorded mass spectra with those reported in catalogues.

Sixty C_1–C_{10} organic compounds were identified in

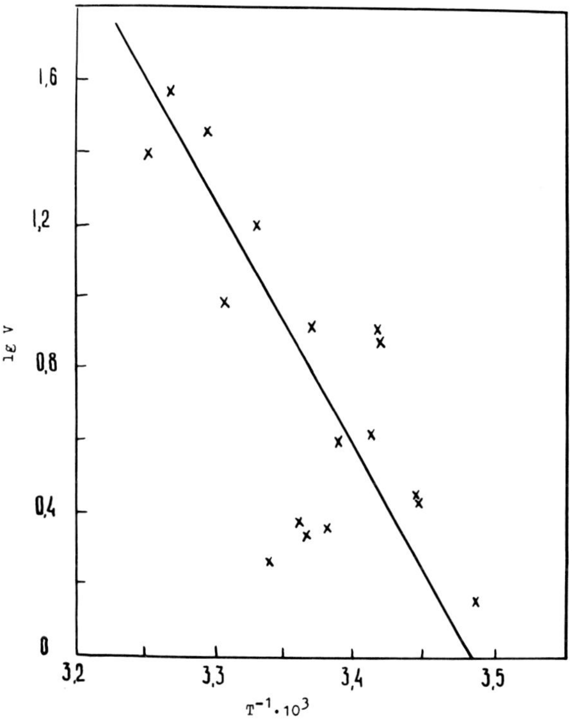

Figure 1. Temperature dependence of the rate of terpene emission by pine needles.

volatile emissions of 14 species of trees, bushes, mosses and ferns. In fact, the number of these compounds is much greater. Many coniferous plants emit c. 100 compounds, but some of them could not be identified because of difficulties encountered with the separation of complex mixtures.

The list in Table 1 includes paraffins and olefins, dienes and terpenes, alcohols and esters, aldehydes and ketones, furans and halocarbons. Most of them with the exception of isoprene and terpenes have been previously considered anthropogenic atmospheric pollutants.

The emission rates of terpene hydrocarbons (main components of the VOC of coniferous plants) and isoprene were determined by using a flow tube (Isidorov et al. 1985). The end of the tree branch was carefully introduced into a glass tube (52×230 mm) and the tube was closed with a plastic foam plug. The tube outlet was connected with a sorption tube packed with Tenax GC or with carboneous sorbent Carbochrome. The air passing through an inlet in the plug was drawn through the tube with foliage and the sorbent at a rate 0.2–0.5 l · min^{-1}. The total sample volume was 1.5 l.

Quantitative analysis of hydrocarbons was done on a gas chromatograph with a flame ionization detector. Table 2 gives the results of measurements of VOC emission rates for nine boreal forest plant species (unpubl. data, Armalis and Nika 1984). I have also listed in the same table, the emission rates of terpenes from some subtropical plants comparison. The emission rates of terpenes range from 1 to 20 μg · g^{-1} · h^{-1} at temperatures 11–30°C. Isoprene emission rates attained 54 μg · g^{-1} · h^{-1} and are profoundly affected by lighting conditions.

Regional evaluation of terpene emission rates

Figure 1 shows the temperature dependence of terpene emission rate by pine needles (*Pinus sylvestris*). This relationship and a similar relationship for spruce (*Picea exelsa*) were used to estimate terpene emissions from the coniferous forests of the North European part of Russia. I used the following data in the calculations: 1) area of pine and spruce forests, 2) the amount of green biomass in forest of different types (absolutely dry needle weight is usually 3–6 t · ha^{-1}), 3) the length of the growing season and 4) the number of days with temperatures higher than 5, 10, 15 and 20°C were used in calculations. The results given in Table 3 show that coniferous forests with an area of 62.6×10^6 ha (c. 10% of the total area of coniferous forests of Russia and

Table 3. Emission rate of terpene hydrocarbons from coniferous forest of the taiga zone in European regions of Russia.

Region	Forested area	Area of coniferous forests	Weight of green needles (pine/fir)	Acative respiration period	Annual emission from forests of	
	10^6 ha	10^6 ha	Tg	d yr^{-1}	Pine Tg yr^{-1}	Fir Tg yr^{-1}
Archangelsk region	20.1	18.3	38/169	120–155	0.15–0.59	0.09–0.16
Vologda region	9.2	5.7	20/42	153–163	0.12–0.34	0.04–0.06
Komi ASSR	27.0	21.6	58/181	120–151	0.22–0.95	0.09–0.18
Karelian ASSR	8.2	7.3	21/60	126–163	0.07–0.33	0.03–0.05
Perm region	8.6	6.2	10/61	130–168	0.07–0.29	0.02–0.06
Kirov region	6.3	3.6	12/26	150–175	0.08–0.35	0.08–0.13
Total	83.9	62.6	160/539	–	0.71–2.85	0.35–0.64

* Mean number of days per year with the mean diurnal temperature above 5°C in the North of the region according to the data of 1880–1969 observations.

Tabel 4 Photocatalytic oxidation of C_5H_8 and $C_{10}H_{16}$ on the surface of various materials.

Hydrocarbon	Photocatalyst	Catalyst weight (g)	Hydrocarbon quantity ($\mu mole^{-1}$)	Irradiation time (min)	$k \cdot 10^5$ (s^{-1})
Isoprene	ZnO	8.8	100	45	200
Isoprene	TiO_2	6.5	100	30	200
Isoprene	Al_2O_3	5.9	100	1000	0.2
Isoprene	sand	7.6	200	1000	0.2
Isoprene	volcanic ash	8.6	100	810	1.0
Isoprene	sea salt	20.0	100	710	2.0
Myrcene	ZnO	4.8	90	240	15
Myrcene	TiO_2	3.6	90	60	40
Myrcene	$CaCO_3$	9.4	90	1020	5.0
Myrcene	volcanic ash	18.4	90	1850	1.5
α-Pinene	ZnO	7.4	250	480	30
α-Pinene	sand	6.0	120	2530	20
α-Pinene	$CaCO_3$	9.9	120	810	20
β-Pinene	ZnO	9.5	90	120	37
Sabinene	ZnO	11.0	90	265	40
α-Phellandrene	ZnO	8.9	90	400	50
Limonene	ZnO	7.5	90	120	80
Limonene	sand	9.0	90	2330	0.6
γ-Terpinene	sand	5.8	90	300	0.5
γ-Terpinene	$CaCO_3$	3.7	90	100	22
γ-Terpinene	TiO_2	6.2	90	45	100
γ-Terpinene	ZnO	9.3	90	40	120

4.8% of those of the Earth) emit from 1.0 to 3.5 Tg of terpenes annually.

This estimate is lower than estimates for forests of the USA (Lamb et al. 1987). They estimated that 6.6 Tg of terpenes are emitted by coniferous forests with an area of 96.4×10^6 ha. These figures imply that 20% more terpene is emitted per unit area of forests in the USA than in forests of Russia. The low productivity, the low mean annual temperature, and the fact that much of the Russian coniferous forest lies north of 55°N latitude explain the difference between the estimates made for the USA and Russia. It may be assumed that the scale of VOC emission by highly productive forests of the subtropical and equatorial zone is higher than it has been observed for the forests of the taiga zone. This assumption is confirmed by high emission rates of terpene and isoprene by tropical plants (see Table 2 and data given by Cronn and Nutmugul 1982).

Photocatalytic oxidation: an additional sink for phytogeneous hydrocarbons?

Many papers deal with the investigation of homogeneous gasphase oxidation of isoprene and terpenes (Brewer et al. 1984, Killus and Whitten 1984, Atkinson et al. 1985, Gu Chee-Ilang et al. 1985). However, the possibility of the occurence of heterogeneous processes with the participation of natural aerosols (the emissions of which into the troposphere are evaluated as 2.3×10^3 Tg · yr^{-1}) has not yet been taken into account. It is noteworthy that metal oxides contained in aerosols (Fe_2O_3, FeO, ZnO, TiO_2, etc.) exibit semiconductor properties and can take part in the photostimulated oxidation of organic substances adsorbed on their surface (Behnke et al. 1987, Isidorov 1990).

The photocatalytic oxidation of isoprene and terpenes was carried out in glass reactor (Isidorov et al. 1990). A mercury lamp (250 W) placed in a quartz finger served as a light source. The lamp was cooled with water running in the space between the quartz finger and pyrex jacket. The reactor design provides for the possibility of using liquid filters for selective light absorption in certain spectran region. All experiments were carried out at a temperature of 21±2°C.

Oxides (ZnO, TiO_2, Fe_2O_3 and Al_2O_3) and natural materials (sand, volcanic ash, sea salt and $CaCO_2$) previously purified from organic impurities were used as photocatalysts. They were applied to the reactor wall from an aqueous suspension and dried. Liquid reagent samples were introduced into the installation mounted for the experiment (reaction volume 430 cm^3). The change in their content and the appearance of reaction products were checked by GC and GC–MS. The measurements of the light flux with the aid of a chemical actinometer showed that in the range of 300–400 nm the catalyst particles absorbed c. 10^{17} quanta · s^{-1}.

When isoprene and terpenes were irradiated with a light at ≥300 nm in the presence of solid particles, rapid disappearance of these hydrocarbons from the gas phase of the reactor was observed. The decrease in hydrocarbon concentration is represented by a straight line in the coordinate system of a firstorder reaction the constant for which are given in Table 4. In the case of isoprene, CO_2 was the main reaction product with a small acetone admixures. However, upon irradiation with light at ≥400 nm, marked changes were observed: apart from CO_2 remaining the main product, seven other compounds were detected in the gas phase of the reactor:

$$C_5H_8 \xrightarrow[\geq 400\ nm]{h\nu,\ ZnO} CO_2 + CH_3COCH_3 + CH_3CHO + CH_3COC_2H_5 + CH_2O + CH_3COC(CH_3){=}CH_2 + CH_3COCH(CH_3)_2 + CH_2{=}C(CH_3)CHO + C{\equiv}C\text{-}CH_3$$

Terpene hydrocarbons have much higher boiling temperatures than isoprene and a greater tendency to adsorp on the surface of solid aerosol particles. Therefore, it may be assumed that photocatalytic reactions play a considerable role in the atmospheric chemistry of terpenes.

The results of measurements given in Table 4 show that the rates of catalytic transformations of monoterpenes on ZnO are lower and those on sand particles are higher than it was in the case of C_5H_8. It is also interesting that on passing from acyclic myrcene to bicyclic pinenes and, further, to monocyclic terpenes with two double bonds, the reaction constants increase in the case of ZnO and decrease when sand is used. However, the oxidation of various terpenes on the surface of $CaCO_3$ proceeds at approximately the same rates.

In experiments with terpenes, the actual CO_2 yield was less than 30% of theoretical (maximum calculated for complete portion oxidation of $C_{10}H_{16}$ molecules). This suggests that a considerable portion of the carbon contained in initial terpenes is bonded to photocatalyst particles. To check this suggestion, I conducted special experiments on the thermal desorption of organic components and analyses by the GC–MS method were carried out. After experiment on photooxidation, catalyst particles were taken from the reactor walls and transferred into a glass tube. Thermal desorption was carried out directly into a capillary column in a helium flow with a gradual (25 min) increase in temperature from 20 to 200°C. The results of analyses showed that particles of the various materials used as photocatalysts contained CO_2, H_2O and a large number of C_2–C_{10} organic compounds, the products of incomplete terpene oxidation. The products included 15 carbonyl compounds, furan and its derivatives, saturated and unsaturated C_2–C_{10} alcohols and aromatic hydrocarbons C_6–C_{10}. CO was not detected among the degradation products of terpenes and isoprene. However, the literatures suggest that CO is one of the main products of homogeneous gas-phase oxidation of these hydrocarbons. As a result, the oxidation of isoprene and terpenes has been postulated to be a large natural source of CO in troposphere (Zimmerman et al. 1978, Graedel 1979, Brewer et al. 1984)

It was not my purpose to elucidate the formation mechanism of the reaction products. It is essential that the oxidation of C_5H_8 and $C_{10}H_{16}$ hydrocarbons occurs on solid particles contained in natural aerosols irradiated with light in the near UV and visible spectral regions. Moreover, the rate of photocatalytic oxidation was found to be comparable to that of a homogeneous gasphase reaction with a HO· radical. For example, the rate constant for C_5H_8 and HO· radical is 7.8×10^{-11} $cm^3 \cdot molecule^{-1} \cdot s^{-1}$. If the concentration of radicals is taken to be 0.5×10^6 cm^{-3}, we will obtain the value of the specific rate of gas-phase reaction $HO + C_5H_8$ (3.9×10^{-5} s^{-1}) close to that given in Table 4 for photocatalytic processes with the participation of components of natural aerosols.

Summary

In summary, photocatalytic oxidation may be a significant atmospheric sink for phytogeneous hydrocarbons emitted to the atmosphere. However, these studies raise a number of new questions. First, what is the subsequent fate and ecological role of the products of incomplete oxidation of hydrocarbons located on the surface of the aerosols? Second, we need to reexamine the idea that phytogeneous hydrocarbon oxidation is a significant source of CO in the atmosphere.

References

Armalis, S. Y. and Nika, A. K. 1984. Some results of the measurements of the hydrocarbons emission by the plants. – Fisika atmos (Vilnus) 9: 74–77.

Atkinson, R., Aschmann, S. M., Winer, A. M. and Pitts, J. N. Jr. 1985. Kinetics and atmospheric implications of the gas-phase reactions of NO_3 radicals with series of monoterpenes and related organics at 294±2 K. – Environ. Sci. Technol. 19: 159–163.

Behnke, W., Nolting, F. and Zetzsch, L. 1987. A smog chamber study on the impact of aerosols on the photodegradation of chemicals in the troposphere. – J. Aerosol Sci. 18: 65–71.

Brewer, D. A., Ogliaruso, M., Augustssen, T. and Levine, J. S. 1984. The oxidation of isoprene in the troposphere. Mechanism and model calculations. – Atmos. Environ. 18: 2723–2744.

Cronn, D. and Nutmagul, W. 1982. Analysis of atmospheric hydrocarbons during Winter MONEX. – Tellus 34: 159–165.

Graedel, T. E. 1979. Terpenoids in the atmosphere. – Rev. Geophys. Space Phys. 18: 937–947.

Gu Chee-Ilang, Rynard, C. M., Hendry, D. G. and Mill, T. 1985. Hydroxyl radical oxidation of isoprene. – Environ. Sci. Technol. 19: 151–152.

Ioffe, B. V. and Isidorov, V. A. 1982. Organic substances in the Earth's atmosphere. – Priroda 7: 68–77.

Isidorov, V. A. 1990. Organic chemistry of the Earth's atmosphere. – Springer, Berlin.

– , Zenkevich, I. G. and Ioffe, B. V. 1985. Volatile organic compounds in the atmosphere of forests. – Atmos. Environ. 19: 1–8.

– , Klockova, E. M. and Zgonnik, P. V. 1990. About photostimulated oxidation of the halogenated organic compounds

on the surface of solid particles. – Vestnik Leningr. Univ. Ser. 4 No. 1, pp. 71–77.
Kholodnii, N. G. 1944. Volatile emissions of flowers and foliage as a nutrition source to the microorganisms. – Dokl. AN SSSR 43: 75–78.
Killus, J. P. and Whitten, G. Z. 1984. Isoprene: a photochemical kinetic mechanism. – Environ. Sci. Technol. 18: 124–148.
Lamb, B., Guenther, A., Gay, B. and Westborg, H. 1987. A national inventory of biogenic hydrocarbon emissions. – Atmos Environ. 21: 1695–1705.
Nilov, V. I. 1928. About quantitative determination of essential oiles emitted by living plants in air. – Nauchno-agron. Zn. (J. Landwirtschaft. Wissenshaft) 11: 810–818.
Rasmussen, R. A. and Went, F. W. 1965. Volatile organic material of plant origin in the atmosphere. – Proc. Natl. Acad. Sci. 53: 215–220.
Zimmerman, P. R., Chatfield, R. B., Fishman, J., Crutzen, P. J. and Hanst, P. L. 1978. Estimates of the production of CO and H_2 from the oxidation of hydrocarbon emissions from vegetation. – Geophys. Res. Lett. 5: 679–682.

Ecological Bulletins 42: 77–85. Copenhagen 1992

Northern Wetlands Study (NOWES): an assessment of the role of northern wetlands in the exchange of atmospheric trace gases

N. T. Roulet, L. Barrie, W. A. Glooschenko, R. C. Harriss, J. I. MacPherson, T. R. Moore, H. Neumann and H. I. Schiff

Roulet, N. T., Barrie, L., Glooschenko, W. A., Harriss, R. C., MacPherson, J. I., Moore, T. R., Neumann, H. and Schiff, H. I. 1992. Northern Wetlands Study (NOWES): an assessment of the role of northern wetlands in the exchange of atmospheric trace gases. – Ecol. Bull. (Copenhagen) 42: 77–85.

The Northern Wetlands Study (NOWES) is a multidisciplinary investigation designed to assess the importance of northern wetlands as a source/sink of biogenic gases to the atmosphere. The primary emphasis of the project is on the exchange of CH_4 between wetland ecosystems and the atmosphere, but the fluxes of N_2O, NO_x, VOC, DMS and CO_2 will also be examined. The Canadian Institute for Research in Atmospheric Chemistry, York Univ., Canada (CIRAC) and the National Aeronautics and Space Administration (NASA) are the lead agencies co-ordinating the project. Eight universities and eleven government agencies from Canada and ten universities and three government agencies from the United States are collaborating on the study. The principal field site for the study is the Hudson Bay Lowland, the second largest continuous wetland in the world. It offers a unique environment because the biological succession over the last 8,000 years is represented along a 100 km transect inland from the coast of James Bay. A second field site in the subarctic/boreal region of Quebec and Labrador where the wetland coverage is discontinuous (27% coverage), will be examined by the NASA Atmospheric Boundary Layer Experiment (ABLE-3B) project. Research activities at both sites include flux measurement programmes using chambers, micrometeorological towers and aircraft. These measurements will be suplimented by studies of the factors that control the gas flux. Land characterization by remote sensing and ecology will be used to develop a classification for the spatial extrapolation of the results. Preliminary studies and reconnaissance were undertaken in 1989 and the main experiment will take place summer 1990, with the modelling effort and extrapolation in 1991.

N. T. Roulet, Dept of Geography, York Univ., 4700 Keele St., North York, Ontario, Canada, M3J 1P3. – L. Barrie and H. Neumann, Atmospheric Environ. Services, Environ. Canada, 4905 Dufferin St., Downsview, Ontario, Canada, M3H 5T4. – W. A. Glooschenko, Nat. Water Res. Inst., P.O. Box 5050, Burlington, Ontario, Canada, L7R 4A6r, – R. C. Harriss, EOS, Univ. of New Hampshire, Durham, NH 03824, USA. – J. I. MacPherson, Flight Res. Lab., Nat. Res. Council, Blging U-61, Montreal Rd., Ottawa, Ontario, Canada, K1A 0Rs. – T. R. Moore, Dept of Geography, McGill Univ. 805 Sherbrooke St. W., Montreal, Quebec, Canada, H3A 2K6. – H. I. Schiff, Dept of Chemistry, York Univ. 4700 Keele St. North York, Ontario, Canada, M3J 1P3.

Introduction

The concentration of atmospheric methane (CH_4) has doubled over the last 200 years (Pearman and Fraser 1988). It is currently 1620 ppb(v) (Steele et al. 1987) and is increasing at c. 1% yr^{-1} (Khalil et al. 1989). Current estimates of total annual emissions to the atmosphere from various sources are very uncertain. Until recently it was estimated 540 Tg of CH_4 is released to the atmosphere annually (Cicerone and Oremland 1988), but this has been adjusted downward to 505 Tg yr^{-1} (Crutzen 1991) because of recent studies on the removal rate of CH_4 with OH (Vaghjiani and Ravishankara 1991). Approximately 20% of the annual contribution is from each of natural wetlands, rice paddies, ruminants and termites and anthropogenic activities such as coal and

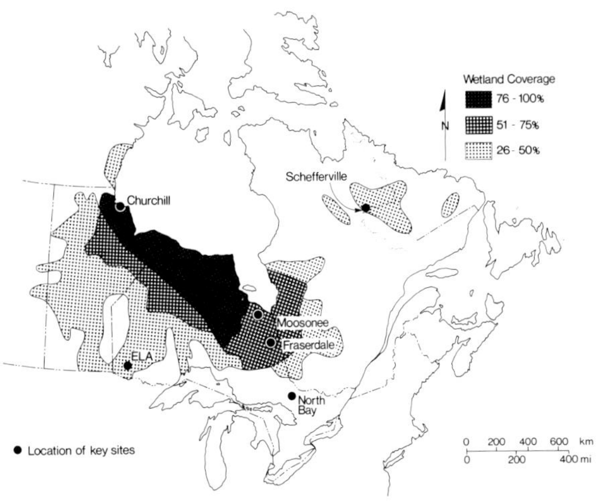

Fig. 1. Location of the Hudson Bay Lowland and centres of NWS/ABLE research activity.

natural gas mining, biomass burning and landfills. Minor sources include the oceans and freshwater lakes.

While estimates vary, wetlands are estimated to contribute between 80 (Aselmann and Crutzen 1989) and 115 Tg (Matthews and Fung 1988) of the annual atmospheric burden of CH_4. However, global extrapolations of the role of wetlands are limited by relatively few measurements and inadequate data on the areal extent of wetlands. In particular most CH_4 flux measurements have been made in small, relatively isolated wetlands in the temperate, south boreal region (e.g. Crill et al. 1988) and the arctic and subartic region (e.g. Whalen and Reeburgh 1988, Moore et al. 1990). The largest concentration of wetlands is in the high latitudes (> 40°N). These wetlands may be significant producers of CH_4 because they possess soils of a relative high fraction of organic matter (National Wetlands Working Group 1988) and are generally water saturated, leading to reducing anaerobic conditions in near-surface environments. These environments may also produce other gases such as N_2O, NO_x, DMS, CMS and volatile organic carbon (VOC). In most climate modelling scenarios (cf. Mitchell 1989), the high latitudes are predicted to experience a greater change in temperature than lower latitudes, so a possible change in trace gas exchange from northern wetlands could be an important biospheric-atmospheric feedback (Robertson 1989).

Of the limited research that has been undertaken on trace gas exchange from wetlands, none has been done in the two largest continuous wetland regions of the world: the Hudson Bay Lowland of Canada (3.2×10^5 km^2) (National Wetlands Working Group 1988) and the Siberian (Vasuugan) peatland of the USSR (5.4×10^5 km^2) (Botch and Masing 1983). The Hudson Bay Lowland (Fig. 1) offers a unique environment to study the exchange of biogenic trace gases for a number of reasons. The Lowlands have evolved in an area which has rebounded isostatically (0.7 to 1.0 m 100 yr^{-1}) for the last 8,000 years following the retreat of continental glaciation (Cowell 1982). A transect from the coast of Hudson Bay inland to the southern perimeter of the Lowland (50 to 250 km) represents 8,000 years of peatland succession. As a consequence there is a relatively consistent set of biogeochemical, ecological and hydrological gradients running inland from the coast. This allows for the comparison of the relative roles of water, heat, chemistry and edaphic controls of CH_4 production, consumption and emission, as wetlands ecosytems evolve. In the Hudson Bay Lowland, the linear changes inland can be used as an analog for changes in time.

A joint Canadian-American (Canadian Institute for Research in Atmospheric Chemistry and NASA Global Tropospheric Experiment ABLE-3B) project was designed to determine the role of northern wetlands and particularly the role of the Hudson Bay Lowland, in trace gas exchange. The project is called the Northern Wetlands Study (NOWES). Over 50 universities, government institutions and agencies from Canada, the United States and other countries participated in the NOWES. The NOWES will examine the trace gas exchange from two different northern wetland regions of Canada. The first is the continuous wetland region Hudson Bay Lowland between the Moose and Albany Rivers (see Fig. 2) and the subarctic/boreal wetland region (27% wetland coverage) centered around Schefferville, Quebec (Fig. 1).

In this paper, we present the scientific design for the NOWES including the objectives, the structure of the research programmes, the location and description of the areas where the NOWES will take place and details of the research activities.

Objectives

The objective of the Northern Wetlands Study is to assess the importance of northern wetlands as a source/sink of biogenic gases to the atmosphere under current climate conditions. The characterization of variations in trace gas source/sink dynamics as a function of wetland hydrology and soil temperature will also provide a basis for modelling the effects of future climate change scenarios on regional feedbacks. This requires: 1) investigat-

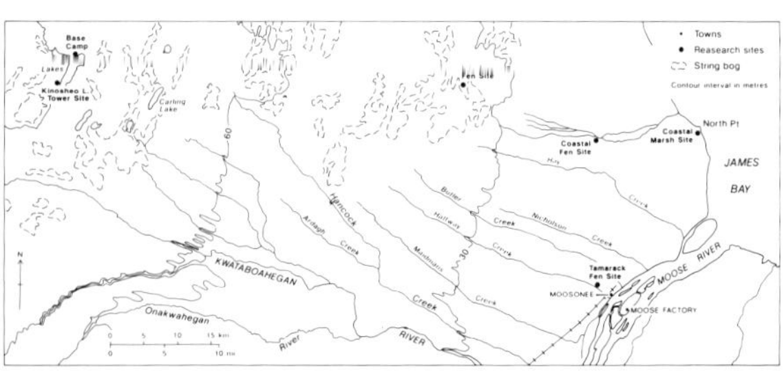

Fig. 2. The location of the field sites near Moosonee, Ontario.

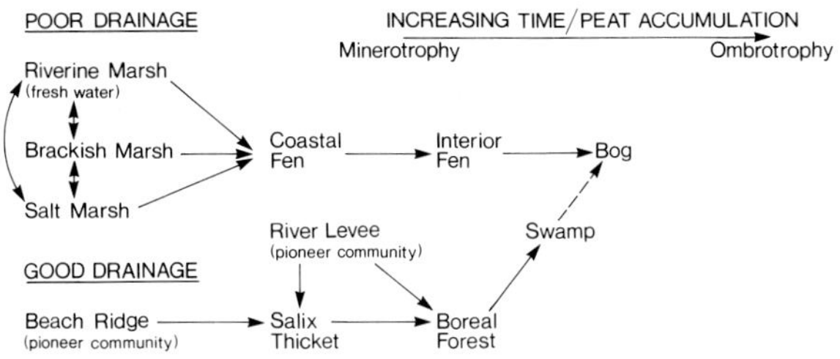

Fig. 3. The sequence of ecological succession across the Hudson Bay Lowland from the coast inland (from Glooschenko and Martini 1983).

ing the atmospheric chemistry of biogenic gases and related compounds in northern wetlands with emphasis upon carbon gases; 2) developing an understanding of the physical, chemical and biological processes influencing biogenic gas production and exchange and the sensitivity of the source/sink to climate change; 3) determining the significance of northern wetlands as a source of biogenic gases to the atmosphere globally; and, 4) characterizing the past and present ecology of the Hudson Bay Lowland.

Study location and field sites

The primary research area for the NOWES is a portion of the Hudson Bay Lowland between the Moose and Albany Rivers (Fig. 2). This section, however, represents a very small portion of the Hudson Bay Lowland which extends from northwest of Churchill, Manitoba (58°15′N, 96°00′W) in the low arctic region of Manitoba to east of the Ontario-Quebec border (51°00′N, 78°15′W) in the boreal forest region (Fig. 1). At its widest point, the Lowland extends c. 250 km inland. The northern section (near Churchill) is underlain by continuous permafrost, but most of the Lowland is in the discontinuous permafrost zone and in the southern region, where the study area is located, permafrost is very sporadic (Cowell 1982).

Several broad-scale generalization can be made about the structure of the Lowland. The average topographic gradient from the Lowland's southern perimeter to the coast is 0.5 to 1.0 m km^{-1}. There has been considerable work done on the peat composition and ecology of the wetlands in a band within 50 km of the coast for the James Bay portion of the Lowland (Jeglum and Cowell 1982, Riley 1982, Sims et al. 1982). These indicate a succession of wetlands from the coast inland (Fig. 3: from Glooschenko and Martini 1983). There are few radiocarbon dates from the Lowland, but to the east of Moosonee the peatlands are 2,000 and 5,000 years old, 20 and 70 km inland respectively (Protz et al. 1988).

Five sites, four of which are located along a 100 km transect running perpendicular to the James Bay coast, have been chosen for detailed trace gas studies (Fig. 2). These sites represent wetland types in relation to the successional sequence shown in Fig. 3. The sequence begins with the establishment of coastal marshes (Fig. 2). Salt marshes are located in the area or adjacent to river discharge plumes and along the coast. Peat begins to accumulate in the inland parts of the marshes that are away from the influence of tidal flows or upper margin of salt laden groundwater (Price and Woo 1989a, b). When 40 cm of peat accumulates and the influence of salt dissipates, the system is classified as a fen. Two fens representative of the coastal portion of James Bay will be sampled (Coastal Fen and Tamarack Fen in Fig. 2). The tamarack fen are characteristic of the area adjacent to the Moose River which probably was an old embayment of the Tyrrell Sea and one of the last portions of the region to rise isostatically. The coastal fen is more common in the areas away from the influence of the large rivers crossing the Lowland.

The coastal fens transform into interior fens which are somewhat poorer in nutrients. As more peat accumulates, the surface of the wetlands become isolated from mineral-rich groundwater and become more ombrotrophic (i.e. only receives atmospheric input of water and nutrients). This transition is represented by the Interior Fen site (Fig. 2). Further inland the ecosystems are more bog-like, characterized by low concentrations of nutrients and higher acidity and small scale topographic variations of ponds, vegetated (flarks) and coalescing lakes being common. The final site on the transect, Kinosheo Lake (Fig. 2), represents the bog like wetland complex.

The hydrology and hydrogeochemistry of the coastal marsh site is dominated by an annual cycle of salinity and saturation controlled by spring snowmelt, summer storms and a small groundwater input (Price and Woo 1989a, b). Throughout the whole of the Hudson Bay Lowland, groundwater is believed to be of little significance in the mass flux of water because of the small hydraulic gradient, but it may be significant in providing a minimal level of nutrients in the fen areas. Inland, the fens appear to remain saturated through the year. The fens have a vascular plant cover in addition to mosses, therefore spatial changes in evapotranspiration will be important (Lafleur and Rouse 1989). These fens are like the fens examined by Moore et al. (1990) and Roulet (1991) in northern Quebec. Further inland, the hydrology is dominated by numerous open ponds, separated by raised beach ridges and oval bog islands. The ponds are permanent, while the bogs and poor fens of the area probably experience a seasonal moisture deficit. Based on ecological and hydrological considerations, the largest CH_4 flux is expected when then the peatlands are saturated (e.g. Harriss et al. 1984, Moore and Knowles 1990) and in areas that remain wet for most of the growing season (Sebacher et al. 1986, Whalen Reeburgh 1988, Crill et al. 1988, Moore et al. 1990).

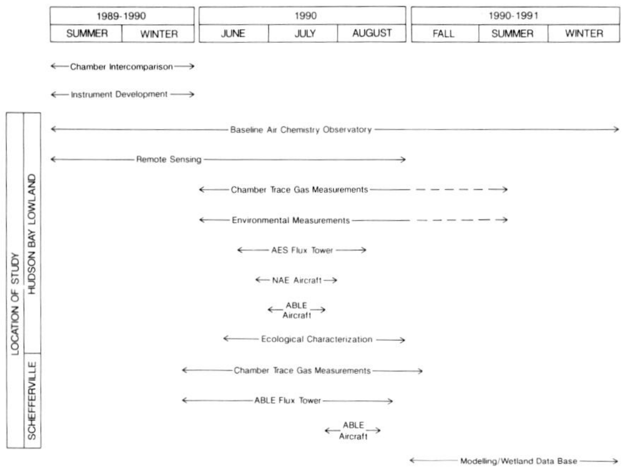

Fig. 4. Schedule for NWS/ABLE-3B.

Research activities

The research activities of the NOWES are divided into four main areas: 1) field studies in the Hudson Bay Lowland, which include chamber, tower and aircraft trace gas fluxes and supporting ground and airborne environmental measurements; 2) a regional atmospheric chemistry observatory located on the southern edge of the Lowland at Fraserdale, Ontario (Fig. 1); 3) field studies at Schefferville, Quebec, which are being co-ordinated by NASA ABLE-3B and NASA Biospheric Research: Emissions from Wetlands (BREW); and 4) modelling activities.

The research activities are taking place over a three year period, 1989 to 1992 (Fig. 4).

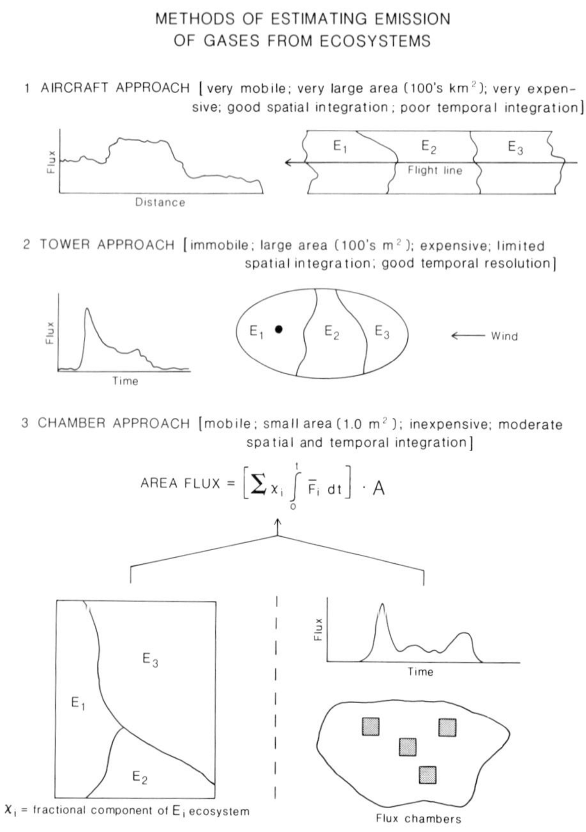

Fig. 5. Different approaches to the measurement of trace gas emissions.

Hudson Bay Lowland field studies

The 1990 Hudson Bay Lowland field programme consists of flux chamber studies, fluxes measured using a meteorological tower and aircraft measurements. In support of the flux measurements, a series of environmental measurements are being made at each site and as part of a regional survey to provide a ground truth for the extrapolation to be undertaken by the modelling group. While the aircraft has the ability to provide good spatial estimates of the trace gas flux, it is expensive to operate and therefore can only be used for a limited time (Fig. 5). The tower flux measurements are less expensive and hence they can be made continuously over a longer study period, but the tower itself is immobile. In a landscape where heterogeneity is high, either several tower installations are required (e.g. at each of the five sites) or an alternative, inexpensive and mobile approach is needed. Flux chambers will be used at all five sites and in a regional survey to be carried out by the National Center for Atmospheric Research (NCAR) team. Kinosheo Lake has been chosen as the site for the tower installation. This site is representative of much of the interior of the Lowland.

Flux chamber studies

The flux chamber studies are listed in Table 1. CH_4 will receive the most attention, while other trace gases, N_2O, VOC, DMS and CO_2 will be measured less frequently or at only a few sites. The isotopic composition of the CH_4 emitted from two of the sites will be determined several times during the summer. In addition to the flux measurements at the five permanent sites, a regional survey will be conducted in the month of July. In support of the flux chamber measurements a series of studies examining the physical, chemical and biological environment at the five sites will be undertaken (Table 2). These studies have been chosen to gain an understanding of the environmental conditions that favour higher gas fluxes and to provide the context for an ecological basis of extrapolating the chamber measurements. These studies are more extensive than the minimum set of measurements recommended to accompany trace gas flux measurements (Schimel 1989). Cores will

Table 1. Surface trace gas flux measurements of the NOWES.

Trace gas	Site	Frequency	PI	Institution
CH_4 – peatlands	all	weekly	T. Moore	McGill Univ.
			N. Roulet	York Univ.
	Churchill	weekly	W. Rouse	McMaster Univ.
CH_4 – ponds/lakes	all	weekly	C. Kelly	Univ. of Manitoba
CO_2			J. Rudd	Freshwater Inst.
DFO				
N_2O – peatlands	KL	int. prd.	D. Hastie	York Univ.
CO_2 – peatlands	all	3 int. prds.	G. Whiting	NASA/BREW
DMS – water	James Bay	int. prd.	J. Dacey	WHOI
CO_2, VOC, NO_y NCO, NO_2, O_3	Schefferville	continuous	S. Wofsy	Harvard
^{12}C, ^{13}C, ^{14}C	CF, KL	occasional	M. Whalen	N.Y. Dept of Health
CH_4, CO_2, NMHC	regional	int. prd.	P. Zimmerman	NCAR
^{12}C, ^{13}C, ^{14}C	sites		L. Klinger	NCAR
N_2O	lab		D. McKeeny	Windsor
CH_4	lab		D. Schmiel	CSU
			D. Valintine	
CH_4 production	all		T. Moore	McDonald
and consumption			R. Knowles	Coll.

DFO = Dept of Fisheries and Oceans Canada; NASA/BREW = Nat. Aeronautics and Space Adm./Biospheric Res. of Exchange from Wetlands; NCAR = Nat. Center for Atmospheric Res., Boulder, CO.; WHOI = Woods Hole Oceanographic Inst.

be obtained from all sites and returned to various laboratories for analysis and examination of the controls of N_2O and CH_4 emissions.

Flux tower studies

The flux tower study will be a collaborative effort between the Atmospheric Environment Service, Environment Canada and the Agrometeorology Section, Univ. Guelph. Micrometeorological measurements will be made, weather permitting, from a 20 m scaffold tower located near the southwest corner of Kinosheo Lake throughout July, 1990. A combination of eddy correlation and Bowen ratio techniques will be used to measure fluxes of heat, water vapour, momentum, CH_4, CO_2, total hydrocarbons (THC) and O_3 at a time resolution of 0.5 h. Spatial resolution will be on the scale of several hundred of metres and consequently the tower measurements are expected to provide a link between the chamber measurements and the regional scale aircraft measurements.

Fast response turbulence and air temperature measurements will be obtained by 3-axis anemometers at two levels on the tower. Eddy correlation fluxes of water vapour, CH_4, CO_2 and THC will be obtained by Lyman-alpha hygrometers, a fastresponse tunable diode laser (TDL), a fast-response infra-red gas analyzer (IRGA), a fast-response flame ionization detector (FID), respectively. A reversing boom with thermocouples and IGRA inlets for water vapour and CO_2 will be employed in the Bowen ratio apparatus. A infra-red radiation thermometer will be used for surface temperature measurements.

Table. Environmental and ecological sampling at the HBL field sites.

Variables	Site	Frequency	PI	Institution
Moisture and thermal regime	all	continuous	N. Roulet	York Univ.
Energy balance	KL, IF, CF	continuous	N. Roulet	York Univ.
			P. Lafleur	Trent Univ.
Radiation balance	KL, IF, CF	continuous	N. Roulet	
			P. Lafleur	
Micromet.	Schefferville	continuous	D. Fitzjarrald	SUNY (Albany)
Water chemistry	all	occasional	R. McCrae	IWD
			W. Glooschenko	NWRI
			N. Roulet	
Peat chemistry	all	na	W. Glooschenko	NWRI
Peat properties	all	na	R. Protz	Univ. Guelph

IWD = Inland Water Directorate Ontario Region; SUNY = State Univ. of New York, NWRI = Nat. Water Resources Inst., Environ. Canada.

Table 3. Measurements conducted routinely at the NOWES baseline air chemistry observatory at Fraserdale, Canada from 1990 onward.

Measurement	Type of observation
1. Wind speed at 3 levels to 40 m	Continuous
2. Wind direction at 3 levels	Continuous
3. Humidity at 3 levels	Continuous
4. Black carbon	Continuous
5. Carbon dioxide	Continuous
6. Methane	Continuous
7. Condensation nuclei	Continuous
8. Carbon monoxide	Continuous
9. Ozone	Continuous
10. Light hydrocarbons	Continuous
11. Wet deposition of nutrients	Weekly, warm season
12. Aerosol trace element chemistry	Daily July 1990
13. Sulphur dioxide, sulphates and methane sulphonic acid	Daily May to Oct 1990

Aircraft Studies

Two atmospheric research aircraft will fly co-ordinated missions in the Lowland during July, 1990. They are the Canadian Twin Otter operated by the National Aeronautical Establishment (NAE) of the National Research Council of Canada (NRCC) and the NASA ABLE Electra. The research flights have two objectives: 1) to quantify the areal average CH_4 flux from the study area and correlate the fluxes to environmental variables measured from the aircraft and on the ground; and 2) to make flux measurements of heat, momentum, CH_4, CO_2, H_2O and ozone at a scale which will allow for the testing of extrapolations from the chamber and tower flux measurements using remote sensing.

The Twin Otter will be configured for flux measurements as during the recent First ISLSCP (International Satellite Land Surface Climatology Project) Field Experiment (FIFE). The instrumentation includes a system to make accurate, realtime wind and gust computations for use with the eddy correlation technique for trace gas flux measurement. The ESRI Gas Analyzer, developed by Agriculture Canada is mounted in a large duct through the rear fuselage and provides CO_2 and water vapour concentration measurements at a frequency response of 20 Hz. A fast response CH_4 analyzer for flux estimates has been developed for the study. The NASA Electra is instrumented to make a significant number of in situ tropospheric chemistry measurements, as well as to remotely sense the distribution of ozone and particulate above and below the flight altitude using a DIAL lidar system (Harriss et al. 1988).

Several different flight patterns will be used to best meet the objectives of the NOWES. With its low altitude and low speed capabilities, the Twin Otter will concentrate between altitudes of the tower height (20 m) and c. 300 m with flights legs of c. 15 km. The faster Electra is restricted to operations above 150 m and will focus on regional scale measurements with flux runs of c. 60 km in length. During each co-ordinated mission (there will be missions where each aircraft is doing a separate experiment), both aircraft will fly a fixed 100 km track (the North Point to Kinosheo Lake transect) at the same altitude and collect data to be used in comparisons of gas concentration and flux estimates. The Twin Otter will also perform routine soundings to 3,000 m to document the thermal and dynamic structure of the atmosphere in the Kinosheo Lake area. Experimental night flights will be performed with the Twin Otter during overcast, windy conditions when the mixed layer may retain considerable depth after dark.

The NASA Electra will also conduct regional scale measurements of trace gas distributions and CH_4, CO and 0_3 flux from the Kinosheo Lake area to Churchill and over the Schefferville region (Fig. 1). In the Hudson Bay Lowland these studies will be particularly important for characterizing the role of permafrost as a factor influencing CH_4 emissions. Flights will also be conducted in an upwind/downwind sampling pattern when Arctic air penetrates across the Lowland to characterize the vertical and horizontal influence of biogenic CH_4 on the regional distribution on atmospheric CH_4.

Land characterization, remote sensing and ecology

Several studies have examined the feasibility of using remote sensing to clarify the wetland ecosystems of the Hudson Bay Lowland (Pala and Boissonneau 1982). Twenty-seven wetland systems are recognized. These correspond to the ecological classification of Riley (1982) which is broadly based on nutrient availability, acidity and moisture regime: three factors which are also important in determining the magnitude of trace gas fluxes. The Ontario Centre of Remote Sensing (OCRS) and the Canada Centre of Remote Sensing (CCRS) will produce LANDSAT-TM (Thematic Mapper) at a 28 m resolution for all the field sites and flight transects. This data will be used in conjunction with the ecological surveys of the Lowland conducted by Riley (1982) to extrapolate the gas flux measurements (see the Modelling studies). CCRS will also use an airborne multispectred electro-optical imaging scanner (MEIS) to determine the vegetative composition of the satellite image pixels. The variability of the method surface temperature and hopefully depth of water table below the surface will be obtained with the MEIS thermal imager and airborne synthetic operative radar (SAR).

To place the present day gas fluxes in the context of the evolution of the Hudson Bay Lowland, several studies will examine the paleao-ecological and present day ecology of wetlands representative of the gas flux study site.

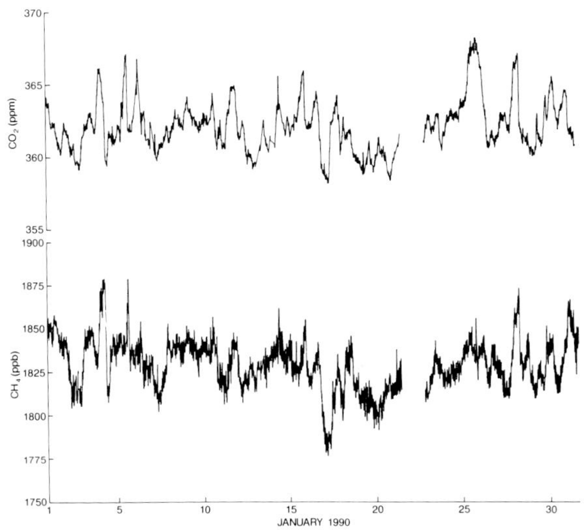

Fig. 6. The CO_2 and CH_4 concentrations measured at the Fraserdale Baseline Chemistry Observatory, Jan 1990.

Regional Atmospheric Chemistry Programme

Although atmospheric CH_4 and CO_2 are measured at many locations around the world the sites have been deliberately located away from the major sources or sinks (Steele et al. 1987). There is a need for an observatory in the northern, midcontinental, wetland region (Tan et al. 1989). Therefore the NOWES has erected an atmospheric observatory at Fraserdale, Ontario (Fig. 1), located on the southern perimeter of the Hudson Bay Lowland (50°N, 87°W). Annual and seasonal variations of the concentration of climate-related atmospheric trace constituents and other substances that are potentially released to the atmosphere by northern wetlands and boreal forests will be measured (Table 3). An example of the observatory's product is shown in Fig. 6. When the winds are from the sector 300 through 50° the Lowland will be a major source of trace gases. When the winds are from 210 to 300° and 50 to 100° the air will have passed over the boreal forest and the remaining sectors will likely bring in air that has received a significant input of anthropogenic emissions.

Schefferville Field Studies

Schefferville, Quebec (54°48′N, 66°49′W) is located in the subarctic/boreal forest ecotone (Waterway et al. 1984). Deglaciated about 5,000–6,000 years ago, the Schefferville area possesses a distinct topographic alignment of ridges and troughs, associated with the sedimentary rocks of the Labrador Trough. On the ridge tops, underlain by permafrost, the dominant vegetation is composed of alpine tundra, with graminoids, ericaceous plants and lichens and mosses covering a thin layer of soil. This grades into subalpine heath, with greater coverage of shrubs and the occurrence of tree species, such as black and white spruce, in a stunted form. On the lower, sheltered slopes, this passes into spruce-lichen woodlands, a very common vegetation association across subarctic Canada and composed of scattered spruce and tamarack trees with a thick ground cover of lichens and a variable shrub cover. In the wetter sites, denser spruce-moss forests similar to boreal forest stands further south, are found.

Wetlands cover c. 25% of the land surface of the Schefferville area and lie in the Low subarctic wetland region of Canada (National Wetlands Working Group 1988). Fens dominate and differences in hydrology and trophic status of the peat water (associated with variations in substrate from quartzite to dolomite) produce a variety of forms and vegetation assemblages. Forested margins grade into open water in the central sections and rich fens dominated by sedges to poor fens with *Sphagnum* spp. form the trophic gradient. Patterned fens are common, with distinctive ridge – pool topography and differences in hydrology and species composition and strong variations in methane emission rates (Moore and Knowles 1987).

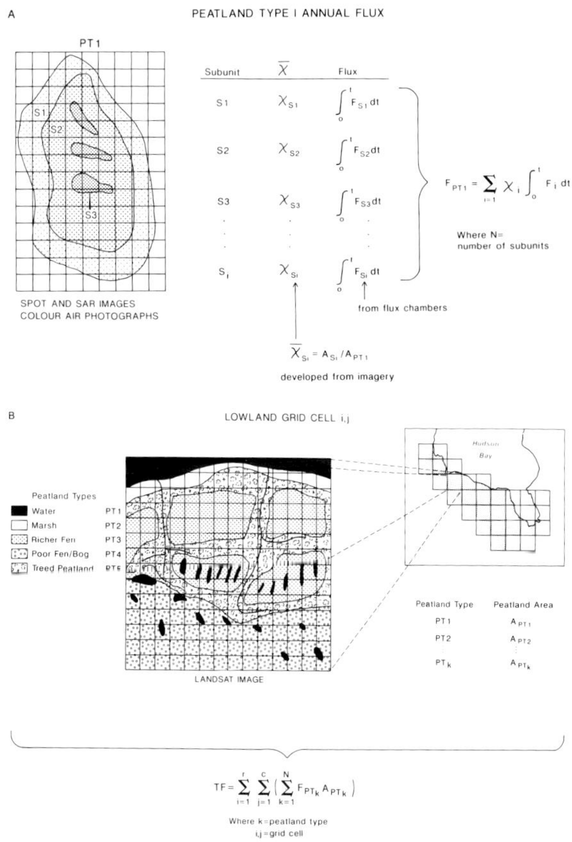

Fig. 7. The approach for the spatial extrapolation of chamber measurements to estimate the role of the Hudson Bay Lowland in global CH_4 emissions.

The ABLE 3B studies in the Schefferville area will follow the same type of experimental design as discussed above for the Hudson Bay Lowland studies. Two micro meteorological towers will continuously characterize fluxes of CH_4, CO_2, O_3, selected nitrogen gases and meteorological parameters. One tower will be located in a woodland environment and the other in a wetland. The NASA Electra aircraft will conduct regional air chemistry and flux measurement studies with the Schefferville towers as critical reference points. These measurements combined with previous studies of wetland hydrology and CH_4 emissions in the Schefferville area (Roulet 1991) will provide the basis for an important comparison of continental interior ecosystems to the Hudson Bay Lowland which have a significant maritime influence.

Modelling studies

The modelling programme of the NOWES consists of two components. The first is the extrapolation of spot measurements to regional and latitudinal estimates of the role of northern wetlands as a source of atmospheric CH_4. The second is to attempt to develop a model for trace gas emissions based on relationships between trace gas flux and several environmental and biological variables. Since the actual production of trace gases is a microbiological process, the variables used in these relationships are surrogates, which themselves correlate with the microbiological activity.

Spatial extrapolation

The approach to be used in the extrapolation is outlined in Fig. 7. The landscape input is based on a hierarchical classification system. Gas flux varies at a scale much smaller than can be remotely sensed (i.e. <5 m^2), so several types of detailed imagery, flight videos and aerial photography will be used in target areas to develop regional statistical distributions of sub-peatland spatial characteristics. These distributions will be combined with the gas fluxes derived from seasonal, chamber measurements to yield an ecosystem weighted total gas flux based on eco-units (Fig. 7). Total emissions from the Lowland will be derived by prorating the peatland weighted fluxes with the actual area of each peatland type determined by LANDSAT Thematic Mapper. The peatlands will be differentiated using a combination of the Canadian Wetland Classification System and the Ontario System (Jeglum and Cowell 1982). In the Hudson Bay Lowland, two wetland classes predominate, bogs and fens, with several wetland forms being important in each class: domed, flat and string bogs, and basin, floating, net and northern ribbed fens. Peatland types are determined by general physiognomy and vegetation cover. In the Lowland these consist of treed, shrub, graminoid and moss. The highest level of classification (that obtained by LANDSAT) used in the Northern Wetland Study will consist of a combination of form and type: bogs, rich fens, poor fens, treed fens, treed bogs and coastal marshes, while the statistical distribution at the sub-peatland level will be based primarily on form.

At the present time the only attempt to test of the validity of flux measurements extrapolated from chambers has been undertaken by the ABLE-3A mission in Alaska (Bartlett et al. 1989). However, a comparison among chamber, aircraft and tower derived fluxes will be made. A problem associated with this kind of test is determining the surface of origin of the gas flux measured by the tower or aircraft. Two experiments have been designed. At the tower site, detailed remote sensing (SPOT and colour aerial photography) will be used to derive an areal, subpeatland weighting, to extrapolate the chamber measured fluxes to the footprint area of the tower. The second experiment will involve spatially extrapolating chamber measurements along the flight transect from North Point to Kinosheo Lake using the full extrapolation model and comparing this with aircraft measured fluxes.

Sensitivity of source/sink strength to change

Many studies (Harris et al. 1984, Sebacher et al. 1986, Moore and Knowles 1987, Crill et al. 1988, Whalen and Reeburgh 1988, Moore and Knowles 1990) have shown that the CH_4 flux from peatlands is correlated with peat temperature and soil moisture. In addition, there appears to be a broad relationship between wetland types and the ratio of net primary productivity and the flux of CH_4 from each wetland group (Aselmann and Crutzen 1989). However, since these relationships are empirically derived, and are not causal, they need to be developed for each major peatland region and each type of peatland. Observations of the variables necessary to derive these relationships, from several peatlands in the Lowland, are being undertaken as part of the summer field campaign. In addition, a simple hydrological and thermal model for each wetland type will be developed. Various climate scenarios will be used to provide inputs to the physical models and the resultant change in peatland environment will provide the input to a gas flux model.

Acknowledgements – This is Scientific Contribution Number 90–1 of the Canadian Institute for Research in Atmospheric Chemistry (CIRAC). The work described herein was undertaken as part of the CIRAC Northern Wetlands Study which was funded in part by Collaborative Grant from NSERC Canada and Atmospheric Environment Service, Environment Canada.

References

Aselmann, I. and Crutzen, P. 1989. Global distribution of natural freshwater wetlands and rice paddies, their net primary productivity, seasonality and possible methane emissions. – J. Atmos. Chem. 8: 307–358.

Bartlett, D. S., Ritter, J., Sache, G., Bakwin, P., Wofsy, S., Fan, S., Bartlett, K. and Fitzjarrald, D. 1989. Regional flux estimates from a synthesis of enclosure, tower, and aircraft measurements. – EOS 70: 286.

Botch, M. S. and Masing, V. V. 1983. Mire ecosystems in the U.S.S.R. – In: Gore, A. P. (ed.), Ecosystems of the worlds 48-mires: swamp, bog, fen, and moor regional studies. Elsevier, Amsterdam, pp. 95–152.

Cicerone, R. J. and Oremland, R. S. 1988. Biogeochemical aspects of atmospheric methane. – Global Biogeochem. Cycles 2: 299–327.

Cowell, D. W. 1982. Earth sciences of the Hudson Bay Lowland: literature review and annotated bibliography. – Working paper no. 18. Lands Directorate, Environ. Canada, Ottawa.

Crill, M. P., Bartlett, K. B., Harriss, R. C., Gorham, E., Verry, E. S., Sebacher, D. I., Madzar, L. and Sanner, W. 1988. Methane flux from Minnesota peatlands. – Global Biogeochem. Cycles 2: 371–384.

Crutzen, P. J. 1991. Methane's sink and sources. – Nature (Lond.) 350: 380–381.

Glooschenko, W. A. and Martini, I. P. 1983. Wetlands of the Attawapiskat River mouth, James Bay, Ontario, Canada. – Wetlands 3: 64–76.

Harriss, R. C., Sebacher, D. I. and Day, F. P. 1984. Methane flux from the Great Dismal Swamp. – Nature (Lond.) 297: 673–674.

– Wofsy, S. C., Garstang, M., Browell, E. V., Molion, L. C. B., McNeal, R. J., Hoell, J. M., Bendura, R. J., Beck, S. M., Navarro, R. L., Riley, J. T. and Snell, R. L. 1988. The Amazon boundary layer experiment (ABLE 2A): dry season 1985. – J. Geophys. Res. 93: 1351–1360.

Jeglum, J. K. and Cowell, D. W. 1982. Wetland ecosystems near Kinoje (Kinosheo) Lakes, southern interior Hudson Bay Lowland. – Naturaliste Canadian 109: 621–635.

Khahil, M. A. K., Rasmussen, R. A. and Shearer, M. J. 1989. Trends of atmospheric methane during the 1960s and 1970s. – J. Geophys. Res. 94: 18279–18288.

Lafleur, P. and Rouse, W. R. 1989. The influence of surface cover and climate on energy partitioning and evaporation in a subarctic wetland. – Boundary Layer Meteorol. 44: 327–347.

Matthews, E. and Fung, I. 1988. Methane emission from natural wetlands: global distribution, area, and environmental characteristics of sources. – Global Biogeochem. Cycles 1: 61–86.

Mitchell, J. B. 1989. The "greenhouse" effect and climate change. – Rev. Geophys. 27: 115–140.

Moore, T. R. and Knowles, R. 1987. Methane and carbon dioxide evolution from subarctic fen. – Can. J. Soil Sci. 67: 77–81.

– and Knowles, R. 1990. Methane emissions from fen, bog and swamp peatlands. – Biogeochemistry: 11: 45–61.

– Roulet, N. T. and Knowles, R. 1990. Temporal and spatial variability and environmental controls of methane flux from subarctic fens. – Global Biogeochem. Cycl. 4: 29–46.

National Wetlands Working Group. 1988. Wetlands of Canada. – Polyscience, Montreal.

Pala, S. and Boissonneau, A. 1982. Wetland classification maps for the Hudson Bay Lowland. – Naturaliste Canadien 109: 653–659.

Pearman, G. L. and Fraser, P. J. 1988. Sources of increased methane. – Nature (Lond.) 332: 489–490.

Protz, R., Ross, R. J., Shipitalo, M. J. and Terasmae, J. 1988. Podzolic soil development in the southern James Bay Lowlands, Ontario. – Can. J. Soil Sci. 68: 287–305.

Price, J. S. and Woo, M.-K. 1989a. Studies of a subarctic coastal marsh. I: Hydrology. – J. Hydrol. 103: 275–292.

– and Woo, M.-K. 1989b. Studies of a subarctic coastal marsh. II: Salinity. – J. Hydrol. 103: 293–307.

Riley, J. L. 1982. Hudson Bay Lowland floristic inventory, wetland catalogue and conservation strategy. – Naturaliste Canadian 109: 543–555.

Robertson, G. P. 1989. Trace gas exchange and chemical and physical climate: Critical interactions. – In: Andrae, M. O. and Schimel, D. S. (eds), Exchange of trace gases between terrestrial ecosystems and the atmosphere. – Life Sci. Res. Report 47. Wiley, New York, pp. 303–320.

Roulet, N. T. 1991. Surface adjustment in northern peatlands: the maintenance of effective surface saturation. – Arctic and Alpine Res. 23: 303–310.

Schimel, D. S. 1989. Research priorities for studies on trace gas exchange. – In: Andrae, M. O. and Schimel, D. S. (eds), Exchange of trace gases between terrestrial ecosystems and the atmosphere. – Life Sci. Res. Report 47. Wiley, New York, pp. 321–332.

Sebacher, D. I., Harriss, R. C., Bartlett, K. B., Sebacher, S. M. and Grice, S. S. 1986. Atmospheric methane sources: Alaskan tundra bogs, an Alpine fen and a subarctic boreal marsh. – Tellus 38: 1–10.

Sims, R. A., Cowell, D. W. and Wickware, G. M. 1982. Classification of fens near southern James Bay, Ontario, using vegetation physiognomy. – Can. J. Bot. 60: 2608–2623.

Steele, L. P., Fraser, P. J., Rasmussen, R. A., Khalil, M. A. K., Masarie, K. A. and Thoning, K. W. 1987. The global distribution of methane in the troposphere. – J. Atmos. Chem. 5: 125–171.

Tan, P. P., Conway, T. J. and Nakazawa, T. 1989. Latitudinal distribution of the sources and sinks of atmospheric carbon dioxide derived from surface observations and an atmospheric transport model. – J. Geophys. Res. 94: 5151–5172.

Vaghjiani, G. L. and Ravishankara, A. R. 1991. New measurement of the rate coefficient for the reaction of OH with methane. – Nature (Lond.) 350: 406–409.

Whalen, S. C. and Reeburgh, W. S. 1988. Methane flux time-series for tundra environments. – Global Biogeochem. Cycl. 2: 399–409.

Waterway, M. J., Lechowicz, M. J. and Moore, T. R. 1984. Vegetation of the Schefferville region, Noveau Quebec. – McGill Subarctic Res. Paper 32: 7–20.

Ecological Bulletins 42: 86–97. Copenhagen 1992

High-latitude ecosystems: sources and sinks of trace gases

N. Roulet, W. Reeburgh, M. Alperin, R. Desjardin, V. Galchenko, W. Glooschenko, R. Lassiter, K. Lassey, T. Moore, H. Schiff, B. H. Svensson, M. Wahlen and G. Zarvarzin

Roulet, N., Reeburgh, W., Alperin, M., Desjardin, R., Galchenko, V., Glooschenko, W., Lassiter, R., Lassey, K., Moore, T., Schiff, H., Svensson, B. H., Wahlen, M. and Zarvarzin, G. 1992. High-latitude ecosystems: sources and sinks of trace gases. – Ecol. Bull. (Copenhagen) 42: 86–97.

Corr. author: N. Roulet, Dept of Geography, York Univ., 4700 Keele St., North York, Ontario, Canada M3J 1P3.

Introduction

High latitude terrestrial ecosystems, including freshwater, which for the purposes of this report encompass the boreal forest, subarctic and arctic regions (Fig. 1), are significant sources of atmospheric CH_4 (Sebacher et al. 1986, Matthews and Fung 1987, Aselmann and Crutzen, 1989), and may be important as a source or sink of other trace gases. Biogenic trace gases that may be significant include NMHC, N_2O, NO_x, DMS, COS, and H_2S. High latitude ecosystems can broadly be divided into wet and dry tundra, wetlands (which at high latitudes are predominantly peatlands), conifer forest with some mixed and hardwood component, and lakes. Many of these ecosystems are significant sources of trace gases because they have large stores of organic matter, especially the northern peatlands. Furthermore, soils are generally low in oxygen through the accumulation of surface water due to poor drainage, topography, and shallow soils underlain by impermeable bedrock and, in many cases, permafrost. Finally, it has been projected in various climate modelling scenarios (e.g. Mitchell 1989) that the high latitudes may experience a greater amplitude of change than the temperate and tropical regions. The biological responses to temperature changes at low temperatures tend to be greater than at high temperatures, suggesting that climate change may have its greatest biological response in the higher latitudes.

Objectives

The objective of this report is to outline a research plan for estimating the role of high latitude ecosystems as sources of biogenic trace gases, and sinks for biogenic and anthropogenic trace gases. To attain this goal, three main research activities are required:

1) estimation of the strength of the trace gas sources and sinks in high latitude ecosystems in a global context. Availability of data on land type (Aselmann 1989) and flux measurements (Harriss 1989), and tests of extrapolation models (Matson et al. 1989) need to be addressed;
2) determination of the principal ecological and environmental characteristics that correlate with, or are surrogates for, the major microbiological and environmental controls of ecosystem trace gas production, consumption and transport (Rosswall et al. 1989) in high latitude ecosystems; and
3) estimation of the sensitivity of the high latitude sources and sinks of trace gases to environmental change (Robertson et al. 1989).

Activity 1: Estimation of trace gas emissions from high latitude ecosystems

The modelling efforts of Aselmann and Crutzen (1989) and Matthews and Fung (1987) produced a large difference in the estimation of the northern wetlands as a source of atmospheric CH_4. This difference is attributable to differences in: 1) the data base on ecosystem type and areal extent of each ecosystem, 2) the flux estimates used for each ecosystem type, and 3) annual duration of flux period. In addition, there are no tests of these or other regional areal extrapolations of chamber measurements. Three tasks are proposed to address these issues.

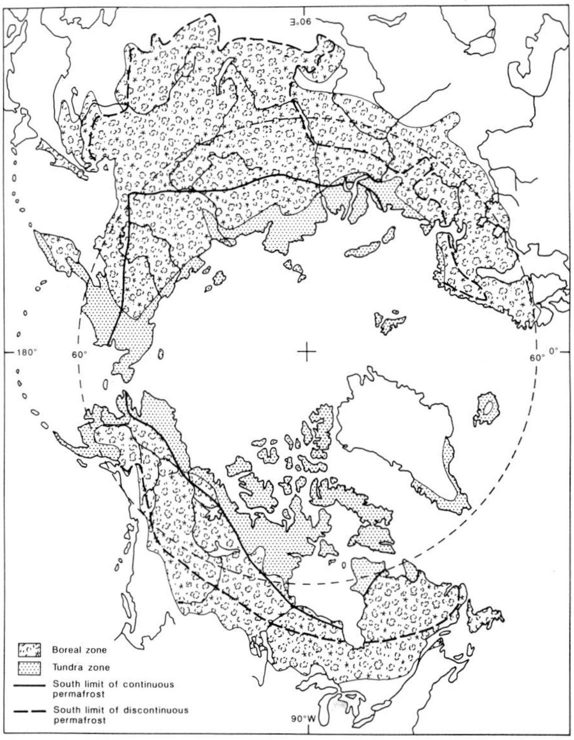

Fig. 1. Location of the boreal forest and arctic regions of the northern hemisphere and the southern extent of continuous and discontinuous permafrost.

Task 1: Standardized high latitude ecosystem data base

There is a need for a standardized classification of high latitude ecosystems to develop a data base that is relevant to the estimation of regional and global trace gas emissions. The classification should be based on vegetation (forested/non-forested; deciduous/conifers/mixed), soils (organic/mineral/permafrost), hydrology (wet/dry/duration of inundation), and biogeochemistry (nutrient status). A hierarchial approach is needed to obtain a spatial resolution broad enough to allow global and regional extrapolation, but fine enough to allow recognition of small scale variations in vegetation and environmental factors which control the gas fluxes. The areal extent of major ecosystems should be determined by remote sensing (e.g. LANDSAT TM), and existing land inventories, and the composition of sub-ecosystems derived from statistical correlations between remotely sensed and ground measured data in targeted areas of each ecosystem. Ecosystem categorization could include bogs, fens, swamps, marshes, lakes, upland forest, lowland forest, and wet and dry tundra. Subecosystem level of classification would entail determining the ecosystem descriptors listed above.

Several classification systems have been developed for wetlands that could be used as a starting point. In Canada an ecologically based classification system for wetlands (Fig. 2) is used (National Wetland Working Group 1988). Through the activity within the Tundra Study (Sweden) a classification has been developed. Classifications have also been developed for forestry purposes, and more recently these have been used to evaluate the areal extent of different peatlands and forests in Sweden.

Objectives
To develop a standardize ecosystem classification for extrapolating point trace gas flux measurements.

Research plan
Similar programmes as that of Canada of wetlands classification should be initiated for areas of North America, Europe and Asia where none exist, and a classification should be developed for non-wetland northern ecosystems. A workshop should be held drawing together land classification specialists and land data base managers from northern countries to determine how to maximize existing data base resources and to determine regions that require primary analysis. Members of the trace gas flux community should also attend.

An institution (e.g. NCAR, Canada Centre for Remote Sensing [CCRS], Climate Research Centre –

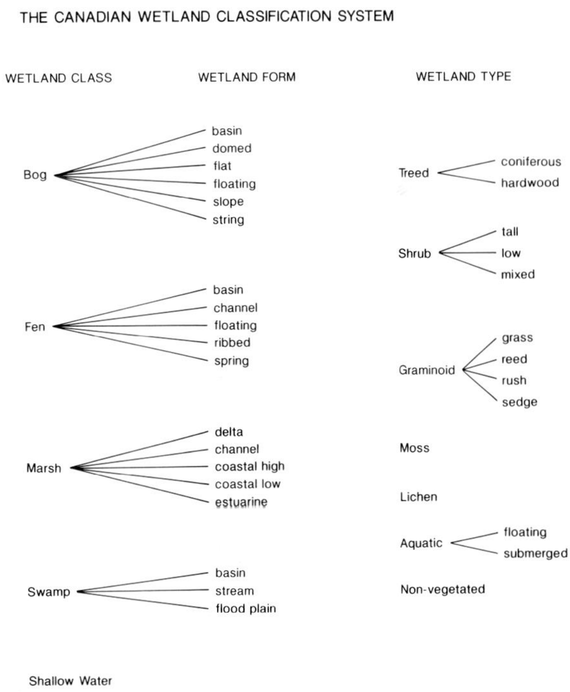

Fig. 2. Example of the Canadian Wetlands classification system (NWWG 1989) in the figure indicating the wetland class and form. A third level of classification, referred to as wetland type, is based on vegetation composition. This may be the most useful form of classification for trace gas exchange work since vegetation is an integrator of many of the variables that influence gas emissions.

Table 1. Examples of existing and ongoing CH_4 flux measurements in northern ecosystems.

Region	Ecosystem	Citation
SA Canada	Mineral poor fens	Moore and Knowles (1987, 1990)
SA Scandinavia	Bogs	Svenson and Rosswall (1984)
A/SA Alaska	Bogs, fens, tundra	
BF USA	Bogs, fens	Crill et al. (1988)
BF Canada	Bogs, fens, swamps, beaver ponds	Vitt et al. (1990) Moore, Roulet and Knowles (unpubl. data)
BF NA.	Beaver ponds	Naiman et al. (1986)
BF Canada	Shield lakes	Rudd and Hamilton (1975), Kelly (1979)
SA/BF Canada	Hudson Bay Lowland	
SA Canada	Mineral poor fens	Hoell et al. (1990)
A Alaska	Wet and dry tundra	Bartlett et al. (1989)

SA = Subarctic, A = Arctic, BF = Boreal Forest

McGill Univ.) is needed to house the data base, as well as a coordinator to oversee the organization and development of the classification. The final data base should be accessible to all interested researchers.

Timetable
1992: The High Latitudes Trace Gas Exchange Group should select a coordinator for the data base and convene a workshop in autumn 1992 to utilize the expertise being developed in this area by such groups as the Northern Wetlands Study and The Boreal Forest Experiment.
1994: Completion of a northern wetland data base.
1995: Completion of a boreal forest and the tundra/polar data base.

Task 2: Ecosystem/geographical distribution of trace gas fluxes

Estimates of trace gas emissions from northern ecosystems require flux measurements using chambers, towers (Chahuneau et al. 1989) and aircraft (Desjardins and MacPherson 1989) which incorporate the temporal and spatial variability in each ecosystem unit and geographical region. Although reliable flux measurements of CH_4 are available for several ecosystems and regions (Table 1) data for many northern ecosystems and regions are missing. Flux measurements of other trace gases are lacking from almost all cases. Measurements should be accompanied by isotope determination where applicable. Biomass burning of northern forested and peatland ecosystems may also be an important source of trace gases but this has received little study.

Table 2: Ecosystem and regions for CH_4 flux measurements.

Ecosystem	Region
Upland boreal forest (wet/dry) (conifer/hardwood/mixed; beaver ponds; coastal marshes; treed wetlands; shallow lakes; large reservoirs	All northern regions
All wetlands (especially peatlands)	Soviet Union (especially western Siberia and Pechora regions), Finland and Scandinavia
Bogs, fens, swamps, and delta wetlands	Canada – Mackenzie Valley, N.W.T., Northern Alberta, Saskatchewan, Manitoba, and Newfoundland

Objectives
To derive estimates of the trace gas fluxes from all major ecosystems and geographical regions of the high latitudes.

Research plan
Annual integrated flux measurements of CH_4 should be obtained from the ecosystems and regions listed in Table 2. These regions and ecosystems have been identified because; 1) they represent a large percentage of the high latitudes (e.g. western Siberia), 2) there have been a few measurements that suggest the ecosystem flux may be large (e.g. beaver ponds), and, 3) they could be a potential large source but no data exists (e.g. reservoirs). Pilot studies should be initiated to determine if other trace gases are emitted in significant quantities to warrant further studies. A priority list for various trace gases is suggested in Table 3.

Research groups are required to carry out flux measurements, especially in the Russia, Finland and Scandinavia. There is a need for the reliable and practical field trace gas measurement techniques that can be used in remote, generally inaccessible, and sometimes hostile environments. When fast response sensors are inadequate (Anderson et al. 1989) eddy accumulation, which is a variant of eddy correlation techniques, need to be further developed (Buckley et al. 1988, Wesely et al. 1989). This technology should be incorporated in aircraft-based systems (Desjardins and MacPherson 1989).

Table 3. Priority list for gas flux measurements for high latitude ecosystems.

Tundra	Boreal forest	Wetlands
CH_4	CO_2	CH_4
CO_2	NMHC	CO_2
N_2O	CH_4	N_2O
NMHC	N_2O	NO_x
NO_x	O_3	
	NO_x	

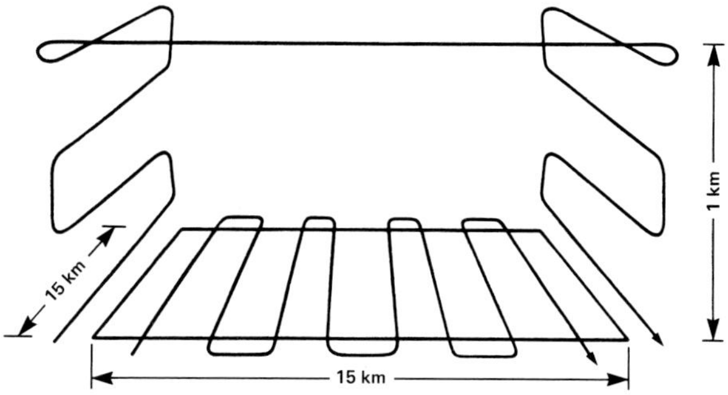

Fig. 3. Flight pattern used for CO_2 flux measurements in the recent FIFE experiment (courtesy of R. Desjardins).

Timetable
1992: Initiate measurement programmes. Development of field techniques for measuring trace gases.
1996: Completion of CH_4 flux measurements from identified ecosystems and regions and initiate measurements of NMHC fluxes.
2000: Completion of measurement of the fluxes on non-methane trace gases from high latitude ecosystems.

Task 3: Extrapolation models and test of extrapolations

Several methods of extrapolating gas fluxes have been delineated by Matson et al. (1989) and Stewart et al. (1989). These include the linear extrapolation based on areal extent of each ecosystem with an associated flux, and extrapolations based on environmental correlates (e.g. temperature, moisture). The first approach is feasible at the present time, but further process studies are required for the second approach to be practical for the high latitudes. However, it must be recognized that both approaches assume that the fluxes are related to ecosystem type or ecosystem variables and these relationships have received little empirical verification.

How reliable are these extrapolations? This can be evaluated in two ways. Schuepp et al. (1989) were first to compare modelled fluxes and regional fluxes estimated using tower and aircraft measurements. Recent progress in estimating the actual area sampled with flux systems should help test this approach (Schuepp et al. 1990). The combination of aircraft based areal flux measurements with satellite information holds much promise (Mack et al. 1990). There is also a requirement for the interchangeability and intercalibration of chamber techniques since chambers are used in both gas emission and consumption studies.

A second test of spatial extrapolations (and of source strengths generally) is based on the quantitative balance (budgets) of constituent isotopes, determined by measurement of isotope ratios of the trace gases in the atmosphere and in their sources. This approach has proven particularly useful in respect of the ^{13}C isotope via measurements of $^{13}C/^{12}C$ ratios which show distinctive differences among some sources (Tyler 1986, Quay et al. 1988, Stevens 1988, Stevens and Engelkemeir 1988) and radiocarbon (^{14}C) abundance (Lowe et al. 1988, Wahlen et al. 1989, Manning et al. 1990) and shows promise also for deuterium balance via D/H ratios. This approach requires a knowledge of the isotopic fractionation in the major tropospheric destruction reaction (CH_4 and OH), and shows that the empirical correlation between D/H in CH_4 from shallow inundated ecosystems and that of local precipitation can provide a basis for reliable flux extrapolation to the global scale. Isotope studies are also helpful in improving our understanding of the important process pathways in bacterial CH_4 production and consumption (e.g. Whiticar et al. 1986, Cicerone and Omerland 1988).

Objectives
Develop methods for extrapolation gas flux measurements and testing the validity of the modelled extrapolation.

Reseach plan
Several comparisons between modelled fluxes (extrapolation) from an ecosystem and fluxes measured from towers and aircraft have been undertaken or are being presently undertaken. Examples of these are FIFE (Sellers et al. 1990), ABLE–3A (Harriss and Wofsy 1989), ABLE–3B (Hoell et al. 1990) and the NOWES (Schiff 1988).

Investigations into the isotope comparison of CH_4 sources and atmospheric composition of CH_4 are being performed (Lowe et al. 1988, Stevens and Engelkemeir 1988, Wahlen et al. 1989). The validity of this approach has been demonstrated and especially the multi-isotope approach is most promising. However, only few data have been obtained from high latitude ecosystems.

To further these investigations, there are three principal research requirements;

1) Model tests by aircraft in the boreal forest regions similar to those of NOWES, ABLE and FIFE: Areal flux estimates of trace gases should be obtained using boundary layer aircraft flights in a designed study on the gas exchange over a particular area. Areal estimates of heat and water vapor have been obtained from a 15×15 km site in FIFE (Betts et al. 1989, Desjardins et al. 1989). This kind of information is required to test various scaling up techniques. For more detailed intercomparison between models, towers and aircraft, a relatively homogeneous area of at least 4×4 km is desirable (Fig. 3). This homogeneity requirement may be difficult to obtain in many high latitude ecosystems.

2) Comparison between extrapolated emissions from high latitude ecosystems and isotope reconstructions of trace gas sources/ sinks: From past results on isotopic composition of biogenic CH_4 sources it seems feasible

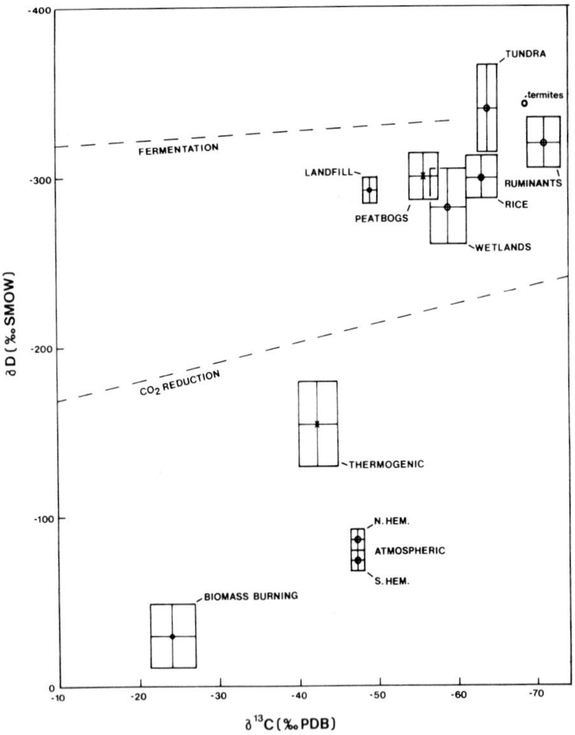

Fig. 4. Delta D versus delta ^{13}C for methane of the various sources and for atmospheric methane (from Wahlen et al. 1990). The boxes give the range encountered in samples collected over the last 3 yr. It demonstrates that the isotopic composition is specific for a given source of methane, and illustrates how a source mix has to be constructed to reproduce the atmospheric methane isotopic composition. Also shown are the boundaries inferred for biogenic methane from acetate fermentation and from CO_2 reduction. The relative contribution from the two processes in different wet ecosystems are very similar, c. 70% from acetate fermentation and c. 30% from CO_2 reduction.

that even subgroups of ecosystems are distinguishable (Fig. 4). These investigations of isotopes in various types of high latitude ecosystems could refine source strength estimates. Since much of the stored carbon is old, ^{14}C measurements on CH_4 could reveal how much "fossil" or aged carbon is released to the atmosphere in this way (Lowe et al. 1988, Wahlen et al. 1989). Isotope analysis and time series measurements of atmospheric CH_4 from source regions would reveal any temporal trends in the CH_4 emissions from high latitude ecosystems (Stevens 1988).

3) Inter-comparison of chambers: Static and dynamic chambers are widely used in estimating fluxes. Chambers have the advantage of yielding inexpensive, rapid and generally intercomparable results (Mosier 1989), but they remain an operationally defined compromise. Dörr and Münnich (1987) have detailed a technique based on soil ^{222}Rn distributions and fluxes, that should be employed as a means of putting the range of flux chamber measurements on a better, and hopefully absolute footing. This method clearly works in dry soils: it requires testing in wetland and waterlogged soils. It also has the advantage of producing non-destructive measurements, and with appropriate corrections for bacterial turnover, and of providing estimates of net primary production. The natural variability in trace gas emissions is so large, even from seemingly homogeneous ecosystems (Fig. 5) that it is doubtful that any meaningful results could be obtained in a natural setting without using the approach as that outlined above. Gas standards should also be addressed for comparative studies (Galbally 1989).

Generally the resources are in place to carry out most of this activity. Plans are well underway for extrapolation experiments as part of the upcoming Boreal Forest Experiment in 1994 in Canada. Preliminary plans have been prepared, but will require refinement once the results of FIFE and the NOWES are analyzed.

Timetable
1990: NOWES Scale up experiment. ABLE–3B Scale up experiment.
1991: Chamber inter-comparison studies.
1994: BOREAS extrapolation study.

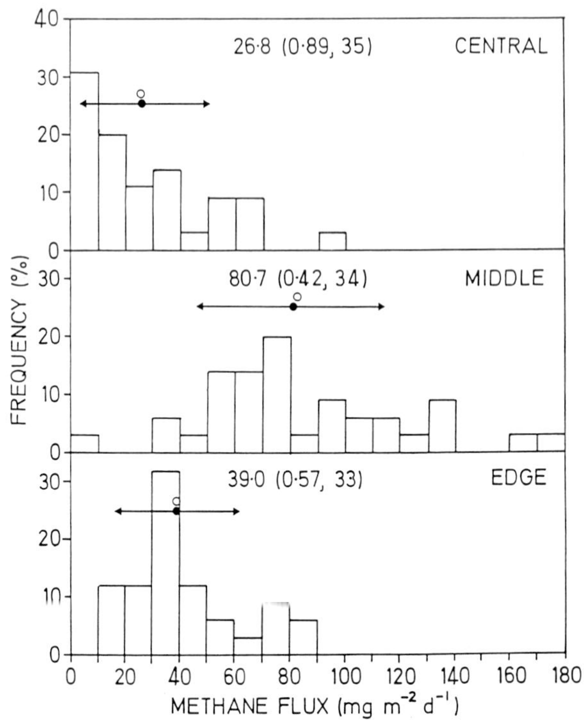

Fig. 5. Histograms for CH_4 flux based on 30 chambers at three different sites (E = edge, M = middle, C = central flooded site) in a mineral-poor subarctic fen (Moore and Roulet unpubl. data). The two dots indicate the arithmetic (solid) and geometric (open) means, and in brackets is the coefficient of variation and the standard deviation.

Activity 2: Ecological and environmental controls and correlates for trace gas source/sink in high latitude ecosystems

Many studies have shown that trace gas emissions, particularly that of CH_4 from wetland ecosystems, are correlated with a number of environmental and biological variables. Examples of the variables that correlate with CH_4 flux are temperature (Svensson 1984, Moore and Knowles 1987, Crill et al. 1988, Whalen and Reeburgh 1988, Harriss 1989), moisture regime (Svensson and Rosswall 1984, Moore and Knowles 1989), and net primary production (Aselmann and Crutzen 1989).

Other aspects of high latitude ecosystems influence trace gas production and consumption such as oxygen content of the substrate, soil acidity, microbial population, quality of litter, soil chemistry, physical properties and water flow. Plant structure affects gas transport. Despite the overall importance of these factors on trace gas exchange, there is little information on how they interact in controlling the net exchange of trace gases in high latitudes.

A second important consideration of the role of high latitudes in trace gas emissions is the sink potential of various ecosystems. Non-wetland areas in Alaska have been shown to consume c. 2 mg m^{-2} d^{-1} of CH_4. Studies in dry temperate forests suggest sinks may be important (Melillo et al. 1989, Steudler et al. 1989). High latitude ecosystems may also be sinks for anthropogenic pollutants.

Objectives

To determine the controls of trace gas production, consumption, and transport in high latitude ecosystems.

Research plan

Many studies are currently examining or have examined aspects of the relationships between environmental conditions and trace gas flux at spatial scales varying from sub-meter to regional and temporal scales of hours to inter-annual (Table 1). Some of these relationships are being tested on a wider range of northern wetlands through the NOWES/ABLE 3B and other studies. To further augment this research effort the following tasks are proposed:

1) Surveys: In survey type studies where the main objective is to derive a flux estimate from ecosystems and geographical areas, a minimal set of additional measurements should be made, such as soil temperature, soil moisture or position of the water table, nutrient status, and vegetation characteristics. The geographical areas and ecosystems in which this work should take place are listed in Table 2.

2) Environmental correlation studies: Specific detailed, process level studies should be undertaken in key ecosystem and geographical regions, particularly those areas listed in Table 2. In these studies, the ecosystem's thermal and soil moisture regime, biogeochemistry (oxygen, C:N, P, Ca, pH), and soil characteristic (available carbon, degree of decomposition), should be examined in conjunction with gas flux measurements. In addition the storage of trace gas in the soils should be measured through time, as well as isotope signatures of stored and emitted CH_4.

3) Mechanistic process studies: It is widely recognized that trace gas fluxes in any given environment are controlled by three processes: a) production, b) consumption, and c) transport. Understanding each process at a mechanistic level is necessary to interpret temporal and spatial variability in the trace gas flux. Process level studies are difficult and not readily amenable to large scale field deployment. However, techniques are available that will allow detailed studies to be conducted in a limited number of key high latitude environments, especially fens, bogs, and forests. General techniques for studying production, consumption, and transport are described below. CH_4 is highlighted because the analytical and experimental methods are currently available but similar experiments should be conducted for other trace gases when techniques are developed.

a) CH_4 production: Ecosystem-level predictive models of CH_4 production rates require an understanding of the vertical distribution of CH_4 production as well as bacterial substrate requirements. Since methanogenic bacteria are capable of CH_4 production from a limited number of substrates (Conrad 1989), it is possible, in principle, to determine the most significant CH_4 precursor(s) for a particular high latitude ecosystem. The isotope composition (^{13}C, D) can be used to quantify CH_4 production by acetate fermentation and CO_2 reduction. In addition the depth distribution of CH_4 concentration provides important circumstantial evidence regarding the role of root exudate, bulk organic matter breakdown, and substrate in regulating CH_4 production.

b) CH_4 oxidation: A limited number of studies have suggested that as much as 50% of the CH_4 produced in rice paddies, and tundra is oxidized and therefore is not released to the atmosphere. Methanotrophic bacteria appear to function as an effective "biofilter" in controlling CH_4 emissions (Galchenko et al. 1989). Studies should be undertaken to delineate this oxidation potential. Techniques are currently available for measuring CH_4 oxidation rates in jar experiments (Conrad 1989) as well as ^{14}C tracer methods (Alperin and Reeburgh 1985). The depth distribution of CH_4 oxidation should be measured along hydrological and ecosystem transects, and the role of water table and root/rhizome oxygen transport of CH_4 oxidation can be assessed. The isotope consequences for CH_4 consumption (fractionation) should be evaluated.

c) Transport mechanisms: There are six transport mechanisms of trace gases. These include: active and passive vascular plants, ebullition, diffusion in gas and liquid phase, and atmospheric pumping. The transport

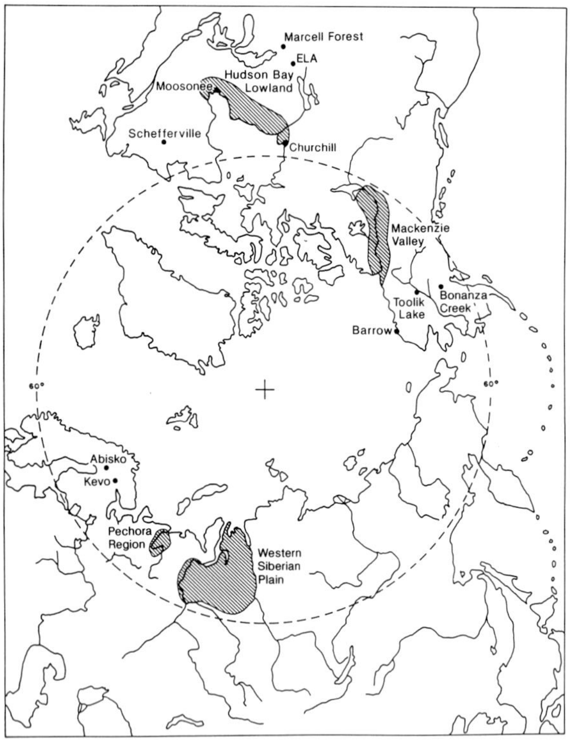

Fig. 6. Location of possible research sites for ecosystem-atmosphere trace gas exchange studies.

in a given ecosystem is regulated by the relationship between the depth of CH_4 production, the water table height, and the abundance of vascular plants. Diffusion can be evaluated by boundary layer experiments, while studies of vascular transport should be conducted in process studies using chambers over ecosystems with and without plants, accounting for differences in CH_4 production and consumption.

Sites for process studies should be chosen from each high latitude ecosystem of large areal extent, or ecosystems with disproportionately high emissions. These studies should wherever possible collaborate with proposed and ongoing programmes such as the Long Term Ecological Research, the NOWES, and old IBP sites. In addition there are a number of research centers where this work could be promoted such as the Experimental Lakes Area, Churchill Centre for Northern Studies, the McGill Subarctic Research Station, Petawawa and Turkey Lakes Forest Research Centre, all in Canada; Abisko, Sweden; and Kevo, Finland (Fig. 6).

Timetable

1992: Continue and initiate further studies in areas where work is required.

Activity 3: Sensitivity of trace gas exchange in high latitudes to ecosystem change

The production and consumption of trace gases in high latitude ecosystems is dependent on several biotic and abiotic variables (e.g. Svensson and Rosswall 1984, Whalen and Reeburgh 1988). The response of trace gas emission from ecosystems to climatic change, or other environmental perturbation, will involve, in the short term, from changes in the abiotic variables, such as temperature and moisture content and, in the longer term, from changes in the plant and microbial communities. Robertson et al. (1989) hypothesized several scenarios of the effect of climatic change on the emission of CH_4 from northern wetlands.

The study of these perturbations, such as climatic

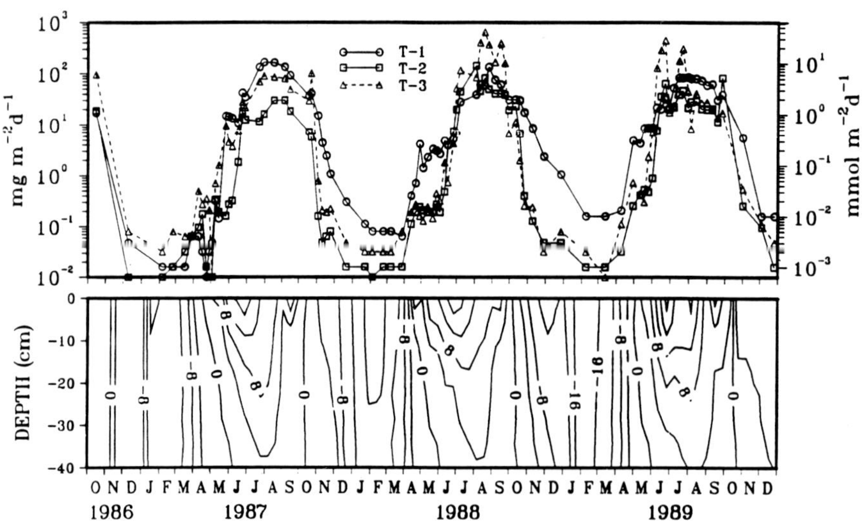

Fig. 7. Time series of the methane gas flux from wetland tussuck sites in Alaska (Whalen and Reeburgh unpubl. data). The lower box indicates the soil temperature profile.

change, on trace gas emissions can be addressed in three main ways (tasks): 1) measurements conducted over several years at the same site can be used to evaluate the effect of year-to-year variations on emissions; 2) measurements from a range of ecosystems can be used to evaluate the impact by changes in community on trace gas emissions; and, 3) manipulation experiments, to simulate the anticipated changes, can be conducted.

A combination of these three approaches, combined with a better understanding of the controlling factors on gas emissions, can be used to address the issue of the sensitivity of high latitude ecosystems.

Task 1: Long term measurements programme

Very few measurement programmes of the flux of CH_4 from high latitude ecosystems have been continued for more than one growing season (cf. Moore and Knowles 1990, Fig. 7 – Whalen and Reeburgh, unpubl. data). Short records are insufficient to determine any relation between climate and trace gas flux. In addition with the large spatial variability of trace gas flux coefficients of variation of 30–50%, even within individual ecosystems (Svensson and Rosswall 1984, Sebacher et al. 1986, Crill et al. 1988, Whalen and Reeburgh 1988), it is difficult to see how temporal variability can be separated from the spatial variability, especially when inter-annual, as well as seasonal variation could be expected. Establishment of atmospheric chemistry observatories near source areas would be one approach to identify long-term changes. Such programmes exist to examine other problems (e.g. GMCC, ALE/GAGE), but these sites are deliberately not situated near major sources of biogenic trace gases.

Objectives

Develop inter-annual and seasonal time series of the concentration and flux of trace gases from the major ecosystems of the high latitudes.

Research plan

The NOWES has established an atmospheric chemistry observatory in Fraserdale, Ontario, Canada, to measure the concentration of several trace gases. The observatory is situated on the boundary of the Hudson Bay Lowland and boreal forest zone. This facility is schedule to operate for approximately 10 years.

Three additional studies are proposed:

1) Long term atmospheric observatories in other source areas: Similar observatories as the Fraserdale site should be established in the tundra region (e.g. Churchill, Manitoba; Schefferville, Quebec; Kevo, Finland; Abisko, Sweden [Fig. 6]). Wherever possible, these sites should tie in with the IGBP Regional Research Centers. A suite of biogenic trace gases should be measured throughout the year. In addition, a subset of samples should be analyzed for isotopes where applicable.

2) A long term chamber programme in key ecosystems at existing research facilities that are part of Long Term Ecological Research (LTER) sites or have been in existence for more than 10 years. Candidate arctic and subarctic sites include Bonanza Creek LTER, Toolik Lake LTER and Barrow, Alaska; Churchill, Manitoba; Schefferville, Quebec; Abisko, Sweden; and Kevo, Finland (Fig. 6). Possible boreal forest sites are the Marcell Forest, Minnesota and the Experimental Lakes Area, Ontario. Both latter sites, however, are at the southern limit of the boreal zone and a more central location would be preferred. The US/Canada BOREAS is currently reviewing potential sites. In Russia, there are several suitable Biosphere Reservations, but no developed facilities for trace gases monitoring.

3) Low altitude north-south aircraft measurements: Fluxes of CH_4, CO_2, H_2O and other radiatively-active important trace gases should be measured as sensors become available in order to obtain flux measurements in correspondence with spectral reflectance data from aircraft and satellite. This information will be helpful to optimize the use of remotely sensed spectral data to parameterize gaseous exchange of radiatively important gases.

Government agencies or institutions such as Atmospheric Environment Services, Environment Canada and NASA should take a lead role in providing infrastructure support for the establishment, operation and maintenance of the long term atmospheric chemistry sites. Research groups are required to take the lead role in the establishment of continuous chamber programmes at the long term ecosystem sites.

Timetable

1990: Establishment of the Northern Wetland and Boreal Forest Atmospheric observatory site at Fraserdale, Ontario.

1992: Establishment of long term measurement programmes at representative ecosystem sites.

1993: Establishment of long term Atmospheric Chemistry observatories in the arctic/subarctic and boreal zones.

Task 2: Experimental analogs of sensitivity of trace gas flux

The sensitivity of trace gas flux to environmental change can be studied using the natural environmental variability along a transect or manipulative experiments within a given ecosystem. Several studies have used transects over different ecosystems and change in latitude to derive estimates of regional fluxes (Sebacher et al. 1986). This approach can be used as an analog for the influence of ecosystem change on trace gas flux. Alternatively, existing ecosystems can be manipulated

Table 4. Priority of ecosystem manipulation for trace gas studies.

Priority	Wetlands	Upland forests	Tundra
1	drainage	temperature	temperature
2	temperature	fertilization	permafrost/thermokarst
3	fertilization	hydrology	hydrology
4	CO_2 enrichment	O_3 enrichment	fertilization
5	fire/forestry	CO_2 enrichment	CO_2 enrichment
6		fire/forestry	

to simulate changes that might occur. Initially, the observed changes will occur in response to physical factors, but if manipulations are run for decades, changes in the biological community that are important to trace gas flux emissions, may be identified. In conducting the manipulative studies, care must be taken to ensure that appropriate pre-manipulation measurements are made to provide a reliable comparison with after manipulation emissions. This opportunity, however, is rarely available and "piggy-backing" on manipulative experiments commonly rely upon contemporary measurements at reference sites with faith that these represent emissions before manipulation. Because of the expense and difficulty of field ecosystem manipulations, laboratory manipulations at microcosm scales should be undertaken. These experiments may give valuable information for planning field studies and allow a clearer identification of the relationships between manipulation factors and gas fluxes.

Objective

Develop manipulation experiments to examine sensitivity of gas exchange to environmental change.

Research plan

Current studies: The NOWES transect from the Hudson Bay coast inland provides an example of successional change in ecosystems, associated with the emergence of the Lowland from Hudson Bay over 5000 years. Major recent disturbances, such as associated with the Alaskan pipeline, also provide opportunities to address changes in trace gas flux along a succession.

Simulated acid rain is being added to a bog at the Experiment Lakes Area, Ontario, and peat water chemistry, plant responses, and CH_4 flux are being monitored (Bayley and Parker 1990). The extremely acid nature of the bog limits the natural production and emission of CH_4, so the response of CH_4 flux to further acidification has not been pronounced.

Moore and Knowles (1990) observed insignificant emission of CH_4 from a drained swamp, compared to an adjacent undrained swamp. Further studies of the effect of drainage on CH_4 flux from boreal peatlands are being planned for several boreal forest sites. The influence of changes in temperature and water table position on CH_4 emission is being investigated at various institutions under phytotron conditions for peat cores representing fens, bogs and swamps.

There are continuous changes imposed on wetlands in Sweden for the exploitation of forestry and peat resources. The range in type and intensity of the drainage could be used to evaluate the effect of changes in water table position. Projects are needed in which initial baseline studies have been made before drainage.

Laboratory studies on the degradation of different parts of peat-forming organic matter are being made in controlled methanogenic systems in Sweden (Svensson 1984). These studies also include the analysis of the substrates behind the methane formation in the peat profile. This type of experiment could easily be redesigned to include the influence of temperature and other changes such as fertilization and N– and S– deposition. Microbial and substrate controls on methane production and consumption are being investigated on a range of peat materials representing bogs, fens and swamps (Knowles pers. comm.).

Isotope analyses have been made on atmospheric CH_4 and CH_4 emitted from the tundra at Churchill, 1986–88 and will be continued.

Proposed studies: Successional transects similar in experimental design to those of the NOWES should be undertaken in the boreal forest and the arctic and subarctic tundra biomes. In these areas, it is difficult to find as clear successional trend as exists in the Hudson Bay Lowland. Much smaller scale transects, across ecosystem boundaries, may be more appropriate in these ecosystems. Manipulation experiments should take place in the major gas producing ecosystems in each of the boreal, subarctic and arctic regions. In each ecosystem type the priority of manipulation is listed in Table 4.

Given the high costs of field manipulations, it is probable that trace gas exchange studies can only be made if contact is made with other projects involving manipulations such as at the Experimental Lakes Area (ELA), Dept of Fisheries and Oceans Canada.

Timetable

1991: Establish field programme and determine control emissions as soon as possible.

Task 3: Process-based trace gas exchange sensitivity models

There are two approaches to the modelling of gas flux from various ecosystems: physically-based mechanistic models and statistical regression models. A biospheric-atmospheric exchange model could be based on the Simple Biospheric (SiB) model (Sellers et al. 1986) or Biospheric Atmospheric Transfer (BATS) model (Dickinson et al. 1986). However, the limiting control on the total emissions of many biogenic trace gases is not turbulent transfer but rather microbial and biogeochemical controls in the soil. Because of the multiplicity of variables, processes and scales involved in the emissions of trace gases, realistic mechanistic models are not attainable at present. The soil environment, however, depends on the exchange of energy and moisture with the atmospheric boundary layer and these processes can be modelled with reasonable success.

The second approach deals with the complexity using empirical correlations between the gas flux and environmental variables. Theses models assume that the relationships remain appropriate when extrapolated out of their validated range by a climate change or other environmental change scenario.

Objective
Develop, in the first instance, empirical and later, mechanistic models of trace gas production, consumption, and exchange to be used in analyzing the sensitivity of trace gas emissions to environmental change.

Research plan
At the present time, there are no simulation models for estimating the sensitivity of high latitude trace gas exchange to environmental change. There have been some models that have succeeded in simulating the loss of water from upland short grass steppe (Parton et al. 1988), but these are restricted to local and regional settings. A very simple conceptual hydrologic model has been developed to estimate the sensitivity of moisture storage in northern fens to changes in precipitation and evaporation. This model has been applied to wetlands in subarctic Quebec and shows that CH_4 may not necessarily increase with increasing atmospheric temperature, but may decrease significantly because of a reduction in moisture storage.

The following research is needed in this area: 1) better characterization of the relationship between the forcing variables and trace gas emission, paying attention to non-linearity and interdependence of variables, 2) improve ecosystem-scale process models for estimating the input of environmental changes in a landscape context, and 3) develop interactive surface exchange functions between climate models and the land surface for the high latitudes.

Timetable
1990: Begin programme.

Additonal considerations

Several of the tasks listed above are not unique to high latitude ecosystems. These include the classification schemes for ecosystems at a level useful for trace gas flux extrapolations, and the development of field techniques to measure trace gases.

This report deals with high latitude ecosystems, and because wetlands are such a prominent landscape type in this region, they have been discussed extensively. However, wetlands are also significant ecosystems in the temperate and tropical regions and should thus form an integral part of trace gas studies there. Similarly, while biomass burning was not discussed, fire is a very important aspect of the boreal forest biome, though the quantity of biomass burned on a unit area basis is probably much smaller than in tropical areas.

Finally, there are three aspects of trace gas fluxes from high latitude environments that may be important, but have not been assessed and no research plan has been suggested. These should be considered, and their significance determined in a global context. The first is the possible source of clathrate hydrates in recently emerged coastal permafrost environment and continental shelf. This could be a potentially large source if the north experiences warming (Kvenolden 1988a, b). Secondly, the high latitudes are receiving an increased anthropogenic input of N, P, S, and heavy metals. There is little information on the effect of this deposition on high latitude ecosystems, and as in temperate regions (Steudler et al. 1989) they may initiate both positive and negative feedbacks for different trace gases. Finally, a large portion of Canada and Russia's high latitude regions, particularly the boreal forest region, are being inundated for the purpose of hydroelectric power generation. Standing vegetation is rarely removed before flooding, so the reservoirs contain a high organic matter content. The effect of inundation on trace gas fluxes from these activities should be determined.

References

Alperin, M. J. and Reeburgh, W. S. 1985. Inhibition experiments on anaerobic methane oxidation. – Appl. Environ. Microbiol 50: 540–945.

Anderson, S. M., Delaney, A. C., Fehsenfeld, F. C., Golden, P. D., Kolt, C. E., Ridley, B. A. and Steele, P. L. 1989. States of chemical sensors for flux measurements. – In: Lenshow, D. H. and Hicks, B. B. (eds), Global tropospheric chemistry, pp. 47–68.

Aselmann, I. 1989. Extrapolation: a critical assessment. – In: Andreae, M. O and Schimel, D. S. (eds), Exchange of trace gases between terrestrial ecosystems and the atmosphere. Wiley, pp. 119–133.

– and Crutzen, P. 1989. Freshwater wetlands: global distribution of natural wetlands and rice paddies, their net primary productivity, seasonality and possible methane emissions. – J. Atmos. Chem. 8: 807–358.

Bartlett, D. S., Ritter, J., Sachee, G., Bakwin, P., Wofsy, S., Fan, S., Bartlett, K. B., Fitzjarrald, D. and Whiting, G. 1989. Regional flux estimates from synthesis of enclosure, tower and aircraft measurements. – EOS 70: 286.

Bayley, S. and Parker, B. 1990. Final report: measurement of methane and carbon dioxide flux from an experimental wetland. – Atmos. Environ. Services, Downsview, Ontario, Canada.

Betts, A. K., Desjardins, R. L., MacPherson, J. I. and Kelly, R. D. 1989. Boundary layer heat and moisture budgets from FIFE. – Boundary Layer Meteorol. 50: 109–138.

Buckley, D. J., Desjardins, R. L., Lalonde, J. L. M. and Brunke, R. 1988. A linearized fast response gas sampling

apparatus for eddy accumulation studies. – Computers and Electronics in Agric. 2: 243–250.
Chahuneau, E., Desjardins, R. L., Brach, E. J. and Verdon, R. 1989. A micrometerological facility for eddy flux measurements of CO_2 and H_2O. – J. Atmos. Oceanic Tech. 6: 193–200.
Cicerone, R. J. and Omerland, R. S. 1988. Biogeochemical aspects of atmospheric methane. – Glob. Biogeochem. Cycl. 2: 299–327.
Conrad, R. 1989. Controls of methane production in terrestrial ecosystems. – In: Andreae, M. O. and Schimel, D. S. (eds), Exchange of trace gases between terrestrial ecosystems and the atmosphere. Wiley, pp. 39–58.
Crill, P. M., Bartlett, K. B., Harriss, R. C., Gorham, E., Verry, E. S., Sebacher, D. I., Modzar, L. and Sanner, W. 1988. Methane flux from Minnesota peatlands. – Glob. Biogeochem. Cycl. 2: 317–384.
Desjardins, R. L. and MacPherson, J. I. 1989. Aircraft-based measurements of Trace gas fluxes. – In: Andreae, M. O. and Schimel, D. S. (eds), Exchange of trace gases between terrestrial ecosystems and the atmosphere. Wiley, pp. 00–00.
Dickinson, R. E., Henderson-Sellers, A., Kennedey, P. J. and Wilson, M.F. 1986. Biosphere/atmosphere transfer scheme (BATS for NCAR community climate model. – NCAR Tech. Note TN-275 + STR, NCAR Boulder, CO, USA.
Dörr, H. and Münnich, K. O. 1987. Annual variation in soil respiration in selected areas of the temperate zone. – Tellus 39B: 114–121.
Galbally, I. E. 1989. The international global atmospheric chemistry (IGAC) programme: a core project of the international geosphere-biosphere programme. – IAMAP Comm. on Atmos. Chem. and Global Pollution.
Galchenko, V. F., Lenin, A. and Ivanov, M. 1989. Biological sinks of methane. – In: Andreae, M. O. and Schimel, P. S. (eds), Exchange of trace gases between terrestrial ecosystems and the atmosphere. Wiley, pp. 59–72.
Harriss, R. C. 1989. Experimental design for studying atmospherebiosphere interactions. – In: Andreae, M. O. and Schimel, D. S. (eds), Exchange of trace gases between terrestrial ecosystems and the atmosphere. Wiley, pp. 291–301.
– and Wofsy, S. 1989. The arctic boundary layer experiment (ABLE-3A). – EOS 70: 282.
Hoell, J. M., Harriss, R. C. and Wofsy, S. C. 1990. The NASA global tropospheric experiment: atmospheric boundary layer expedition (ABLE-3B). – Tropos. Chem. Program, Earth Sci. and Appl. Div. Off. of Space Sci. and Appl., NASA, Washington DC.
Kelly, C. 1979. Quantitative comparison of the significance of methane in the carbon cycles of two small lakes. – Arch. Hydrobiol. Beih. 12: 123–135.
Kvenvolden, K. 1988a. Methane hydrates and global change. – Global Biogeochem. Cycl. 2: 221–229.
– 1988b. Methane hydrates – a major reservoir of carbon in the shallow geosphere? – Chem. Geol. 71: 41–51.
Lowe, D. C., Brenninkmeijer, C. A. M., Manning, M. R., Sparks, R. and Wallace, G. 1988. Radiocarbon determination of atmospheric methane at Baring Head, New Zealand. – Nature (Lond.) 332: 522–525.
Mack, A. R., Desjardins, R. L., MacPherson, J. I. and Schuepp, P. H. 1990. Photosynthetic activity of agricultural lands from airbourne CO_2 and satellite data. – Int. J. of Remote Sensing 11: 237–252.
Matson, P., Vitousek, P. M. and Schimel, D. S. 1989. Regional extrapolation of trace gas flux based on soils and ecosystems. – In: Andreae, M. O. and Schimel, D. S. (eds), Exchange of trace gases between terrestrial ecosystems and the atmosphere. Wiley, pp. 97–108.
Matthews, E. and Fung, I. 1987. Methane emissions from natural wetlands: global distribution, area, and environmental characteristics of sources. – Glob. Biogeochem. Cycl. 1: 61–86.
Melillo, J. M., Steudler, P. A., Aber, J. D. and Bowden, R. D. 1989. Atmospheric deposition and nutrient cycling. – In: Andreae, M. O. and Schimel, D. S. (eds.), Exchange of trace gases between terrestrial ecosystems and the atmosphere. Wiley, pp. 263–280.
Mitchell, J. F. B. 1989. The "greenhouse" effect and climate change. – Rev. of Geophys. 27: 115–140.
Moore, T. R. and Knowles, R. 1987. Methane and carbon dioxide evolution from subarctic fens. – Can. J. Soil Sci. 67: 77–81.
– and Knowles, R. 1989. The influence of water levels on methane and carbon dioxide emissions from peatland soils. – Can J. Soil Sci. 69: 33–38.
– and Knowles, R. 1990. Methane emissions from fen, bog and swamp peatlands in Quebec. – Biogeochem. 11: 45–62.
Mosier, A. R. 1989. Chamber and isotope techniques. – In: Andreae, M. O. and Schimel, D. S. (eds), Exchange of trace gases between terrestrial ecosystems and the atmosphere. Wiley, pp. 175–188.
Naiman, R. J., Melillo, J. M. and Hobbie, J. E. 1986. Ecosystem alteration of boreal forest streams by beaver (*Castor canadensis*). – Ecology 67: 1254–1269.
National Wetlands Working Group. 1988. The wetlands of Canada. – Ecol. Land Classification Ser., No. 24, Sustainable Dev. Branch, Environ. Canada, and Polysci. Publ., Montreal.
Parton, W. J., Mosier, A. R. and Schimel, D. S. 1988. Rates and pathways of nitrous oxide production in shortgrass steppe. – Biogeochem. 6: 45–58.
Quay, P. D., King, S. L., Landsdown, J. M. and Wilbow, D. O. 1988. Isotopic composition of methane released from wetlands: implications for the increase in atmospheric methane. – Global Biogeochem. Cycl. 2: 385–397.
Robertson, G. P., Andreae, M. O., Bingemer, H. G., Crutzen, P. J., Delmas, R. A., Duyzer, J. H., Fung, I., Harriss, R. C., Kanakidou, M., Keller, M., Melillo, J. M. and Zavarin, G. A. 1989. Trace gas exchange and the chemical and physical climate: critical interactions. – In: Andreae, M. O. and Schimel, D. S. (eds), Exchange of trace gases between terrestrial ecosystems and the atmosphere. Wiley, pp. 303–320.
Rosswall, T., Bak, F., Baldocchi, D., Cicerone, R. J., Conrad, R , Ehhalt, D. H., Firestone, M. K., Galbally, I. E., Galchenko, V. F., Groffman, P. M., Papen, H., Reeburgh, W. S. and Sanhueza, E. 1989. What regulates production and consumption of trace gases in ecosystems: biology or physicochemistry? – In: Andreae, M. O. and Schimel, D. S. (eds), Exchange of trace gases between terrestrial ecosystems and the atmosphere. Wiley, pp. 73–95.
Rudd, J. W., and Hamilton, R. D. 1975. Methane oxidation in a eutrophic Canadian shield lake. – Verh. Internat. Verein. Limnol. 19: 2669–2673.
Schiff, H. I. 1988. A proposal for a collaborative grant to the Natural Science and Engineering Research Council of Canada for a Northern Wetlands Study (NOWES). – Can. Inst. for Res. in Atmos. Chem., York Univ., North York, Ontario, Canada.
Schuepp, P. H., Desjardins, R. L., Macpherson, J. I., Boisvert, J. B. and Austin, L. B. 1989. Interpretation of airbourne estimates of evapotranspiration. – In: Estimation of areal evapotranspiration. IAHS Publ. No. 177, pp. 185–196.
– , Leclerc, M. Y., MacPherson, J. I. and Desjardins, R. L. 1990. Footprint prediction of scalar fluxes from analytical solutions of the diffusion equation. – Boundary Layer Meterol., (in press).
Sebacher, D., Harriss, R. C., Bartlett, K. B., Sebacher, S. M. and Grice, S. S. 1986. Atmospheric methane sources: Alaskan tundra, bogs, in alpine fen, and a subarctic boreal marsh. – Tellus 38B: 110.

Sellers, P., Hall, F. G., Strobel, D. E., Kanemasu, E. T., Kelly, R. D., Blad, B. L., Markham, B. J. and Wang, J. R. 1990. Experimental design and operation. Symposium on the first ISLSCP field experiment. – Am. Meteor. Soc., Anaheim, California, pp. 1–5.

Steudler, P. A., Bowdenm, R. D., Melillo, J. M. and Aber, J. D. 1989. Influence of nitrogen fertilization on methane uptake in temperate forest soils. – Nature (Lond.) 341: 314–316.

Stevens, C. M. 1988. Atmospheric methane. – Chem. Geol. 71: 11–21.

Stevens, C. and Engelkemeir, A. 1988. Stable carbon isotopic composition of CH_4 from natural and anthropogenic sources. – J. Geophys. Res. 93: 725–733.

Stewart, J. W. B., Aselmann, I., Bouwan, A. F., Desjardins, R. J., Hicks, B. B., Matson, P. A., Rodhe, H., Schimel, D. S., Svensson, B. H., Wassmann, R., Whiticar, M. J. and Yang, W.-X. 1989. Extrapolation of flux measurements to regional and global scales. – In: Andreae, M. O. and Schimel, D. S. (eds), Exchange of trace gases between terrestrial ecosystems and the atmosphere. Wiley, pp. 303–320.

Svensson, B. H. 1984. Different temperature optima for methane formation when enrichments are supplemented with acetate or hydrogen. – Appl. Environ. Microbiol. 48: 389–394.

– and Rosswall, T. 1984. In situ methane production from acid peat in plant communities with different moisture regimes in a subarctic mire. – Oikos 43: 341–350.

Tyler, S. C. 1986. Stable carbon isotopes in atmospheric methane and some of its sources. – J. Geophys. Res. 91: 13232–13238.

Vitt, D., Bayley, S., Jin, T., Halsey, L., Parker, B. and Craik, R. 1990. Methane and carbon dioxide production from wetlands in boreal Alberta. – Alberta Environ. Contract Rep. No. 90–0270, Edmonton, Alta, Canada.

Wahlen, M., Tanaka, N., Henry, R., Deck, B., Zeglen, J., Vogel, J. S., Southou, J., Shemesh, A., Fairbanks, R. and Broecker, W. 1989. Carbon-14 in methane sources and in atmospheric methane: the contribution from fossil carbon. – Science 245: 286–291.

Wesely, M. L., Lenshow, D. H. and Denmead, O. T. 1989. Flux measurement techniques in global tropospheric chemistry. – In: Lenshow, D. H. and Hicks, B. B. (eds), pp. 31–46.

Whalen, S. C. and Reeburgh, W. S. 1988. A methane time series for tundra environments. – Global Biogeochem. Cycl. 2: 399–409.

Whiticar, M. J., Faber, E. and Schoell, M. 1986. Biogenic methane formation in marine and freshwater environments: CO_2 reduction vs. acetate frementation – isotope evidence. – Geochem. Csomchem. Acta. 50: 693–709.

Ecological Bulletins 42: 98–108. Copenhagen 1992

Anaerobic burial of refuse in landfills: increased atmospheric methane and implications for increased carbon storage

Jean E. Bogner

Bogner, J. E. 1992. Anaerobic burial of refuse in landfills: increased atmospheric methane and implications for increased carbon storage. – Ecol. Bull. (Copenhagen) 42: 98–108.

Anaerobic decomposition of landfilled refuse produces biogenic methane, most of which is vented to the atmosphere. Estimates for landfill methane emissions, based on worldwide solid waste quantities, range from <10 to >50 Tg yr^{-1}. Major controls on rates of landfill methane generation include water content, nutrient availability, and characteristics of interbedded soils. The dynamic process of methane movement through soil materials from the site of gas generation to the atmosphere is controlled by soil gas concentration and pressure gradients, as well as by properties of the porous medium, particularly moisture content. Examining percentages of cellulose, hemicellulose, and lignin in refuse that are not decomposed under optimal laboratory conditions suggests that widespread anaerobic burial of refuse also results in a measurable increase in the amount of organic carbon entering long-term sedimentary storage.

J. E. Bogner, Argonne Nat. Lab., EES-362-2A 9700 S Cass., Ave., Argonne, IL 60439, USA.

Background

Burial of refuse with resulting production of methane and carbon dioxide under anaerobic conditions has been documented at least since the early 1940s (Elliason 1942). However, in most developed countries, aerobic decomposition practices (open burning and informal composting at dump sites) were also widely practiced up to the 1960s. At that time, for better protection of water resources, for suppression of rodents and pathogens, and for aesthetic concerns, controlled burial of refuse in engineered landfills became widespread. Currently, controlled landfilling of refuse is becoming more prevalent in developing countries. In landfills decomposition of organic substrates occurs through a complex series of microbial reactions, predominantly under anaerobic conditions. The terminal reaction is the production of methane by methanogenic bacteria, which are strict anaerobes and function best at near-neutral pH. The gas mixture produced is approximately half methane and half carbon dioxide, with minor amounts of other gases. Because of its methane content, landfill gas may present a fire and explosion hazard from uncontrolled migration and mixing with air. At a small number of sites worldwide, landfill methane is flared or recovered commercially for fuel use. At most sites worldwide, the landfill gas is vented into the atmosphere, where it contributes to the atmospheric methane and carbon dioxide pools. Landfills provide concentrated sources of atmospheric methane, which is of particular concern due to its potential contribution to greenhouse warming. To date, the strength of the landfill methane source has not been well quantified.

Several estimates of worldwide landfill methane emissions exist. Bingemer and Crutzen (1987) estimated that landfills may emit 30–70 Tg yr^{-1}, based on current estimates of solid waste generated; steady-state methane generation from historic quantities of waste equal to quantities currently being landfilled; and the following assumptions: 20% degradable organic carbon content for the waste; conversion of 80% of the degradable organic carbon (a methane-generation rate of 0.1 kg methane kg^{-1} refuse); and all methane being vented to the atmosphere. Recently, Orlich (1990) estimated worldwide landfill methane-emissions as 33 Tg yr^{-1} for 1990, based on assumed per capita refuse-generation rates and net methane-generation/emission rates (reduced to allow for methane oxidation) of 0.086 kg me-

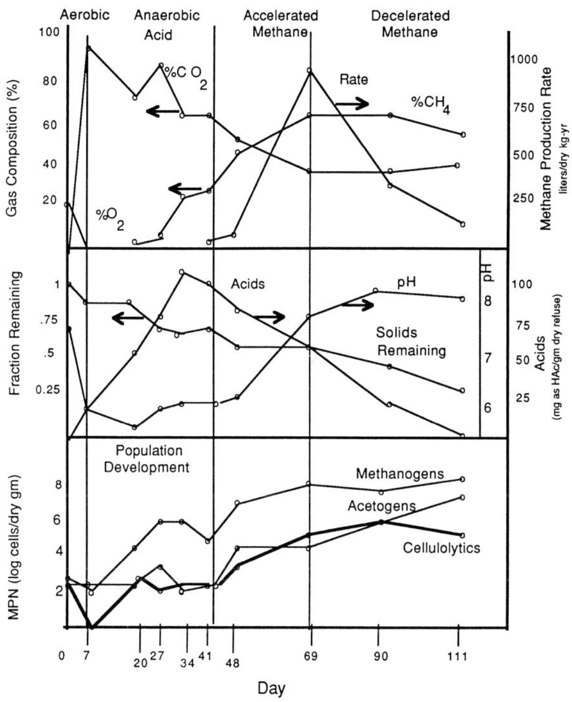

Fig. 1. Observed trends in refuse decomposition with leachate recycle in 2-L laboratory containers. The total carboxylic acids are expressed as acetic acid equivalents. Methanogen MPN data are the log of the average of the acetate– and hydrogen/carbon dioxide-utilizing populations. "Solids remaining" equals the ratio of (cellulose + hemicellulose) removed from a container divided by the weight of (cellulose + hemicellulose) added to the container initially (Barlaz et al. 1989 – used by permission).

thane kg^{-1} refuse for developed countries and 0.030 kg methane kg^{-1} refuse for developing countries. Richards (1990) estimated that landfills emit only 9–18 Tg yr^{-1}, based on the following assumptions: proportionality of worldwide refuse quantities to gross domestic product; 80% of waste being landfilled; a conversion rate of 0.036 kg methane kg^{-1} refuse; and steady-state methane generation.

For the United States, Thorneloe (1990) of the U.S. Environmental Protection Agency (EPA), which is preparing to regulate landfill emissions, estimated that 15–18 Tg yr^{-1} are emitted from U.S. landfills. For comparison, if one considers representative volatile solids (VS) content for the various biodegradable fractions in the total quantity of U.S. refuse landfilled during 1966–1986 (refuse quantities from Franklin Associates 1988); assumes a typical ultimate anaerobic digester conversion of 0.35 kg methane kg^{-1} VS (based on 0.5 m^3 biogas containing 50% methane kg^{-1}; Metcalf et al. 1979); and emits all the resulting methane evenly over the 1966–1986 period – a total of 10.5 Tg yr^{-1} is attributable to U.S. landfills. Ultimate landfill methane generation rates are undoubtedly much lower then digester rates; thus, one might argue that the figure should be less than 10.5 Tg yr^{-1}.

The key to refining these wide ranges for global estimates lies in better understanding what fraction of organic carbon deposited in landfills is converted to methane and carbon dioxide within a few decades and what fraction remains in sedimentary storage into geologic time. Ultimate rates for methane generation from refuse given in the literature, based on assumed refuse composition and stoichiometry or on carefully controlled laboratory experiments conducted under optimal conditions, range from 0.03 to 0.15 kg methane kg^{-1} refuse (Stegman and Ehrig 1982, Halvadakis et al. 1983, Pacey and Dietz 1986, Fabry 1988, Richards 1987). For commercial landfill methane recovery, some percentage below 50% is typically applied to these ultimate rates as the percent recoverable. It is well documented that old landfills (c. 1960–1975) contain widespread undecomposed organic substances, particularly lignocellulosic materials: indeed, old newspapers are often used for absolute dating of refuse strata. Thus, during the short history of widespread controlled sanitary landfilling, observed decomposition of organic materials has not been complete.

The objectives of this paper are, first, to review landfill methane generation and emission processes; and, second, to focus on the long-term role of landfills in the global organic carbon cycle.

Controls on landfill methane generation

In terms of microbiological processes, the decomposition of refuse in landfills is comparable with degradation of organic materials in an anaerobic digester. Fig. 1, based on a laboratory decomposition study by Barlaz et al. (1989), summarizes the evolution of gases, liquids, solids, and major microbial groups. In this study, multiple containers packed with fresh shredded refuse and added water were destructively sampled at various stages in the decomposition process. No microorganisms were added, and leachate was recycled and neutralized during the early stages of decomposition.

Initially in the decomposition process, or soon after burial in an actual landfill, oxygen and oxidized species are rapidly depleted. Hydrolytic and fermentative microorganisms convert complex organic materials (food waste, garden waste, paper and paper products) to intermediate carboxylic acids, alcohols, hydrogen, and carbon dioxide. Acetogenic bacteria produce acetate, hydrogen, and carbon dioxide from longer-chain carboxylic acids and alcohols. Methanogenic bacteria produce methane by either acetate cleavage or carbon dioxide reduction. In Fig. 1, "solids reduction" refers to (cellulose + hemicellulose) degradation; Barlaz et al. (1989) observed 72% reduction during 111 d of incubation at 41°C. They concluded that, initially, acetate

Table 1. Observed ranges, means and standard deviations (SD) for solids analysis of landfill samples from northern Illinois, southern Wisconsin, and eastern Colorado, USA. All values are dry-weight basis, except total solids. N = 18: six samples from each of three sites. Modified from Bogner (1990).

Property	Range	Mean	SD
Total solids (%)	39–86	72	14
Volatile solids (%)	8–67	31	21
Water content (%)	16–159	46	40
Total carbon C (%)	6–39	20	12
Kjeldahl nitrogen (ppm)	805–9120	3080	2320
Total phosphorus (ppm)	260–1650	605	323
"Total" potassium (ppm)	815–3040	1650	706

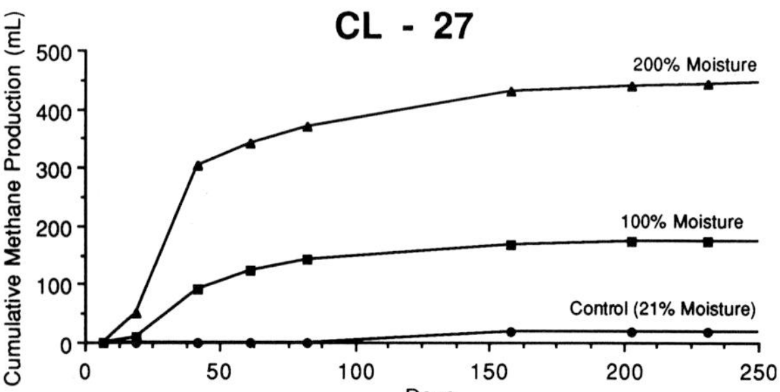

Fig. 2. Effect of moisture addition on biogas production. Typical plot of cumulative methane production (mL) from in-vitro anaerobic incubation of 25-g landfill samples with gravimetric water content adjusted to 100 and 200%.

utilization by developing methanogenic populations limited methane production, whereas later, polymer hydrolysis limited methanogenesis. There was negligible lignin decomposition.

In terms of controlling variables, several differences are apparent between methane production in anaerobic digesters and landfills. Landfills have larger particle size, higher density, lower water content, and poorer mixing, all resulting in lower methane production rates than attainable for digesters. Digesters typically contain <10% total solids (wt/wt) in a mixed system; 10–30% is considered "high solids", and solids content > c. 30% is termed "dry anaerobic digestion" (Jewel et al. 1982). The Barlaz et al. (1989) refuse samples (discussed above) were adjusted to 27% total solids.

The total solids content of landfill samples taken from three sites in Illinois, Wisconsin, and Colorado ranged from 39 to 86% (Table 1). Landfills typically have moisture contents below field capacity; that is, they are not freely draining. In addition, they display wide variations in moisture content both vertically and horizontally due to; 1) variable moisture at the time of burial; and 2) subsequent moisture movement which is predominantly under unsaturated conditions, where interfaces between materials of different hydraulic conductivity are important zones of preferential moisture movement or retardation. For biodegradation, low water contents and highly variable water contents imply low and variable methane production rates due to poor circulation of nutrients and microorganisms to degradable substrates. In vitro anaerobic incubation at 35°C of ground landfill samples from the three sites summarized in Table 1 yielded average biogas production (methane plus carbon dioxide) from triplicate assays ranging from 1.3×10^{-3} to 1.7×10^{-1} m^3 (dry) kg^{-1} yr^{-1} (Bogner et al. 1989a, b, Bogner 1990). The extreme high and low values were from the same depth at the same 17-ha Illinois site, which contained refuse more than 15 years old. For methane alone, the highest rate from these three sites was 5.4×10^{-2} m^3 (dry) kg^{-1} yr^{-1}. Water addition to parallel sets of assays from the same sites reduced the overall variability observed in the controls by raising the extreme low value for total biogas an order of magnitude, to 1.2×10^{-2} m^3 (dry) kg^{-1} yr^{-1}. In this case, the water addition was to a constant gravimetric water content of 200% – or c. 30% total solids, the range of "dry" anaerobic digestion.

The solids analysis given in Table 1, combined with laboratory in vitro assays (Bogner et al. 1989a, b, Bogner 1990), can be used to predict biodegradability of individual landfill samples. A major consideration is that landfills are a three-part system: refuse, soil, and water. Each component may contribute degradable organics, nutrients, and microorganisms. In general, the combination of high water content, low volatile solids content, and high indigenous nutrients was optimal for high rates of gas production in the controls. For the three sites studied, the amount and characteristics of interbedded soil materials influenced the volatile solids loading, nutrient characteristics, and buffering capacity of individual assays.

Some specific examples of these traits are given in Figs 2–4. Fig. 2 indicates the beneficial effect of water addition alone; the sample illustrated (CL-27 from the

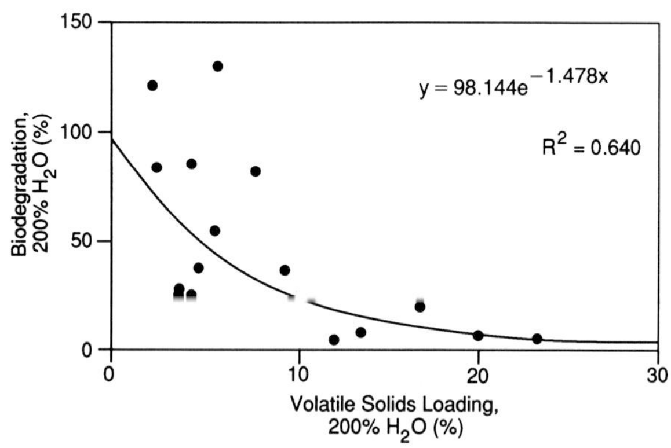

Fig. 3. Relationship between biodegradation and volatile solids (VS) loadings (%) after 200 d incubation. 100% biodegradation based on 0.5 m^3 biogas kg^{-1} VS in assay (Metcalf et al. 1979). Includes assays with gravimetric water content adjusted to 200% from sites in Illinois, Wisconsin, and Colorado, USA. (Modified after Bogner 1990.)

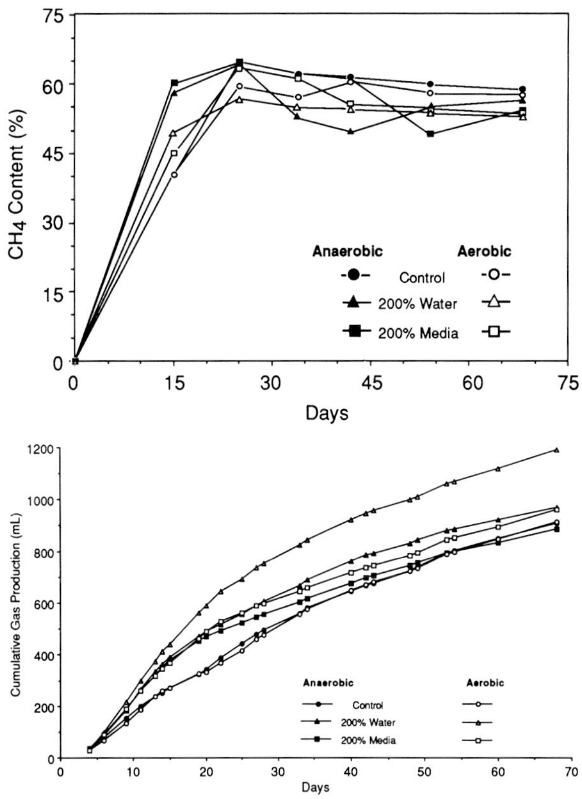

Fig. 4. Historic methane content a) and cumulative biogas production b) from in-vitro anaerobic incubation of northern Illinois landfill sample MN-72. Samples were handled anaerobically in the field and laboratory, with duplicate sample sets being given a controlled aeration before assays were degassed (nitrogen/carbon dioxide) and incubated at 35°C. MN-72 had a water content of 67% and Kjeldahl nitrogen was at 0, 9%; it contained high visible content of calcareous glacial till soil (Bogner et al. 1989b, by permission).

Colorado site) had low natural water content. Fig. 3 indicates the retardation of biodegradation due to higher volatile solids loading; in general, high volatile solids loading promotes high concentrations of intermediate carboxylic acids, which depress the pH and retard development of methanogenic populations. Similar curves were obtained by plotting volatile solids loading vs. percent biodegradation after a given period of incubation for any comparable sets of assays. Fig. 4 indicates historic biogas production and methane (%) for MN-72, which had the maximum gas production rate among the Illinois, Wisconsin, and Colorado samples discussed above. MN-72 was a sample taken from a northern Illinois (USA) landfill in which all the factors discussed in the previous paragraph were optimal, i.e. low volatile solids content, high natural water content, and high content of interbedded calcareous soil with high indigenous nutrients. There were no statistically significant differences between the MN-72 controls and trials with added water or nutrient media, or between the same trials for sample splits handled anaerobically or aerobically. All trials for MN-72 had >50% methane at 25 d.

The controls on landfill decomposition discussed above are directly related to waste composition, site construction practices, and site hydrogeology. Waste composition varies widely worldwide with regard to percent degradable organic materials and the relative proportions of paper, food waste, and garden waste. Water content is related to the pre-filling depth to the top of the zone of saturation and to construction practices. Historically, older landfills in developed countries, which include many closed sites with low maintenance levels, were designed according to less stringent regulations regarding both cover materials and compaction; hence, they receive more infiltration. The amount and characteristics of interbedded soil materials used for daily cover or for final top cover influence decomposition rates by variable supply of nutrients, indigenous microorganisms, and carbonates. Finally, due to poor circulation of liquids within many landfill sites, gas production rates may vary by at least two orders of magnitude within a single site. Thus, within-site variability may be as great as between-site variability. The available literature on biogas production from refuse (including laboratory studies with a wide variety of refuse substrates; anaerobic incubation of landfill samples at various temperatures; and assumed gas production rates derived from field landfill pumping tests) indicates rates ranging from more than 1.0 m^3 (dry) kg^{-1} yr^{-1} to c. 0.007 m^3 (dry) kg^{-1} yr^{-1} (Ramaswami 1970, Rovers and Farquhar 1973, Augenstein et al. 1976, DeWalle et al. 1978, Buivid 1980, Pohland 1980, EMCON Ass. 1981, Klink and Ham 1982, Jenkins and Pettus 1985, Pacey and Dietz 1986, Emberton 1986, Barlaz et al. 1987, 1989). The higher rate resulted from laboratory studies using a high percentage of degradables under optimal conditions and is not representative of landfill decomposition in the field. The low rate was derived from field pumping tests at landfill sites where the rate of biogas generation is typically equated to the sustainable gas recovery from wells drawing from a given mass of refuse.

Controls on landfill methane transport and emissions

The top of the landfill is a dynamic mixing zone for air and landfill gas. In cover soils, the major air gases (oxygen and nitrogen) decrease with depth while methane and carbon dioxide increase with depth (Bogner and Moore 1986, Bogner et al. 1987b, 1988). The major driving forces for soil gas transport are pressure gradients (convection) and gas-specific concentration gradients (diffusion) (Alzaydi et al. 1978, Massman et al. 1981). At most landfill sites, diffusion through the cover

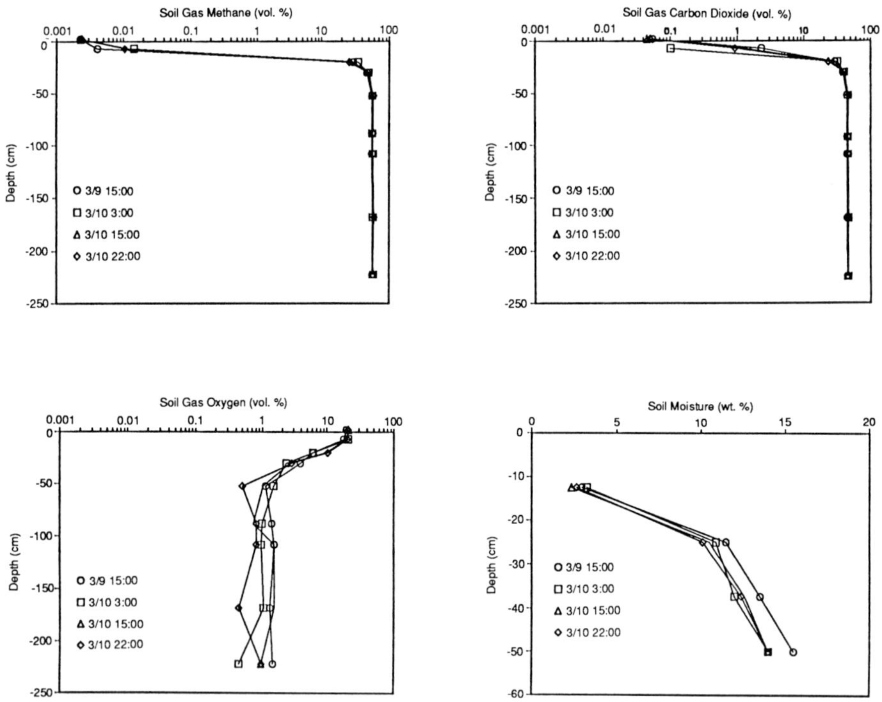

Fig. 5. Soil methane, carbon dioxide, oxygen, and soil moisture at a southern California landfill (Olinda) during March 9–10, 1988. Oxygen values include a small percentage of argon (estimated from expected N_2:Ar to be <5% of given oxygen values). Top of refuse is at 50 cm depth (Bogner et al. 1989b, by permission).

materials or through vertical perforated pipes is the major mechanism for gas transport, resulting in gas venting to the atmosphere. At a small percentage of total sites with pumped gas recovery systems piped to flares or commercial gas recovery operations, convection is the major mechanism. In general, gas transport has both diffusive and convective components.

The physical properties of the porous medium (cover soils and underlying refuse), moisture content, and physical properties of the gas mixture affect convective and diffusive transport. Methane is often monitored in probes placed to various depths at the landfill perimeter; uncontrolled lateral subsurface migration is of concern because 5–15% (volume) methane in air is an explosive mixture. The barometric pressure of the atmosphere and wind-driven convection are external controls on landfill gas emissions; soil moisture is a major internal control on gas transport through a given material. Due to Henry's Law partitioning as a result of increased soil moisture and increased soil gas pressure, carbon dioxide may become part of the aqueous phase, resulting in methane contents of 70% (vol.) or more in the gaseous phase.

Figure 5 shows an extreme case for diffusive methane venting from an unvegetated landfill with sandy silt cover in southern California. There, profiles of soil gases indicate rapid decreases in oxygen from the ground surface downward and rapid decreases in methane and carbon dioxide from the top of the refuse upward. The top of the refuse is at a depth of 50 cm. Methane decreases over several orders of magnitude; the change in slope at c. 30 cm depth coincides with changing soil moisture (also shown in the figure) and a change to more highly compacted cover materials. Even though this is an extreme example of methane venting, the shapes of the three soil gas profiles are typical for unpumped landfill sites in general. Note that measurable oxygen is present in the top of the refuse; this is also typical, indicating that the upper refuse is not a zone of methanogenesis but rather reflects methane transport from deeper zones of gas generation.

Instantaneous diffusive flux of methane was calculated for the data displayed in Fig. 5; diffusive venting was also measured using a closed chamber technique. Calculations for diffusive flux were based on the following equation and on the steep concentration gradient from 20 cm depth to the surface:

$$J = -D\frac{\varphi_A}{\theta}\frac{\Delta C}{\Delta Z}$$

where J = flux (g cm^{-2} S^{-1}); D = molecular diffusion coefficient for CH_4 in given gas mixture (cm^2 S^{-1}), which is 0.153 cm^2 S^{-1} for CH_4 in CO_2 (Satterfield 1970) and 0.220 for CH_4 in air at NTP (Lerman 1979); φ_A = gas-filled porosity (fraction of total porosity); θ = tortuosity (dimensionless); ΔC = vertical methane concentration difference for ΔZ (g cm^{-3}) and ΔZ = depth difference (cm). D was corrected for soil gas pressure and temperature at probe depth according to the following relationship (after Campbell 1985, Rolston 1986, Currie 1960):

Table 2. Comparison of calculated and measured methane flux at the surface of the Olinda Landfill, March 9–10, 1988. θ is taken to be $\varphi_A^{0.333}$ (Millington 1959). Modified from Bogner et al.(1989b).

Date	Time	Depth (cm)	φ_A	D for CH_4 (cm^{-2} S^{-1})	Calculated CH_4 flux[a] (g cm^{-2} S^{-1})	Measured CH_4 flux[b] (g cm^{-2} S^{-1})	Flux ratio: calc. meas.
3/ 9/88	15:00	20–0	0.42	0.25	1.93×10^{-6}	0.63×10^{-6}	3.1
3/10/88	3:00	20–0	0.42	0.24	1.97×10^{-6}	1.97×10^{-6}	1.0
3/10/88	15:00	20–0	0.42	0.26	2.18×10^{-6}	2.21×10^{-6}	1.0
3/10/88	22:00	20–0	0.42	0.24	1.93×10^{-6}	0.37×10^{-6}	5.3
Average	–	–	–	–	$2.0\ \times10^{-6}$	$1.3\ \times10^{-6}$	–
SD	–	–	–	–	$\pm0.1\ \times10^{-6}$	$\pm0.9\ \times10^{-6}$	–

$$D \text{ at } T, P = D_o \left(\frac{T}{T_o}\right)^{1.72} \left(\frac{P_o}{P}\right)$$

where D_o = diffusion coefficient at NTP (273.16 K, 101.3 kPa); T_o = 273.16 K; P_o = 101.3 kPa.

Field-measured variable for various depths included soil gas concentrations, soil gas pressures, soil moisture, soil density, and soil temperature. Detailed methods are given in Bogner et al. (1989b). Low wind speeds (<0.5 m^{-1} s^{-1}), low diurnal soil gas pressure changes (confirmed by continuous monitoring to be under 0.5 kPa), and close conformity of soil gas pressures to atmospheric pressure suggested that convective flux could be neglected in this field setting.

Methane venting was also measured directly using twin closed chambers adjacent to the probes where depth-based variables were measured. The closed chambers each had a 0.25 m^2 area with a volume of 0.3 m^3; six samples were drawn for analysis at 5 min intervals. Samples from both of the twin chambers were drawn at the same time. Results for both calculated and measured methane flux are compared in Table 2. Very high rates of methane flux, probably approaching a maximum venting rate, were observed with both techniques. The mean value for the chamber method (410 kg m^{-2} yr^{-1}), with high standard deviation, was c. 35% lower than the equivalent calculated mean (630 kg m^{-2} yr^{-1}). The variation between techniques and the internal variation in the closed chamber measurements was attributed to the difficulty in seating closed chambers in dry, highly compacted landfill cover materials. For comparison, calculated methane fluxes from the top and sides of a mounded Illinois landfill with vegetated clay cover ranged from one to four orders of magnitude less than the California results presented in Table 2 (Bogner unpubl. data).

Soil moisture changes, based on field data from Illinois and California sites, exert a major influence on soil gas profiles and, in turn, on methane emissions. Fig. 6A shows soil gas methane and oxygen changes due to natural infiltration at an Illinois site during the autumn of 1985; Fig. 6B shows a soil gas profile from the Olinda site after an infiltration experiment in July 1988. At the Illinois site, after a very dry summer, c. 25 cm of rain fell between Sept 15, 1985, and Nov 15, 1985. The profiles reflect relatively slow wetting of cover materials and the top of the refuse (Probe 4A given in Fig. 6A), as evidenced by tensiometer values of 5 cbar or less from early Oct 1985. During the period shown in Fig. 6A, methane increased and oxygen decreased slowly as gas-filled porosity was reduced due to increased soil moisture, effectively sealing the cover soil. In terms of gas diffusion to the atmosphere, gas through-liquid diffusion coefficients are approximately four orders of magnitude less than gas-through-gas diffusion coefficients. At saturation in late Oct 1985, increased soil gas pressures in the upper refuse due to surface sealing were also observed. Soil gas pressures exhibited a diurnal effect in phase with barometric pressure, but were elevated a consistent 2 kPa above atmospheric pressure. Thus, through autumn 1985, diffusive flux of landfill gas was initially dominant but was overshadowed by lateral convective transport as subsurface soil gas pressure increased due to surface saturation.

In contrast, during a rapid sprinkling experiment at Olinda in July 1988, c. 5 cm were evenly infiltrated into the sandy silt cover materials during a 10-h period. There was no measurable runoff. Methane concentrations in soil gas at a depth of 7 cm in the cover materials (Fig. 6B) increased from <200 ppm to c. 35% (vol.) over this 10-h period, indicating rapid "sealing" of shallow gas-filled pore spaces. Because only an area of 12×24 m was sprinkled, no pressure effects could be noted as a result of increased soil moisture. The dynamics of infiltration, whether slow as at the Illinois site, or rapid as at Olinda, are a major influence on changing soil gas profiles, and thus on atmospheric methane emissions.

A mitigating effect on methane emissions may be methane oxidation by methanotrophic bacteria in aerated cover materials. Although field and laboratory work by Hoeks (1972) suggested that virtually all methane in soil gas in the vicinity of a leaking natural gas line can be converted to carbon dioxide by indigenous microbes, laboratory work by Mancinelli and McKay (1985) indicated that a maximum of c. 10% of the methane in a simulated landfill gas mixture was consumed by methanotrophs. Mancinelli et al. (1981) had previously shown that the number of methanol-oxidiz-

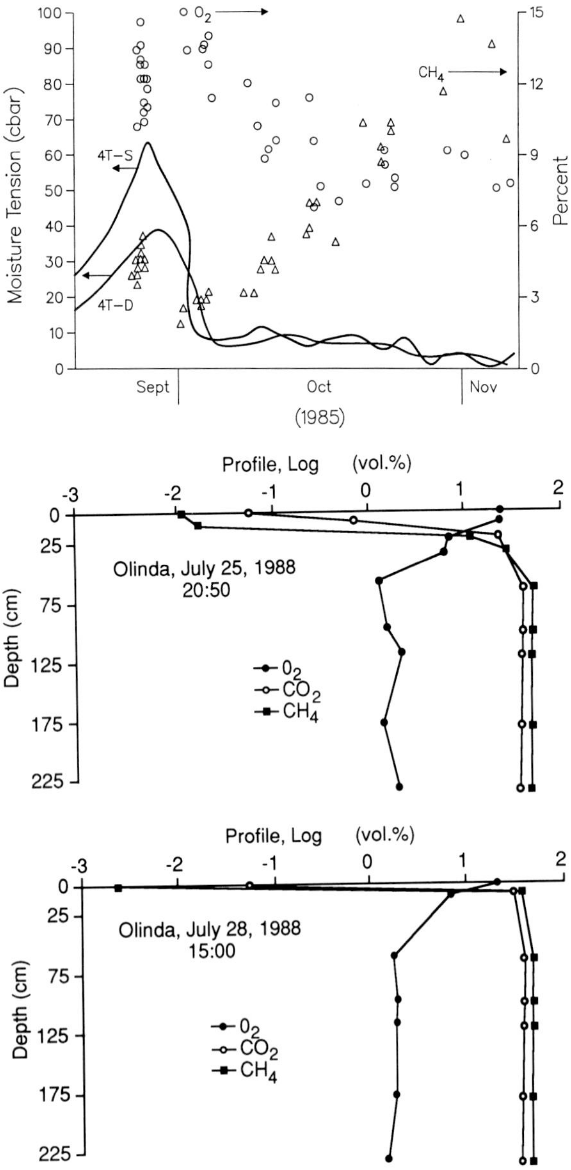

Fig. 6. Effect of increased soil moisture on soil gas concentration profiles. a) Slow infiltration at an Illinois site with clay cover during autumn 1985. Probe in top of refuse at depth of 1.7 m. 4T-S is a shallow tensiometer at 0.8 m depth; 4T-D is a deeper tensiometer at 1.5 m depth. Tensiometer values approaching zero indicate increasing saturation. Oxygen values include argon (6–7% of given values). (Bogner 1986). b) Rapid infiltration at a California site (Olinda) after a sprinkling experiment in July 1988. Profiles of soil gas methane, carbon dioxide, and oxygen before (7/28) 5 cm water infiltrated into sandy silt cover materials. Oxygen values include argon <5% of values shown (Bogner unpubl. data).

ing bacteria in landfill soils were proportional to CH_4 concentration up to c. 50% methane (vol.). Recently, Whalen et al. (1990) reported rapid oxidation of methane in a landfill cover soil gassed with up to 7.7% CH_4 (vol.) in air; they also showed a dependence of methane oxidation on moisture content and temperature at concentrations of 1.05% and 0.95% CH_4, respectively. The role of limiting variables, such as moisture, oxygen, and nutrients, upon methanotrophic activity needs further investigation. The possible role of anaerobic methane oxidation in landfill environments, especially by sulfate-reducing bacteria, also needs further investigation (Dalton and Hocknull 1990).

Organic carbon cycling in landfill environments

Landfill methane-generation rates may vary by several hundred percent, and methane-transport rates to the atmosphere may vary by several orders of magnitude. To a large extent, the major controls are soil moisture content, the relative volume and physical/chemical properties of associated soil materials, and site design features. An evolution has occurred in landfill design from the 1960s to the 1990s. Increasingly strict regulatory requirements have mandated that landfills become tighter hydrogeologic structures to provide greater isolation of refuse from liquid influx and transport. These requirements include bottom and side liners, highly compacted covers to minimize infiltration, and bottom leachate collection systems to collect landfill liquids for alternative disposal. The cumulative effect of these controls is to lower methane generation rates and thus extend the duration of methane generation.

The duration of methane generation is problematical. Old landfills (1960s), for the most part, are still producing methane and contain undecomposed organic materials. Often, for commercial gas-recovery considerations, models are used that embrace a first-order decomposition reaction to match the results of field pumping tests to an assumed gas generation rate function (EMCON 1981). In addition, the refuse mass may be partitioned according to assigned percentages of readily, moderately, and slowly degradable materials. Individual curves are developed for each component over the period of gas generation. Sites may be placed within this framework to estimate future production potential, frequently with the conservative assumption that the total duration of gas generation is 20–30 years.

For purposes of long-term, worldwide climatic considerations, it is well to consider the fate of organic carbon deposited in sanitary landfills. Three major pathways are possible: 1) eventual atmospheric emission as methane, 2) eventual atmospheric emission as carbon dioxide following microbial oxidation or combustion, and 3) long-term sedimentary storage in landfill reservoirs. Taking Bingemer and Crutzen's (1987) estimates of worldwide quantities of landfilled municipal waste (but neglecting their estimated quantities of industrial waste) and the laboratory work on refuse

Table 3. Assumptions and calculated results for sedimentary storage of organic carbon from landfilled refuse (developed countries only, annual basis).

Location	Waste carbon from municipal refuse[a,b] 10^6 T yr^{-1})	Degradable organic carbon in waste (%)	Total waste (10^6 T yr^{-1})
USA, Canada, Australia	37	22	168.2
Other OECD	19	19	100.0
Russia and eastern Europe	13	17.5	74.3
Total			342.5[c]

Fraction	Amount in fresh USA refuse (%)[d]	Estimated annual quantities in refuse from developed countries (10^6 T yr^{-1})[e]	Recalcitrant to landfill decomposition assumed to go to sedimentary storage[f] (%)	(10^6 T yr^{-1})	(% C infraction)	(10^6 T carbon yr^{-1})
Cellulose[h]	45	115.6	18	20.8	45	9.4
Hemicellulose[h]	12	30.8	18	5.5	40	2.2
Lignin[h]	13	33.4	100	33.4	60	20.0
Total		256.9[g]				31.6

Source/location	Total organic carbon productivity (10^9 T yr^{-1})	Organic carbon to long-term sedimentary storage (10^6 T yr^{-1})
Krey 1970 (Marine)	20	20[i]
Fleming 1957 (Marine)	20	
Koblents-Miske et al. 1968 (Marine)	23	
Berger et al. 1987 (Marine)	27	
Sundquist 1985 (Terrestrial)	45–60	
Garrels et al. 1976, Lerman, 1979[j] (Marine)		30
Bolin 1983[k] (Marine)		40
Lasaga et al. 1985[l] (Marine)		48
Holligan 1990[m] (Marine)		22–40

[a] Bingemer and Crutzen 1987.
[b] Excludes industrial waste; excludes municipal waste from developing countries.
[c] C. 20% degradable organic carbon content.
[d] Barlaz et al. 1989, Bookter and Ham 1982; basis: Madison, WI, refuse w/water content of 25% (w/w).
[e] Derived in part from data presented above.
[f] Barlaz et al. 1989.
[g] Based on assumption of 25% water.
[h] Dry-weight basis.
[i] 0.1% of primary productivity.
[j] Using wordwide organic carbon model.
[k] Annual organic C flux to marine sediments.
[l] Calculated rate for last 60 mill yr using geochemical model constrained by stable carbon isotopic content of carbonate rocks.
[m] Estimated range of rates for organic C burial limited to "final burial after decomposition" for shelf carbonates and deep ocean sediments (based on Berner 1982, Whitfield and Watson 1983).

decomposition by Barlaz et al. (1989), one can develop a preliminary estimate of long-term organic carbon partitioning to sedimentary storage. This is a highly conservative estimate, based on idealized laboratory decomposition studies using well-mixed refuse at elevated moisture contents and temperatures. In field settings, particularly at post-1980 sites with high rates of compaction and moisture limitation, the quantities of organic materials recalcitrant to anaerobic digestion and entering sedimentary storage would be expected to be much greater.

Table 3 summarizes assumptions, calculations, and results; it also provides some estimates of current primary productivity of organic carbon (mainly oceanic phytoplankton) and annual flux of organic carbon to sedimentary storage. Barlaz et al. (1989), under idealized laboratory decomposition conditions at 41°C, observed 72% decomposition of cellulose and hemicellulose but negligible decomposition of lignin. On a dry-weight basis, the raw refuse had 40–50% cellulose, 12% hemicellulose, 10-15% lignin, and 4% protein – a total of 64–81% potentially degradable organic materials. If these values are applied to Bingemer and Crutzen's (1987) worldwide estimates of landfilled municipal re-

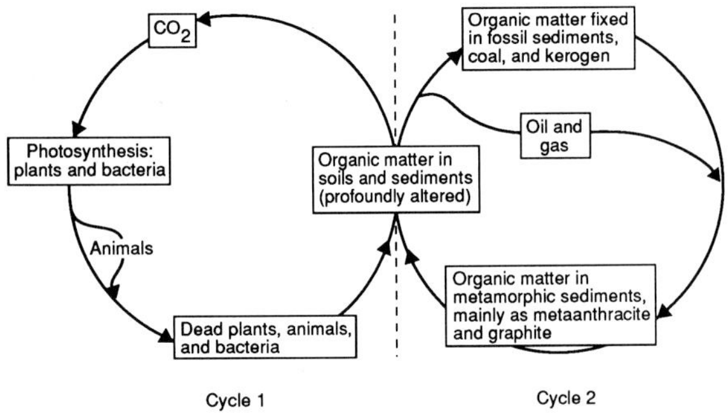

Fig. 7. The two major cycles of organic carbon. Most organic carbon is recycled in Cycle 1. The crossover from Cycle 1 to long-term sedimentary storage in Cycle 2 amounts to 0.1% or less of primary organic productivity. Modified from Welte (1970).

fuse, excluding the uncertain contribution from developing countries, one can develop conservative estimates for landfill organic carbon that enters sedimentary storage. Results shown suggest that the annual amount of recalcitrant organic carbon entering sedimentary storage from landfill deposition, c. 30 mill. tons of carbon yr^{-1}, is approximately equal to previously published estimates of annual organic carbon flux to marine sedimentary storage, the major long-term reservoir for organic C. A number of sources from the literature give a range of 22–48 mill. tons yr^{-1} for organic carbon entering marine sedimentary storage. For comparison, if one considers the base estimates by Bingemer and Crutzen (1987) of annual organic carbon landfilled (top of Table 3) with c. 18% recalcitrant, one still derives c. 12,5 mill. tons of carbon to sedimentary storage annually. Although refuse composition varies greatly worldwide, the sum of degradables (food, waste, yard/garden waste, paper) in municipal solid waste is fairly constant at 55–75% (wt/wt) for the developed countries when data for the European Economic Community, Canada, and the United States (Franklin Associates 1990, Trunick 1991) are compared.

An important implication of the above calculation is that, particularly with modern landfilling practices, we appear to be adding significant organic carbon to terrestrial sedimentary storage, as well as increasing atmospheric methane. In sedimentary environments exclusive of landfills, organic carbon is best preserved under conditions of anaerobic burial. In such environments, too, during early diagenesis, biogenic methane may be produced by microbial degradation of some portion of the original organic matter. In the long-term, i.e. through geologic time, later diagenesis and metagenesis resulting from deeper burial under increased temperatures and pressures transform organic materials to fossil fuel resources. Most organic matter derived from continental higher plants evolves to precursors for natural gas or coal.

Fig. 7 presents a representation of the global organic carbon cycle. Basically, there are two cycles: a short-term cycle, in which carbon is rapidly recycled to atmospheric carbon dioxide, and a long-term cycle, in which organic carbon enters sedimentary storage through geologic time to be transformed into the fossil fuels of the future. Microbial production of methane lies at the cross-over point of the two cycles. After anaerobic burial, methane may be produced from some portion of the organic matter present. By landfilling practices, we are increasing the worldwide production of such biogenic methane, as well as increasing the amount of organic carbon that is stored in sediments.

Acknowledgements – Work supported by the U.S. Dept of Energy, Ass. Secretary for Conservation and Renewable Energy, under Contract W-31-109-Eng-38. Helpful comments provided by M. Barlaz (North Carolina State Univ., Raleigh, NC, USA) are gratefully acknowledged.

References

Alzaydi, A. A., Moore, C. A. and Rai, I. S. 1978. Combined pressure and diffusional transition region flow of gases in porous media. – Am. Inst. Chem. Engin. J. 24: 35–42.

Augenstein, D. C., Cooney, C. L., Wise, D. L. and Wentworth, R. L. 1976. Fuel gas recovery from controlled landfilling of municipal wastes. – Resour. Recov. Conserv. 2: 103–107.

Barlaz, M. A., Milke, M. W. and Ham, R. K. 1987: Gas production parameters in sanitary landfill simulators. – Waste Manag. Res. 5: 27–39.

– , Schaefer, D. M. and Ham, R. K. 1989. Bacterial population development chemical characteristics of refuse decomposition in a simulated sanitary landfill. – Appl. Environ. Microbiol. 55: 55–65.

Berger, W. H., Fischer, K., Lai, C. and Wu, G. 1987. Ocean productivity and organic carbon flux. Part I. Overview and maps of primary production and export production. – In: Berger, W., Smetacek, V. and Wefer, G. (eds), Productivity of the oceans: present and past. Wiley, New York, pp. 429–455.

Berner, R. A. 1982. Burial of organic carbon and pyritic sulfur in the modern ocean. – Am. J. Sci. 282: 415–473.

Bingemer, H. G. and Crutzen, P. J. 1987. The production of methane from solid wastes. – J. Geophys. Res. 92: 2181–2187.

Bogner, J. E. 1986. Understanding natural and induced gas migration through landfill cover materials – the basis for improved landfill gas recovery. – In: Proc. 21st intersoc. energy conversion eng. conf., Am. Chem. Soc., Washington, D.C., pp. 199–204.

– 1990. Controlled study of landfill biodegradation rates using modified BMP assays. – Waste Manag. Res. 8: 329–352.

– and Moore, C. A. 1986. Gas movement through fractured landfill cover materials. – In: Proc. 9th ann. Madison waste conf., Univ. of Wisconsin, Madison, Wis., pp. 288–309.

– , Vogt, M., Moore, C. and Gartman, D. 1987. Gas pressure and concentration gradients at the top of a landfill. – In: Proc. GRCDA 10th int. landfill gas symp., West Palm Beach, FL., Feb 9–13 1987.

– , Rose, C., Vogt, M. and Gartman, D. 1988. Understanding landfill gas generation and migration. – In: Proc. GRCDA 11th int. landfill gas symp., Houston, TE, March 1988, pp. 226–242.

– , Rose, C. and Piorkowski, R. 1989a. Modified biochemical

potential (BMP) assays to assess biodegradation potential of landfilled refuse. – In: Proc. 5th int. conf.: solid wastes, sludges, and residual materials, Rome, Italy, April 26–29 1989, pp. 53–67.

– , Vogt, M. and Piorkowski, R. 1989b. Landfill gas generation and migration, Review of current research II. – In: Proc. anaerobic digestion reviews meeting, Solar Energy Res. Inst., Golden, CO, pp. 140–163.

Bolin, B. 1983. The carbon cycle. – In: Bolin, B. and Cook, R.B. (eds), The major biogeochemical cycles and their interactions, SCOPE Report no. 21. Wiley, Chichester, New York, pp. 41–45.

Bookter, T. J. and Ham, R. K. 1982. Stabilization of solid waste in landfills. – J. Environ. Eng. Div. ASCE 108: 1089–1100.

Buivid, M. G. 1980. Laboratory simulation of fuel gas production enhancement from municipal solid waste landfills. – Dynatech R&D Co., Cambridge, MA.

Campbell, G. S. 1985. Soil physics with basic: transport models for soil-plant systems. – Elsevier, New York.

Currie, J. A. 1960. Gaseous diffusion in porous media, Part 1. – non-steady state method. – Brit. J. Appl. Phys. 11: 314–317.

Dalton, H. and Hocknull, M. D. 1990. Methane oxidation: is anaerobic methane oxidation possible (in landfills)?. – In: Lawson, P. and Alston, Y. R. (eds), Landfill microbiology, R&D Workshop. Harwell Lab., Oxfordshire, U.K., pp. 102–114.

DeWalle, F. B., Chian, E. S. K. and Hammerberg, E. 1978. Gas production from solid waste in landfills. – J. Environ. Eng. Div. ASCE 104: 415432.

Elliason, R. 1942, Decomposition of land-fills. – Am. J. Publ. Health 32: 1029–1037.

Emberton, J. R. 1986. The biological and chemical characterization of landfills. – In: Emberton, J. R. and Emberton, R. F. (eds), Energy from landfill gas. Harwell Lab., Oxfordshire, U.K., pp. 150–163.

EMCON Associates 1981. State of the art of methane gas enhancement in landfills. – Argonne Nat. Lab. Report ANL/CNSV-23.

Fabry, R. 1988. Landfill gas exploitation within the European economic community. – In: Alston, Y. R. and Richards, G. E. (eds), Landfill gas and anaerobic digestion of solid waste. Harwell Lab., Oxfordshire, U.K., pp. 58–68.

Fleming R. H. 1957. General features of the ocean. – In: Hedpeth, J. W. (ed.), Treatise on marine ecology and paleoecoology. Geological Society of America Memoir 67: 87–107, cited in W.H. Berger 1989, Global Maps of Ocean Productivity. – In: Berger, W, Smetacek, V. and Wefer, G. (eds), Productivity of the oceans: present and past. Wiley-Interscience, New York, pp. 429–455.

Franklin Associates. 1988. Characterization of municipal solid waste in the United States, 1960–2000. – (Update 1988) U.S. Environ. Protection Agency Report EPA/53O-SW-88–033.

1990. Characterization of municipal solid waste in the United States: 1990. – Update, U.S. EPA report EPA/ l53O-SW-90–042.

Garrels, R. M., Lerman, A. and MacKenzie, F. T. 1976. Controls of atmospheric O_2 and CO_2: past, present, and future. – Am. Sci. 64: 306–314.

Halvadakis, C. P., Robertson, A. P. and Leckie, J. O. 1983. Landfill methanogenesis: literature review and critique. – Tech. Report no. 271, Dep. of Civil Eng., Stanford Univ., Stanford, CA.

Hoeks, J. 1972. Changes in composition of soil air near leaks in natural gas mains. – Soil Sci. 113: 46–54.

Holligan, P. 1990. Coastal ocean fluxes and resources, Report of a CP2 Ad Hoc Workshop, Tokyo, Japan, Sept 1989, IGBP Report no. 14. – Int. Council of Sci. Unions, Stockholm, Sweden, pp. 11.

Jenkins, R. L. and Pettus, J. A. 1985. The use of in-vitro anaerobic landfill samples for estimating landfill gas generation rates. – In: Antonopoulos, A. (ed.), Proc. first. symp. on biotech. advances in processing municipal wastes for fuels and chemicals. Argonne Nat. Lab. Report ANL/ CNSV-TM-167, pp. 419–435.

Jewell, W. J., Dell'Orto, S., Fanfoni, K. J., Fast, S. J., Gotling, E. J., Jackson, D. A. and Kabrick, R. M. 1982. Agricultural and high strength wastes. – In: Hughes, D. E. et al. (eds), Anaerobic digestion 1981. Elsevier Biomed. Press, Amsterdam, pp. 151–168.

Klink, R. E. and Ham, R. K. 1982. Effects of moisture movement on methane production in solid waste landfill samples. – Resour. Conserv. 8: 29–41

Koblents-Mishke, O. I., Volkovinshiy, V. and Kabanova, Y. 1968. Noviie danniie or velichine pervichnoi producktsii mirovogo okeana. – Doklady Akad. Nuak SSSR 183: 1189–1192, cited in Berger, W.H., 1989, Global Maps of Ocean Productivity. – In: Berger, W., Smetacek, V. and Wefer, G. (eds), Productivity of the oceans: present and past. Wiley-Interscience, New York, pp. 429–455.

Krey, J. 1970. Die UrProduktion des Meeres. – In: Dietrich, G. (ed.), Erforschung des Meeres. Umschau, Frankfurt, p. 183.

Lasaga, A. C., Berner, R. A. and Garrels, R. M. 1985. An improved geochemical model of atmospheric carbon dioxide fluctuations over the past 100 million years. – In: Sundquist, E. T. and Broecker, W. S. (eds), The carbon cycle and atmospheric carbon dioxide: natural variations Archean to present. Geophys. Monograph 32, Am. Geophys. Union, Washington, D.C., pp. 397–411.

Lerman, A. 1979. Geochemical processes – water and sediment environments. – Wiley-Intersci., New York.

Mancinelli, R. L. and McKay, C. P. 1985. Methane-oxidizing bacteria in sanitary landfills. – In: Antonopoulos, A. (ed.), Biotech. advances in processing municipal wastes for fuels and chemicals. Argonne Nat. Lab. Report ANL/CNSV-TM-167.

Mancinelli, R. L., Shulls, W. A. and McKay, C. P. 1981. Methanol-oxidizing bacteria used as an index of soil methane content. – Appl. Environ. Microbiol. 42: 70–73.

Massman, J. W., Moore, C. A. and Sykes, R. M. 1981. Development of computer simulations for landfill methane recovery. – Argonne Nat. Lab. Report ANL/CNSV-26.

Metcalf and Eddy Inc. and Tchbanoglous, G. 1979. Wastewater engineering: treatment, disposal, reuse. 2nd Edition. – McGraw-Hill, New York.

Millington, R. J. 1959. Gas diffusion in porous media. – Science 130: 100–102.

Orlich, J. 1990. Methane emissions from landfill sites and waste water lagoons. – In: Proc. int. workshop on methane emissions from natural gas systems, coal mining, and waste management systems, Washington, D.C., April 9–13 1990. U.S. Environ. Protection Agency, pp. 465–472.

Pacey, J. G. and Dietz, A. M. 1986. Gas production enhancement techniques. – In: Emberton, J. R. and Emberton, R. F. (eds), Energy from landfill gas. Harwell Lab., Oxfordshire, U.K., pp. 361–374.

Pohland, F. G. 1980. Leachate recycle as landfill management option. – J. Environ. Eng. Div. ASCE 106: 1057–1069.

Ramaswami, J. N. 1970. Nutritional effects on acid and gas production in sanitary landfills. – Ph.D. Thesis, West Virginia Univ., Morgantown, WV.

Richards, K. 1988. Landfill gas – a global review. – In: Houghton, D. R. (eds), Biodeterioration 7. Elsevier Appl. Sci., London, pp. 774–790.

– 1990. Landfill gas: working with gaia. – In: Proc. int. workshop on methane emissions from natural gas systems, coal mining, and waste management systems, U.S. Environ. Protection Agency, pp. 525–540.

Rolston D. E. 1986. Gas diffusity and gas flux. – In: Klute, A. (ed.), Methods of soil analysis, Part I, physical and mineralogical methods, Agronomy Monograph No. 9, 2nd. ed.

Am. Soc. of Agronomy, Soil Sci. Soc. of America, Madison, WI., pp. 1089–1119.
Rovers, F. A. and Farquhar, G. J. 1973. Infiltration and landfill behavior. – J. Environ. Eng. Div. ASCE 9: 671–690.
Satterfield, C. N. 1970. Mass transfer in heterogeneous catalysis. – M.I.T. Press, Cambridge, MA.
Stegman, R. and Ehrig, H.-J. 1982. Enhancement of gas production in sanitary landfill sites: experiences in West Germany. – In: Senugpta, S. and Wong, K.-F. V. (eds), Proc. Resource Recovery from Solid Wastes. Pergamon Press, New York, pp. 425–434.
Sundquist, E. 1985. Geological perspectives on carbon dioxide and the carbon cycle. – In: Sundquist, E. and Broecker, W. (eds), The carbon cycle and atmospheric CO_2: natural variations archean to present. Am. Geophys. Union Monograph 32, Washington, D.C., pp. 5–60.
Thorneloe, S. 1990. Methane emissions from landfills and other waste management facilities. – In: Proc. int. workshop on methane emissions from natural gas systems, coal mining, and waste management systems, Washington, D.C., April 9–13 1990. U.S. Environ. Protection Agency, pp. 559–572.
Trunick, P. 1991. Sharing MWS planning in the European community. – MWS Management 1: 46–49.
Welte, D. H. 1970. Organischer Kohlenstoff und die Entwicklung der Photosynsthese auf der Erde. – Naturwissenschaften 57: 17.
Whalen, S. C., Reeburgh, W. S. and Sandbeck, K. A. 1990. Rapid methane oxidation in a landfill cover soil. – Appl. Environ. Microbiol. 56: 3405–3411.
Whitfield, M. and Watson, A. J. 1983. The influence of biomineralization on the composition of seawater. – In: Westbroek, P. and De Jong, E. W. (eds), Biomineralization and biological metal accumulation. Reidel Pub. Co., pp. 57–72.

Ecological Bulletins 42: 109–115. Copenhagen 1992

Denitrification in a wheat cropped field – comparison of methods

K. Haider and U. Schneider

Haider, K. and Schneider, U. 1992. Denitrification in a wheat cropped field – comparison of methods. – Ecol. Bull. (Copenhagen) 42: 109–115.

Gaseous N-losses were measured by three independent methods from a fertilized wheat field within an experimental water shed. They varied from 12 to 24 kg N ha^{-1} yr^{-1}. Denitrifying enzyme activity and soil respiration were measured during the year. Enzyme activity was generally high and decreased only after prolonged dry or cold periods, and rapidly reestablished at more favorable soil conditions. Furthermore, gaseous and leaching losses as well as plant uptake of fertilizer- and soil-N were determined by a N-balance upon application of highly ^{15}N enriched fertilizer. Leaching losses of nitrate from soil- and fertilizer-N occurred only between fall and spring, in particular when the field was barren. Most of the leached NO_3^- came from mineralized soil-N while little came from previously immobilized fertilizer-^{15}N.

K. Haider, BFA Landwirtschaft (FAL); Bundesallee 50, D-3300 Braunschweig, Germany. – U. Schneider, Bayer AG., Pflanzenschutzzentrum Monheim, D-5090 Leverkusen, Germany.

Introduction

Denitrification is considered an important process for the removal of surplus fertilizer nitrogen not taken up by plants or removed by crops. Reported amounts of N-losses by this process differ widely and range from 20 to 40% or more of the applied fertilizer-N (Jagnow and Söchtig 1981, Hauck and Weaver 1986, Hauck 1988). Denitrification losses varying between 6 and 165 kg N ha^{-1} within 8 weeks were recently reported by Becker et al. (1990) Other reports show that however, in well drained and tilled soil of the humid temperate climates denitrification losses did not exceed 15 kg N ha^{-1} per cropping period (Aulakh et al. 1982, Germon 1985, Benckiser et al. 1986, 1987). In several studies, N-loss determined by the ^{15}N-balance method considerably exceeded those measured as gaseous losses. N-losses as low as 10% or as high as 50% of the applied ^{15}N-fertilizer were compiled by Hauck (1988) from lysimeter studies using the ^{15}N-balance method. On the other hand several studies indicated a good agreement between gaseous and ^{15}N-balance losses (Mosier et al. 1986, Heinemeyer et al. 1988, Schneider et al. 1990). Gaseous losses from highly enriched ^{15}N nitrate were comparable to those of simultaneously determined balance losses and amounted 7–10 kg N ha^{-1} per cropping period. Any measurements of denitrification in the field, however, have great spatial and temporal variability in common (Parkin et al. 1987).

Western Europe and USA have a highly intensified agriculture where input of nitrogen from fertilizer and wet or dry depositions frequently exceeds uptake by plants and removal by crops (Newbould 1989). A survey by Bach (1985) indicated for Germany that N-addition exceeds removal by an average of c. 100 kg N ha^{-1} yr^{-1}. Similarly, Keeney (1982) reported for the corn belt in Illinois, USA, 60 to 80 kg N ha^{-1} was added above its removal by crops.

If denitrification losses do not compensate the surplus fertilizer inputs, leaching of nitrate into ground and surface waters may be expected (Keeney 1982, Bach 1987). Therefore, a study was conducted in the scope of the "Sonderforschungsbereich 179" (Special Cooperative Group) at Braunschweig on the input and output of N-fertilizer in an intensively cropped watershed, focusing on estimation of denitrification losses in the investigated area by different analytical methods. Nitrate leaching losses were also followed using lysimeters. Preliminary results of this study were published by Schneider et al. (1990).

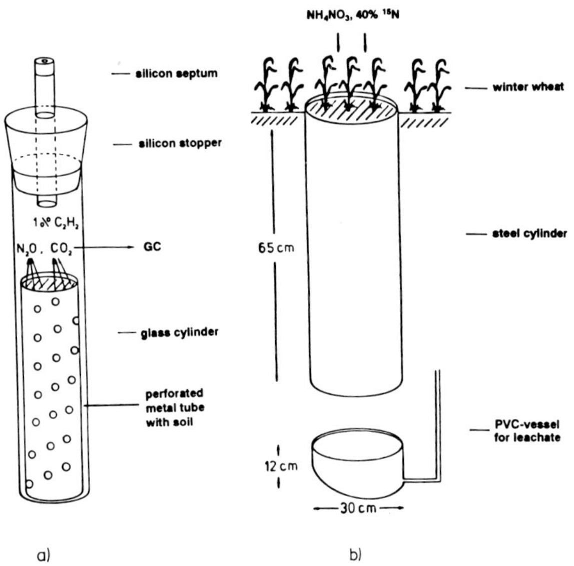

Fig. 1. a) soil cores in glass cylinders for determination of the N_2O release, b) soil cylinders with lysimeter bottles for determination of gaseous N- and leaching-losses, and for plant uptake of fertilizer- and soil-N.

Materials and methods

The experimental site has a humid temperate climate and is located in a watershed near Schladen, south of Braunschweig (Lower Saxony, Germany). The soil, a silty loam overlaying Emscher-Marl (Gleyic Luvisol, Othmer and Bork 1989), was cropped to winter wheat during two growing seasons (1988, 1989), then remained fallow until sugar-beets were seeded in April 1990. The application of mineral fertilizer-N amounted to 200 and 220 kg N ha^{-1} in 1988 and 1989, respectively.

Denitrification was measured by the "open cover method" in the field as described by Ryden et al. (1978) or by Benckiser et al. (1986). Six parallel chambers (50 cm long, 10 cm wide, 15 cm high) were put at weekly intervals into the soil between the rows. C_2H_2 was diffused into the soil at 60 cm depths for 2 h. After one more hour the actual measurements began. Air at a rate of 30 l per h was sucked through each chamber for 6 h and after removal of CO_2 and moisture from the air, N_2O was trapped in molecular sieve (5 Å, Merck, Darmstadt). The adsorbed N_2O was released and determined by a GC column system with backflush to remove contaminants (CO_2, C_2H_2, H_2O) coupled with an ECD heated to 330°C (Mosier and Mack 1980).

Another method used measurements of the N_2O release from undisturbed soil cores after C_2H_2 addition (Rice and Smith 1982). Thirty cores (Ø 3.2 cm, length 15 cm) were taken randomly at weekly intervals from a 400 m^2 plot at 0–15 and 15–30 cm soil depths, respectively. The cores were taken with stainless steel metal tubings (1 mm wall thickness) with perforated walls to allow gas exchange. They were inserted into glas cylinders (500 cm^3) closed with a rubber stopper fitted with a glass tube capped with a serum stopper (Fig. 1a) and amended with C_2H_2 (1% vv). The tubes were immediately incubated after sampling at actual field temperatures. The N_2O-released after 24 h was quantified by GC as described. Soil respiration (CO_2-production) was determined in gas samples taken from the glass cylinders by means of a thermal-conductivity detector using He as a carrier gas. Production of both gases was calculated from the difference between concentration in the cylinders after 24 h and time zero. Denitrifying enzyme activity was determined using methods described by Smith and Tiedje (1979) and Parkin et al. (1987). Soil from 10 cores were mixed and 3 replicate 20 g soil samples (fresh weight) from each layer were placed with 20 ml phosphate buffer (0.05 M, pH 7) containing 5 mM KNO_3, 10 mM glucose and 1 g L-1 chloramphenicol into serum flasks (120 ml) and incubated under N_2 and 10.1 kPa C_2H_2 under shaking at 50 rpm. At different time intervals up to 120 min, 1 ml of headspace gas was analyzed for N_2O.

Fertilizer losses were determined by ^{15}N-balances. Fourteen steel cylinders (Ø 30 cm, 65 cm length) were inserted into the soil of the wheat field at early spring before fertilization. The soil surface was fertilized with an aqueous solution of 40% ^{15}N enriched $(NH_4)_2SO_4$ and KNO_3 (75% NH_4–N, 25% NO_3–N) in split doses at the same amounts and times as the wheatfield was fertilized. Three parallel cylinders were removed at different stages of plant maturity. Five separate cylinders of the same size, but equipped with water collection bottles at the bottom were used as lysimeters to quantify leaching of nitrate (see Fig. 1b).

After removal of the cylinders, plants and roots were separated and the soil was divided into layers of 10 cm depths. Total and mineral N were separately analyzed for ^{15}N contents by mass spectrometry or optical emission (SIRA 12, VG Instruments, England; NIA–1 ^{15}N Analyzer, Jasco, Japan). Each analysis was made in triplicate and the means and standard deviations were calculated and used for statistical tests (Doerffel 1987). Precipitation, soil temperature at different depths, water and nitrate contents in soil were independently and continuously monitored at the experimental site.

Results

Gaseous N-losses

Gaseous losses determined by the "open cover method" were only continuously measured in 1988 and ranged from 6 to 12 kg N ha^{-1} during the growing season. They were therefore, at the same magnitude as formerly reported by Benckiser et al. (1986, 1987). Fig. 2a–d presents the denitrification losses obtained from the undisturbed soil cores, nitrate contents in the soil, water

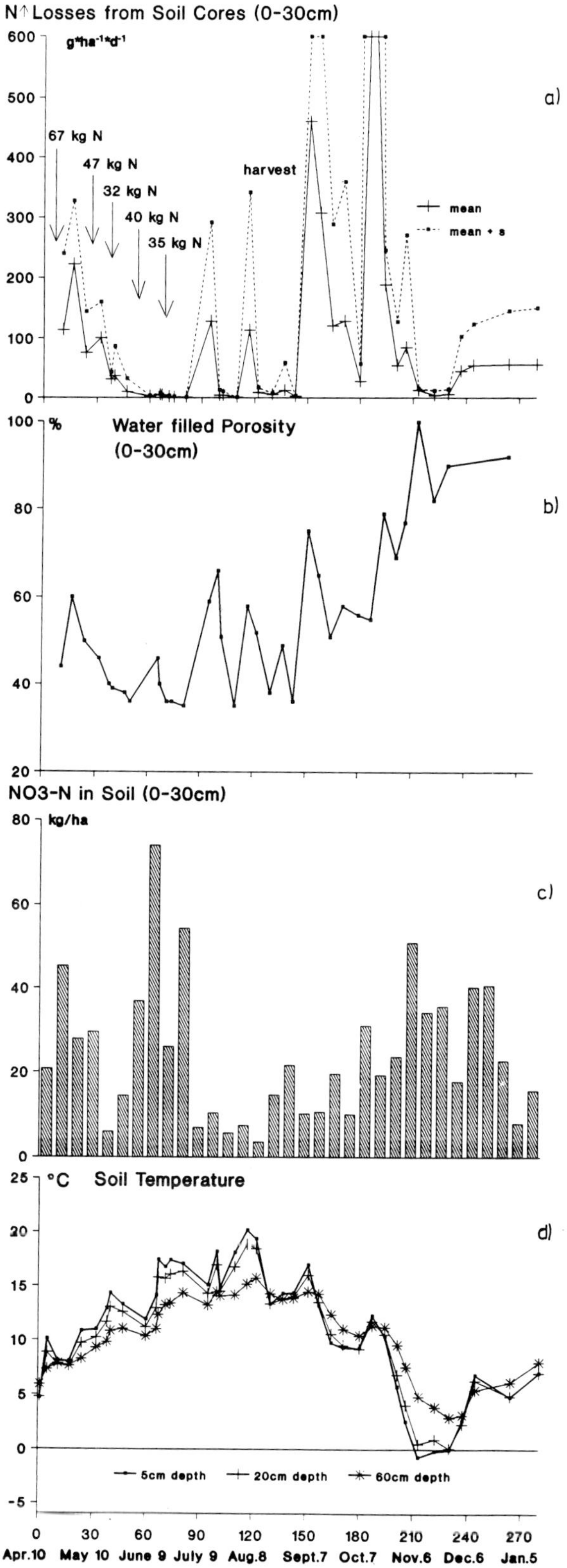

Fig. 2. a) N-losses from soil cores, b) % water filled porosity, c) NO_3-N contents, 0–30 cm soil layer, d) soil temperatures in 5, 20 and 60 cm soil depths. Period from April 10 1989 until Jan 15 1990.

filled porosity, and soil temperatures in 1989 at three different depths. Similar measurements for 1988 have been published by Schneider et al. (1990).

The time course of the N_2O releases during the growing season for both the "open cover" and the "soil core" methods followed approximately the precipitation events and the percentages of water filled soil porosity. The N_2O release from the surface soil (0–15 cm) was 3 to 4 times higher than that of the deeper soil layer. During late spring and the following main growth period, precipitation was uncommonly low. Denitrification between planting and harvest equalled to only 12 (CV%=50) kg N ha^{-1}. At harvest the soil was nearly void of nitrate. After tillage the field remained fallow until spring 1990 when sugar beets were seeded. With high precipitation and soil temperatures of 10–15°C from Sept until the end of March 1990 denitrification was vigorous and equalled 19 (CV%=42) kg N ha^{-1}. End of Nov soil temperatures decreased to ≥8°C and denitrification was small even at high moisture and nitrate contents. The total N_2O–N losses from the beginning of the cropping period 1989 until April 1990 equalled 31 (CV%=32) kg N ha^{-1}.

Denitrifying enzyme activity in the soil layer of 0–15 and 15–30 cm depths, respectively was determined each week and is shown in Fig. 3 together with the released N_2O–N from soil cores and the soil respiration (CO_2 release) from May 1989 until May 1990. During the dry period from May to the end of Aug 1989, both the denitrification losses and the denitrifying enzyme activity were low, but both increased after precipitations in Sept and decreased after a prolonged cold and dry period at end of 1989 which lasted until April 1990. Generally, enzyme activity in the upper soil layer was higher than in the lower. During prolonged dry or cold periods the activity decreased, but increased rapidly again in response to more favorable soil conditions. Similar to the enzyme activity, the CO_2 release from the soil increased at higher temperatures and moisture contents in the soil. C. 15 kg C ha^{-1} d^{-1} were released as a maximum and release was never lower than 4 kg C ha^{-1} d^{-1}. It was therefore, 1000 to 100,000 times higher than the gaseous N-release.

^{15}N-balance losses and N-uptake by plants

^{15}N balances were analyzed at three different stages of the plant development. Fig. 4 shows the distribution of fertilizer-N in plants and different soil layers from 0 to 90 cm depth for 1989, using shooting and harvest. In May at shooting stage (Fig. 4a), 25% of the applied fertilizer-N remained in soil and was mostly located in the upper (0–20 cm) layer. In Aug at harvest (Fig. 4b), most of the residual fertilizer-N remaining in soil was still concentrated in the surface layer. Mineral-N portion was then much smaller than in May, and corresponded only to 1.3 and 3.3 kg N ha^{-1} of fertilizer and

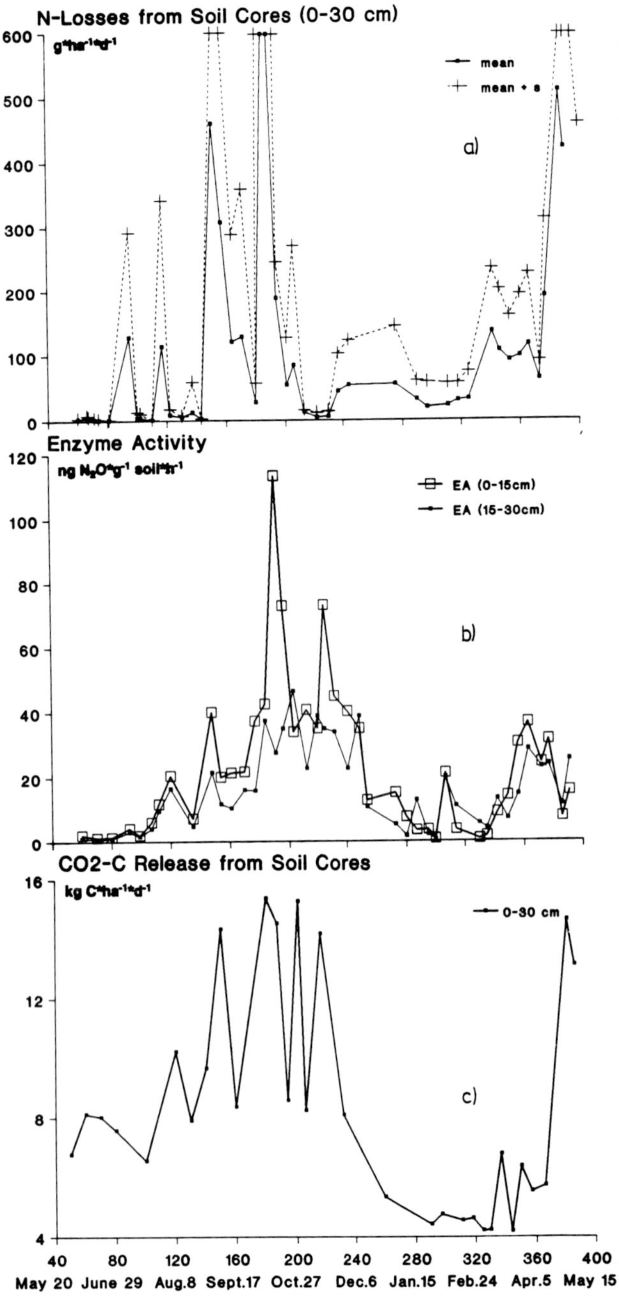

Fig. 3. a) N-losses from soil cores, b) denitrifying enzyme activity, c) CO_2 release (respiration) from soil cores, from May 1989 until May 1990, 0–30 cm soil depths, each.

total mineral-N, respectively (Fig. 2c, 4b). More than 70% of the fertilizer-N was concentrated in the plants and there preponderantly in the ears.

Table 1 shows the total recovery and losses of fertilizer-N from day zero until harvest in Aug 1989. The recovery was 85 to 90% and losses 11 to 22 kg N ha^{-1}. They did not regularly increase with time; the losses at day 90 were higher than those at day 120. The high standard deviations and the applied Scheffé test revealed analytical errors together with spatial variability of the randomly distributed soil cylinders. The fertilizer N balance losses, however, agreed within the order of magnitude to the gaseous N-losses obtained from the soil cores as well as from measurements by the "open cover method" (see also Fig. 2a). Fertilizer N-losses in both 1988 (Schneider et al. 1990) and 1989 were similar and totaled c. 20 kg N ha^{-1}. The uptake of total- and fertilizer-^{15}N by the wheat plants indicates that 50 to 60% of the N-uptake originated from mineralized soil-N and the smaller portion from the applied fertilizer-N. At harvest, N in the grain consisted of c. 60% from fertilizer-N. Total N removal by the grain crop amounted to c. 180 kg N ha^{-1} with 110 kg N ha^{-1} originating from fertilizer. The N removal by the grain crop in the surrounding field corresponded to 160 kg N ha^{-1}. Due to the dry soil conditions in 1989 during the growing season, total N removal by the grain crop was lower than in 1988 when 200 kg N ha^{-1} was removed (200 kg N ha^{-1} applied as fertilizer). Together with wet and dry N deposition of 20 to 25 kg N ha^{-1} (Dämmgen et al. 1990), N input in 1989 exceeded the N output in form of the grain crop (180 kg N ha^{-1}) together with denitrification (20 to 25 kg N ha^{-1}) by 60 to 70 kg N ha^{-1}.

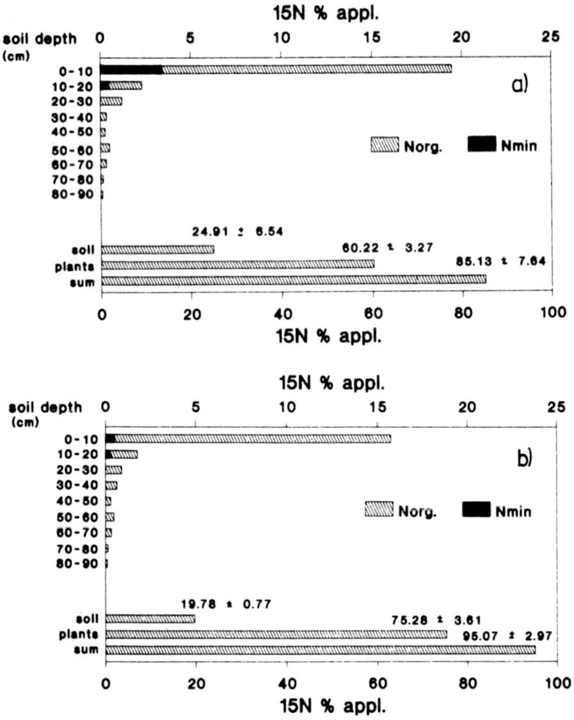

Fig. 4. Distribution and standard deviation of ^{15}N fertilizer in different soil depths, and in plants, a) at shooting stage (May 10, 114 kg N ha^{-1} appl.), b) at harvest (Aug 15, 220 kg N ha^{-1} appl.).

Table 1. ^{15}N balance in plant and soil, and fertilizer losses at different stages during the growing season 1989, mean values and standard deviation from three parallel cylinders each.

Days after first fertilization (April 10)	N-fertilizer applied	N-uptake by plants total[1)]	N-uptake by plants fertilizer	Fertilizer-N remaining in soil	Fertilizer-N loss
			kg N ha^{-1}		
60 shooting	114	164 c[2)] ±12 7	69 a ±3.6	28 a ±4.5	17 a ±6.8
90 ear-emergence	178	263 b ±41.5	117 b ±10.6	40 ab ±8.7	22 a ±13.7
120 harvest	220	343 ab ±39.6	166 ab ±7.9	43 b ±10.8	11 a ±13.4
				N-content in grain total-N	N-content in grain fertilizer-N
				182 ±27.4	109 ±10.1

[1)] Soil-N + fertilizer-N
[2)] Scheffé Grouping: Figures followed by different letters differ on the $p_{0.05}$ level.

Leaching losses of soil- and fertilizer-N

During the growing season the water balance between precipitation and evapotranspiration was equal in 1988 and negative in 1989. Therefore, no leaching occurred during this period. During fall, winter and spring with precipitation mostly in the form of rain, water leached into the collection bottles of the lysimeters. Due to reseeded winter wheat in fall 1988, leaching losses were low during the winter period and amounted to 14 kg NO_3–N ha^{-1}. In fall 1989 the field remained barren until spring 1990 when sugar beets were seeded.

Mineral N content in soil at harvest in 1989 was low (Fig. 2c) and most of the remaining fertilizer N was immobilized in organic form. As can be noticed from Fig. 5a, the mineral N-contents, however, increased considerably during fall in the different soil layers. In Nov, this mineral-N consisted mostly of mineralized organic soil-N and less of fertilizer-^{15}N. Nitrate-N in the leachates amounted to 7.5 kg N ha^{-1} with 1 kg N ha^{-1} derived from fertilizer. During the winter period until end of March 1990, 140 mm from a total precipitation of 350 mm was leached into the lysimeter bottles and contained 26 kg NO_3-N ha^{-1}. ^{15}N-analyses of the soil in the lysimeters indicated that even at the end of the investigated period, most of the fertilizer N remained im mobilized in the upper layers. (Fig. 5b). Therefore, losses from fertilizer-N were small during the winter period. Apparently mineralization and nitrification occurred preferentially on organic soil nitrogen and less on recently immobilized fertilizer. Nitrate contents in the leachates always exceeded the limits for drinking water, set by the EC and the German Federal Department of Health of 50 mg nitrate L^{-1} and amounted on the average to 79 mg L^{-1} in 1989/1990.

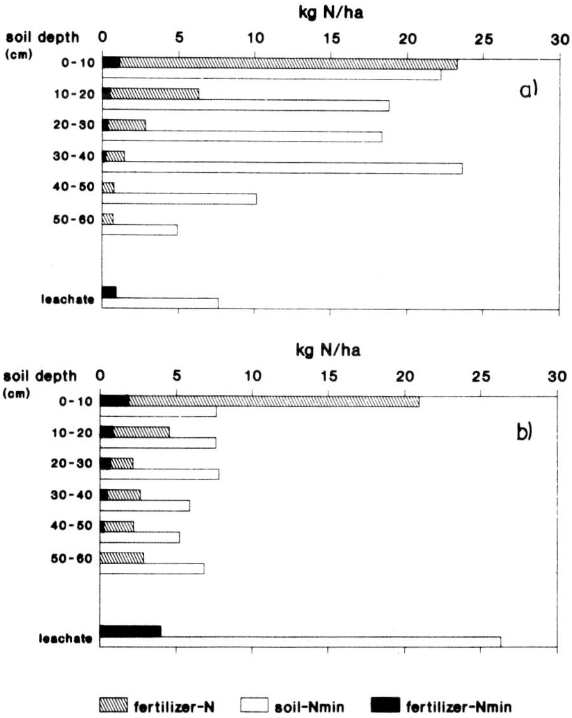

Fig. 5. Distribution of fertilizer- and soil-N in different soil depths, and in leachates, a) at Nov 20 1989, b) at March 20 1990.

Discussion

Compared to other data in the literature, denitrification losses determined by three independent methods during the growth period were low and ranged from 10 to 12% of the applied fertilizer-N. Although the balance losses comprise only fertilizer-N and those measured by the N_2O release from C_2H_2 treated soil include both fertilizer and soil N, this agreement was reasonable. During the cropping season, nitrate in the upper soil layer, where most of the denitrification occurs, consisted mainly of fertilizer N, and should therefore dominantly contribute to gaseous N-losses. After harvest, the soil was nearly emptied of mineral nitrogen. As already reported by Lochmann et al. (1989) mineralization and nitrification of organic soil-N then proceeded and mineral-N is nearly exclusively present as nitrate. This becomes denitrified or is leached down after high precipitation. Denitrification in the present study occurred until soil temperatures dropped below 8°C. This agrees with other observations of the influence of soil temperatures on denitrification which indicated that denitrification ceases at temperatures <5°C (Frede et al. 1975, Firestone 1982, Benckiser et al. 1986). Parsons et al. (1991) however, recently reported that denitrification in soil still occurred at temperatures <5°C. Denitrifying enzyme activity only decreased after longer periods of drought or frost, but increased rapidly with increases in temperature or moisture. Therefore, considerable denitrification losses of available nitrate can be expected at suitable soil temperature and moisture conditions even during the winter.

Nitrate in the soil and the leachates consisted of ≥ 80% mineralized soil-N and little of freshly immobilized fertilizer-^{15}N. This agrees with observations by e.g. Legg and Meisinger (1982) that fertilizer-N once immobilized, does not become preferentially mineralized compared to the bulk of the soil-nitrogen.

Acknowledgements – This work was supported by the Deutsche Forschungsgemeinschaft within the scope of the "Sonderforschungsbereich 179" (Special Cooperative Group) at Braunschweig, Germany. The authors are grateful to A. R. Mosier for baluable advice and to A. Lorenz, B. Simon and H. Lemke as well as D. Maßmann for excellent technical assistance.

References

Aulakh, M. S., Rennie, D. A. and Paul, E. A. 1982. Gaseous nitrogen losses from cropped and summerfallowed soils. – Can. J. Soil Sci. 62: 187–196.

Bach, M. 1985. Stickstoffbilanzen der Kreise der Bundesrepublik Deutschland als Grundlage einer Abschätzung der möglichen Nitratbelastung des Grundwassers durch die Landwirtschaft. – Mitt. Dtsch. Bodenkdl. Ges. 43: 625–630.

– 1987. Regional differenzierte Abschätzungen des möglichen Beitrags der Landwirtschaft zur Nitrat-Belastung des Sickerwassers in der Bundesrepublik Deutschland. – Mitt. Dtsch. Bodenkdl. Ges. 55: 561–566.

Becker, K. W., Janssen, E. and Meyer, B. 1990. The ^{15}N-balance method for calculating denitrification losses in arable fields and its verification by direct measurement of gaseous ^{15}N – losses. Int. workshop on denitrification soil, rhizosphere and aquifer, Gießen, 1989. – Mitt. Dtsch. Bodenkdl. Ges., 60: 59–64.

Benckiser, G., Haider, K. and Sauerbeck, D. 1986. Field measurements of gaseous nitrogen losses from an Alfisol planted with sugar-beets. – Z. Pflanzenernähr. Bodenk. 149: 249–261.

– , Gaus, G., Syring, K.-M., Haider, K. and Sauerbeck, D. 1987. Denitrification losses from an Inceptisol field treated with mineral fertilizer or sewage sludge. – Z. Pflanzenernähr. Bodenk. 150: 241–248.

Dämmgen, U., Grünhage, L., Küsters, A. und Jäger, H.-J. 1990. Messungen von Stoff-Flüssen in der bodennahen Atmosphäre in Agrar-Ökosystemen. – VDI-Bericht 837: 565–583.

Doerffel, K. 1987. Statistik in der analytischen Chemie. – VEB Deutscher Verlag für Grundstoffindustrie, Leipzig.

Firestone, M. K. 1982. Biological denitrification. – In: Stevenson, F. J. (ed.), Nitrogen in agricultural soils. Agronomy, Am. Soc. Agron., Madison, WI, pp. 289–326.

Frede, H., Gebhardt, G. H. und Meyer, B. 1975. Größe, Ursachen und Bedingungen von Boden- und Dünger-N-Verlusten durch Denitrifikation aus dem Ap-Horizont einer Acker-Parabraunerde aus Löß. – Gött. Bodenkundl. Ber. 34: 69–213.

Germon, J. C. 1985. Denitrification in cropped soils. – Fertilizer and Agric. 89: 3–14.

Hauck, R. D. and Weaver, R. W. 1986. Field measurements of dinitrogen fixation and denitrification. – SSSA Special Publ. No 18 Madison.

– 1988. A human ecosphere perspective of agricultural nitrogen cycling. – In: Wilson, J. R. (ed.), Advances in nitrogen cycling in agricultural ecosystems. CAB Int., pp. 3–19.

Heinemeyer, O., Haider, K. and Mosier, A. R. 1988. Phytotron studies to compare nitrogen losses from with corn planted soil by ^{15}N-balance or direct dinitrogen and nitrous oxide measurements. – Biol. and Fert. of Soils 6: 73–77.

Jagnow, G. und Söchtig, H. 1981. Stickstoffverluste aus dem Boden in die Atmosphäre und das Grundwasser – Möglichkeiten ihrer Begrenzung. – Ber. Landw. Sh. 197: 223–234.

Keeney, D. R. 1982. Nitrogen management for maximum efficiency and minimum pollution. – In: Stevenson, F. J. (ed.), Nitrogen in agricultural soils. Agronomy, Am. Soc. Agron., Madison, WI, pp. 605–649.

Legg, J. O. and Meisinger, J. J. 1982. Soil nitrogen budgets. – In: Stevenson, F. J. (ed.), Nitrogen in agricultural soils. Agronomy, Am. Soc. Agron., Madison, WI, pp. 503–566.

Lochmann, R., van der Ploeg, R. Z. and Huwe, B. 1989. Zur Parametrisierung der Stickstoff-Mineralisierung in einem Ackerboden unter Feldbedingungen. Z. Pflanzenernähr. Bodenk. 152: 319–324.

Mosier, A. R. and Mack, L. 1980. Gas chromatographic systems for precise, rapid analysis of N_2O. – Soil Sci. Soc. Am. J. 44: 1121–1123.

– Guenzi, W. D. and Schweizer, E. E. 1986. Soil losses of dinitrogen and nitrous oxide from irrigated crops in northeastern Colorado. – Soil. Sci. Soc. Am. J. 50: 344–348.

Newbould, P. 1989. The use of nitrogen fertilizer in agriculture. Where do we go practically and ecologically? – Plant Soil 115: 297–311.

Othmer, H. and Bork, H. R. 1989. Characterization of the soils at the investigation sites. – Landschaftsgenese und Landschaftsökol. 16: 73–86.

Parkin, T. B. 1987. Soil microsites as a source of denitrification variability. – Soil Sci. Soc. Am. J. 51: 1194–1199.

– Starr, J. L. and Meisinger, J. J. 1987. Influence of sample size on measurement of soil denitrification. – Soil Sci. Soc. Am. J. 51: 1492–1501.

Parsons, L. L., Murray, R. E. and Smith, M. S. 1991. Soil

denitrification dynamics: spatial and temporal variations of enzyme activity, populations, and nitrogen gas loss. – Soil Sci. Soc. Am. J. 55: 90–95.

Ryden, J. C., Lund, L. J. and Focht, D. D. 1978. Direct in-field measurement of nitrous oxide flux from soils. – Soil Sci. Soc. Am. J. 42: 731–737.

Rice, W. and Smith, M. 1982. Denitrification in no tilled and plowed soils. – Soil Sci. Soc. Am. J. 46: 1168–1173.

Schneider, U., Haider, K. and Mahmood, T. 1990. Preliminary results by comparing gaseous denitrification losses with ^{15}N-balance losses in a wheat and a barley cropped field. Int. Workshop on Denitrification Soil, Rhizosphere and Aquifer, Gießen, 1989. – Mitt. Dtsch. Bodenkdl. Ges. 60: 37–44.

Smith, M. S. and Tiedje, J. M. 1979. Phases of denitrification following oxygen depletion in soils. – Soil Biol. Biochem. 11: 262–267.

Ecological Bulletins 42: 116–123. Copenhagen 1992

Measurement and modelling of nitrous oxide emissions from soils

Keith A. Smith and Jonathan R. M. Arah

Smith, K. A. and Arah, J. R. M. 1992. Measurement and modelling of nitrous oxide emissions from soils. – Ecol. Bull. (Copenhagen) 42: 116–123.

Closed flux chamber/gas chromatographic methods used in one particular study of emissions of nitrous oxide from fertilised arable and grassland soils are described, and the principal results are outlined. Fluxes varied greatly in time and space. The highest observed exceeded 500 g N_2O-N ha^{-1} d^{-1}. In most experiments attempts were made also to measure rates of denitrification, using the acetylene inhibition technique. The method worked satisfactorily in reasonably permeable soils (although not in impermeable clays), making it possible to calculate the fractions of the total gaseous products of denitrification emitted as N_2O. Procedures developed for modelling the processes of N_2O and N_2 production by denitrification in soils are outlined, and the possible extension of the model to field and regional scales, and the inclusion of N_2O emission due to nitrification are discussed.

K. A. Smith and J. R. M. Arah, The Edinburgh School of Agriculture, West Mains Rd, Edinburgh EH9 3JG, U. K.

Introduction

Increasing concentrations of nitrous oxide, N_2O, in the atmosphere are contributing both to global warming and to depletion of the stratospheric ozone layer. Perhaps 90% of the N_2O being released into the atmosphere is derived from soils and land use (Bouwman 1990). Significant contributions to these N_2O emissions come from denitrification, which is the main source process in poorly-aerated, nitrate-rich soils, and from nitrification, which predominates in well-aerated conditions.

Better prediction of the likely extent of global warming, and of the responses of important soil processes to predicted changes in climate, requires a) better and more representative measurements of gaseous emissions, including those of N_2O, which characteristically exhibit great spatial and temporal variability, and b) improved models of the emission processes, as an essential prerequisite for prediction of emissions from large areas of land.

In this paper we describe, as an illustrative case study, the methods we have used to estimate N_2O fluxes from, and denitrification rates in, arable and grassland soils in the cool temperate environment of southeast Scotland. We also outline our procedures for mathematical modelling of the processes involved in N_2O production, indicate some of the predictions of our model, and suggest possible directions for future work.

We wish to define two terms that we believe will prove useful for the quantitative description of N_2O fluxes and associated processes. The first is the "N_2O fraction" (Arah and Smith 1990), which is the fraction of the gaseous products of denitrification emitted as N_2O, and which can have values between 0 and 1 (or 0 and 100%). The second is the "N_2O ratio", which is the ratio of N_2O emitted in the absence of acetylene (which inhibits both nitrification and the terminal reduction step of denitrification – N_2O to N_2) to the N_2O emitted in its presence. Where denitrification is the only significant mechanism of N_2O emission, the N_2O ratio becomes equal to the N_2O fraction; where nitrification is the dominant source, the N_2O ratio can exceed unity.

Measurement of N_2O emissions

Methods

The principal method for measuring emissions of N_2O from land surfaces is the so-called chamber method, which involves placing a cover a few cm high and typ-

Table 1. Description of experimental sites.

Site	Bush House	Cowpark	Croft Field	Glencorse Mains	South Road A	South Road B
Date	1986	1989	1988	1987–88	1985–88	1985–88
Soil series	Darvel	Winton	Alluvium[a)]	Winton	Macmerry	Winton
Texture	SL	CL	SL	CL	L	CL
sand (%)	66.5	34.0	59.9	43.0	49.5	47.5
silt (%)	23.7	44.0	28.1	33.0	35.2	35.3
clay (%)	9.8	22.0	12.0	24.0	15.2	17.4
O.M (%)	3.3	4.3	4.6	4.8–6.0	4.4	3.3
pH	5.9	6.3	6.0	6.2–6.7	6.6	6.5
Crop	spring barley	ryegrass	none	spring barley	winter barley	winter barley

[a)] unclassified alluvial deposit

ically 0.05–0.5 m^2 in area over the ground surface to entrap the gas evolving from the soil surface. Chamber methods are relatively easy to apply, and may be cheap and simple, or much more elaborate and correspondingly more expensive. Closed chambers usually take the form of an inverted box, the rim of which is buried a few centimetre into the soil. The N_2O emitted from the soil surface accumulates in the chamber, whence samples are taken for gas chromatographic analysis. Open chambers have an airstream continuously pumped through them, sweeping the emitted gas into an analyser or a trapping system.

The closed chamber has been favoured because of its simplicity and low cost (Matthias et al. 1980). One disadvantage is that accumulation of N_2O inside the chamber reduces the diffusion gradient and thus the rate of emission, and an allowance should be made for this in the flux calculation (Hutchinson and Mosier 1981). The open system was preferred by Ryden et al. (1978) and Denmead (1979) on the grounds that N_2O does not accumulate in an open chamber, but they expressed reservations about the effect of a constant mass flow over the soil surface. The spatial variability of N_2O fluxes is great (Folorunso and Rolston 1984); their study, therefore, demands the maximum possible replication, even at the cost of some loss of precision in each individual measurement; this is an argument in favour of the simpler closed system.

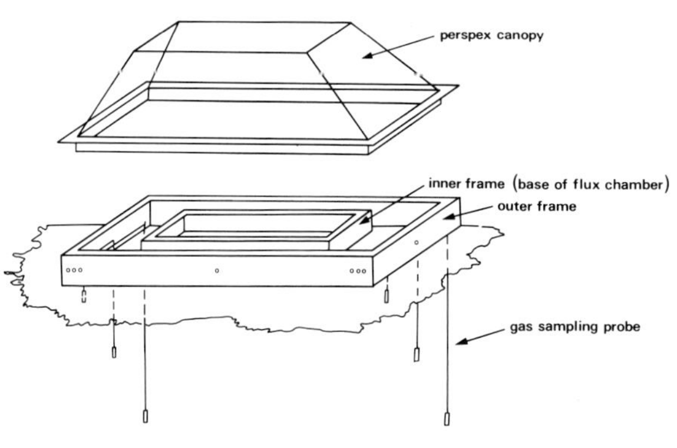

Fig. 1. Apparatus for field application of acetylene and measurement of gaseous emissions from soil (Arah et al. 1991).

In our work we used closed chambers, usually 70 cm square, made up of steel frames 15 cm high inserted into the soil to a depth of 5 cm, with tight-fitting removable steel covers (Arah et al. 1991). These chambers are larger than those used in many N_2O emission/denitrification studies (e.g. Ryden et al. 1978, Ryden and Dawson 1982); the size was increased in order to minimise spatial variability. Our measurements were all made on the Edinburgh School of Agriculture's farms at Bush, c. 10 km south of Edinburgh. The soils ranged from a freely drained sandy loam to poorly drained clay loams developed on glacial till (Table 1).

Our chamber technique proved generally satisfactory for studies of N_2O emission. Attempts to extend the technique to denitrification measurements, by a modification (vertical rather than lateral diffusion of acetylene) of the method of Ryden et al. (1979) in which acetylene is applied to the soil to inhibit the reduction of N_2O to N_2 (Fig. 1), were successful on the relatively coarse soils at Bush House and Croft Field, and the technique also appeared to work on the grassland site at Cowpark immediately following surface application of nitrate fertiliser. Similar attempts at the South Road and Glencorse Mains sites foundered, however, because of the excessively slow diffusion rate of acetylene into (and N_2O out of) the wet clay soils (Arah et al. 1991).

Field application of the acetylene inhibition method thus proved to be unsuitable for investigating conditions which are particularly conducive to the development of anaerobism and hence denitrification. It appears that, in heavy soils, satisfactory measurements of the total amount of denitrification with which N_2O fluxes are associated will have to involve the determination of ^{15}N-labelled N_2 by mass spectrometry. Even this approach, however, has its drawbacks, prominent among them the cost of highly enriched fertiliser and the requirement of a uniform denitrifying pool.

We attempted to estimate N_2O fluxes by measuring

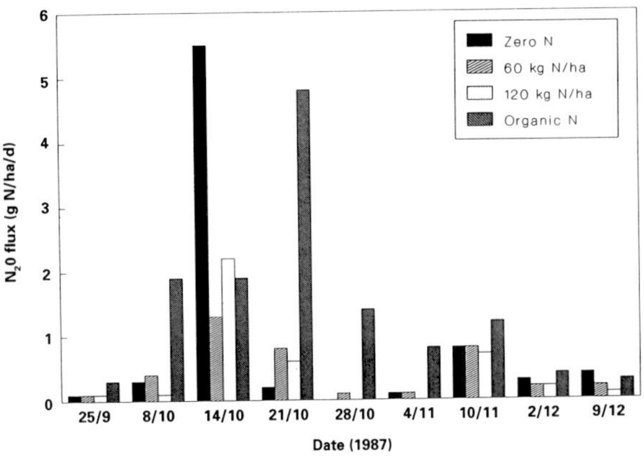

Fig. 2. Surface N_2O flux measured using closed chambers, Glencorse Mains, autumn 1987 (Arah et al. 1991).

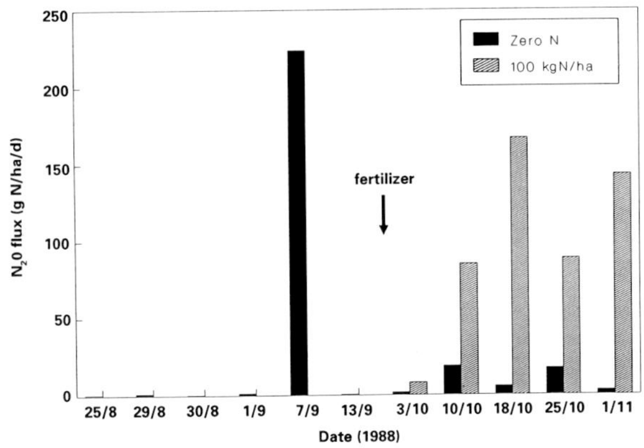

Fig. 3. Surface N_2O flux measured using closed chambers, Croft Field, autumn 1988.

diffusion constants and N_2O concentrations in the soil profile, and applying Fick's law of diffusion. We concluded that this was not a satisfactory technique in impermeable soils, largely because of the presence of sinks as well as sources of N_2O in the profile (Arah et al. 1991).

We also measured the N_2O released from incubated soil cores, but concluded that while this technique is useful for measuring N_2O production rates or total denitrification rates in the presence of acetylene (Aulakh et al. 1984, Ryden et al. 1987), it is unsatisfactory for estimating field emission rates. This is because some of the N_2O produced in the soil may be reduced to N_2 before it can escape to the atmosphere. Escape from a core is easier than escape from the undisturbed soil in the field because the surface-to-volume ratio in the former environment is so much greater and diffusion path lengths much shorter. Core incubation, therefore, is very likely to produce an overestimate of the true N_2O flux to the atmosphere.

Attempts have been made to discriminate between N_2O fluxes originating from denitrification and those from nitrification by applying different concentrations of acetylene and oxygen (Robertson and Tiedje 1987). Interpretation of the results of such measurements is complicated by the spatial heterogeneity inevitable even within a relatively small soil core; whilst it is possible to ensure a uniform concentration of oxygen or acetylene in the atmosphere surrounding an incubated soil core it by no means follows that the concentration of the relevant gas throughout the core is similarly uniform. The measurements provide suggestive, rather than definitive, data. Alternative approaches include the use of nitrification inhibitors such as dicyandiamide (DCD) and nitrapyrin (N-Serve) in conjunction with additions of mineral N, and mass spectrometric techniques depending on the differential isotopic fractionation rates of the two metabolic pathways. We have measured N_2O ratios (fluxes in the absence of acetylene divided by those following its application) to indicate dominant source processes.

Results

In an experiment on a poorly drained Winton soil at the Glencorse Mains site, N_2O fluxes were measured in the autumn of 1987, following the harvest of spring barley and the incorporation into some plots of a green manure crop (forage peas). Except on one occasion, the highest fluxes were observed from the green manured plots, but values were generally low, amounting to no more than a few g N ha^{-1} d^{-1} (Fig. 2). In contrast, on the uncropped alluvial soil at the Croft Field site, an N_2O flux of over 200 g N ha^{-1} d^{-1} was recorded the following September in the absence of fertiliser; and following application of nitrate, values in the order of 100–150 g N ha^{-1} d^{-1} were observed for several weeks (Fig. 3). This was slightly surprising; we expected the higher denitrification rates

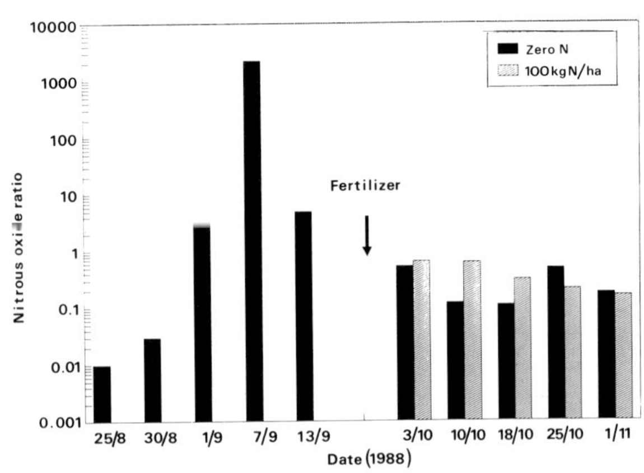

Fig. 4. Measured N_2O ratio (flux in the absence of acetylene divided by flux in its presence), Croft Field, autumn 1988.

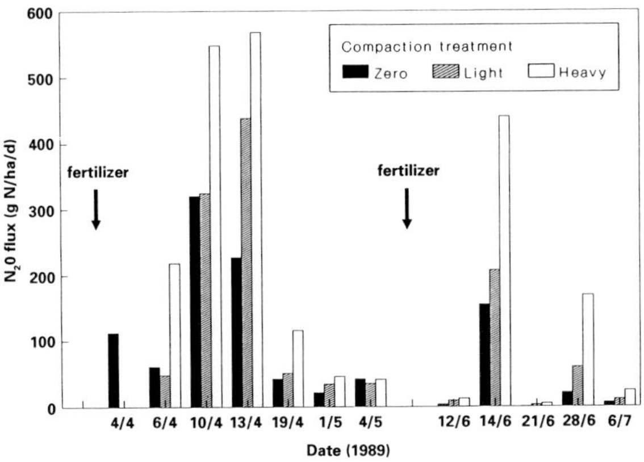

Fig. 5. Surface N_2O flux measured using closed chambers, Cowpark, 1989.

to occur in the heavier (Winton) soil. It is probable that significant reduction of N_2O to N_2 occurred in this soil; the geometric mean N_2O ratio over the period (not shown) was c. 0.1. Fig. 4 suggests another factor: N_2O ratios very much greater than unity were observed on occasions at Croft Field, implying a significant contribution from nitrification.

Fluxes of N_2O from a grassland compaction experiment at the Cowpark site in the spring and summer of 1989 are shown in Fig. 5. Soon after the first application of fertiliser the fluxes rose, to reach peak values in excess of 500 g N ha^{-1} d^{-1}. The second peak maximum was slightly lower and much shorter in duration. The N_2O ratios observed at the Cowpark site, and also in earlier work on a nearby grassland site (Egginton and Smith 1986) are shown in Fig. 6. They varied from 100% (unity) at low rates of denitrification to <5% at the highest rate – which was in excess of 2 kg N ha^{-1} d^{-1}. All values were less than unity, suggesting that denitrification was the dominant mechanism giving rise to N_2O emission at these sites.

Measurements of N_2O fluxes from agricultural soils following addition of nitrifiable fertilisers (principally anhydrous or aqueous ammonia and urea) show major effects of fertiliser form (Breitenbeck and Bremner 1986, Eichner 1990). The greatest fluxes followed application of anhydrous ammonia (Fig. 7), these fluxes are smaller than the maximum denitrification rates reported above, but the peaks last longer than those typical of denitrification.

Modelling of N_2O emissions from soil

Emissions of N_2O can be modelled in two fundamentally different ways: empirically, by regression of measured N_2O fluxes on measured controlling parameters; and at a process level, by synthesizing what is known about the factors governing the kinetics of the two major source processes (nitrification and denitrification) into a mathematical model of the soil. Most published models are in fact hybrids of the two approaches, but the distinction is nevertheless useful. Empirical models are by their nature site-specific (transfer to different soil

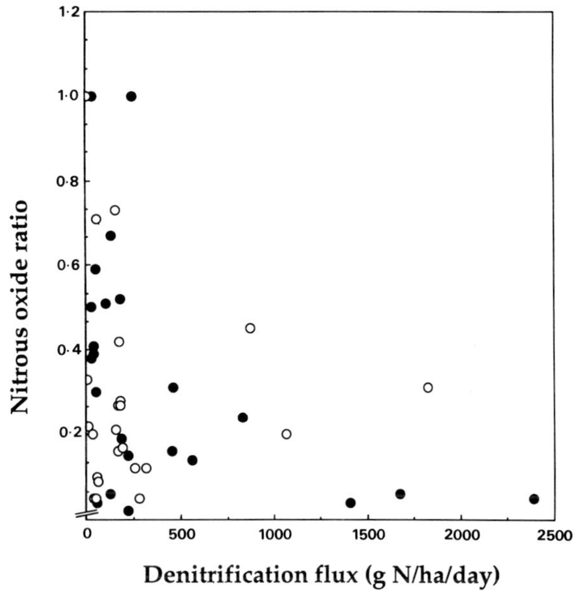

Fig. 6. Relationship between measured N_2O ratio (flux in the absence of acetylene divided by flux in its presence) and denitrification flux at two grassland sites. ●: data from Egginton and Smith (1986), ○: Cowpark, 1989 (Arah and Smith 1990).

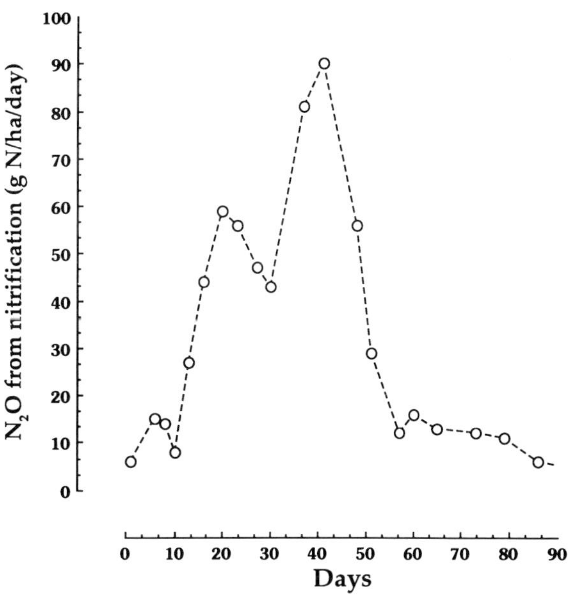

Fig. 7. N_2O flux resulting from nitrification following application of 180 kg N ha^{-1} as NH_3 to soil on day 0 (redrawn from Breitenbeck and Bremner 1986).

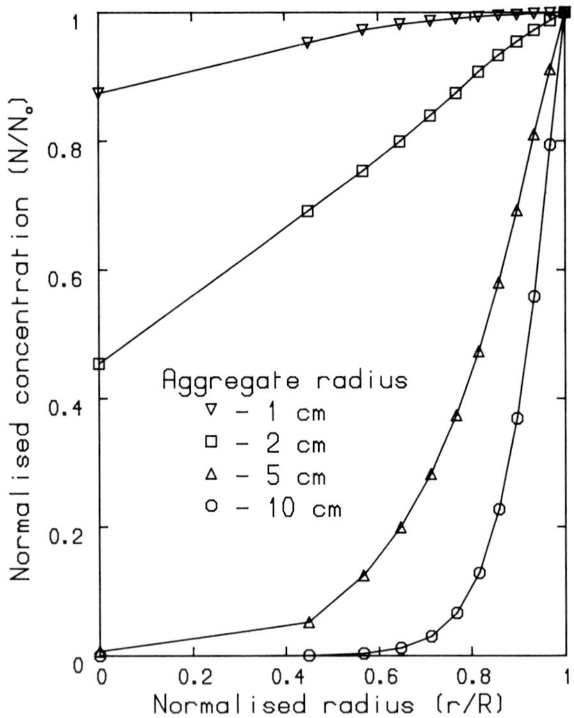

Fig. 8. Normalized intra-aggregate nitrate concentration profiles in uniform spherical aggregates of different radius (Arah and Smith 1989).

types or ecosystems requires a new curve-fitting exercise, essentially the development of a new model); process-based models are theoretically transferable, but they tend to require more (and more detailed) information than is commonly available.

A review of all the published models of N_2O emission is beyond the scope of this paper. Instead, we propose briefly to discuss one or two examples of empirically-based models and then to describe our attempts to develop a process-based model of denitrification. Finally, we shall touch on the prospects for scaling up such small-scale models and incorporating the process of nitrification.

Empirical modelling

The empirical approach to modelling involves the accumulation of a large body of data: measurements of the gas flux of interest, with simultaneous measurements of as many as possible of the soil properties (e.g. temperature, moisture content, nitrate concentration) thought to affect the flux. A best-fit equation (the form of which must be decided in advance) relating the measured flux to the controlling parameters is then derived by multifactorial regression analysis. Mosier et al. (1983), adopting such an approach, were able to explain 50% of the variation in surface N_2O flux from native and cropped soils in terms of variations in water content and the concentrations of nitrate and ammonium. Their inability to account for the other 50% may be attributable to the fact that N_2O is actually produced by two separate processes, which respond in quite different ways to changes in soil conditions. Recognising this, and introducing separate terms to allow for the differential responses of the two processes, Mosier and Parton (1985) and subsequently Parton et al. (1988) were able to account for some 80% of the observed variation in N_2O emissions from a shortgrass steppe, albeit in a uniformly nitrification-dominated system. Benckiser et al. (1987) obtained similar multiple regression coefficients relating surface N_2O flux to moisture content (the dominant controlling parameter), soil temperature, and the concentrations of organic carbon, nitrate and ammonium in an agricultural system favouring denitrification.

Process modelling – denitrification

Denitrification, the reduction of nitrate via nitrite and nitric oxide to N_2O and N_2, occurs when anaerobic conditions develop, whether in entire horizons, as occurs in flooded soils, or in microsites. In all cases, this condition is brought about because the microbial demand for oxygen exceeds the supply, which is primarily controlled by diffusion. Anaerobic microsites above the water table may occur at the centres of soil aggregates (Currie 1961, Greenwood and Berry 1962, Smith 1980), or in small saturated regions within a structureless soil (Arah 1988), or wherever the local oxygen demand is exceptionally high (Parkin 1987) – a so-called "hotspot" of activity.

Denitrification in a microsite can be modelled as a process of simultaneous diffusion and reaction of O_2, nitrate, and N_2O. Full details are provided in Arah and Smith (1989); here we seek merely to give an outline of the approach. It is assumed that enzyme-mediated microbial reactions follow dual-substrate Michaelis-Menten kinetics, and that nitrate reduction is inhibited if any O_2 is present. Intra-microsite concentration profiles of O_2, nitrate and N_2O are calculated by iterative finite-difference approximation, using a microcomputer. The only requirement for convenient mathematical simulation is the assumption of some degree of symmetry. Both hotspots of decaying organic matter, and soil aggregates, are taken to be spherical; the diffusion zone around a root or old root channel is assumed to be cylindrical. Fig. 8 shows calculated concentration profiles for nitrate in spherical aggregates of different radius.

Denitrification rates (and N_2O fractions) can be calculated for volumes of soil large enough to contain many microsites differing in size, water content and oxygen demand, by integrating over appropriate probability distributions (e.g. log-normal distributions of ag-

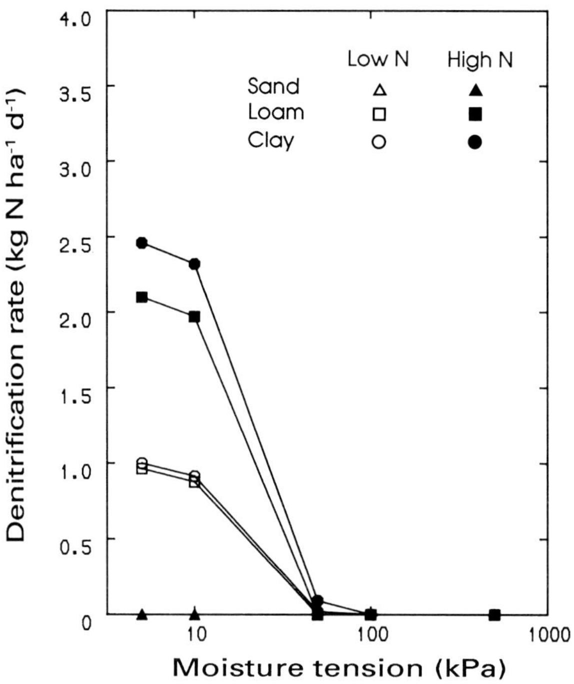

Fig. 9. Calculated whole-soil denitrification rates for three model soils at two nitrate concentrations and a range of moisture tensions (Arah and Smith 1989).

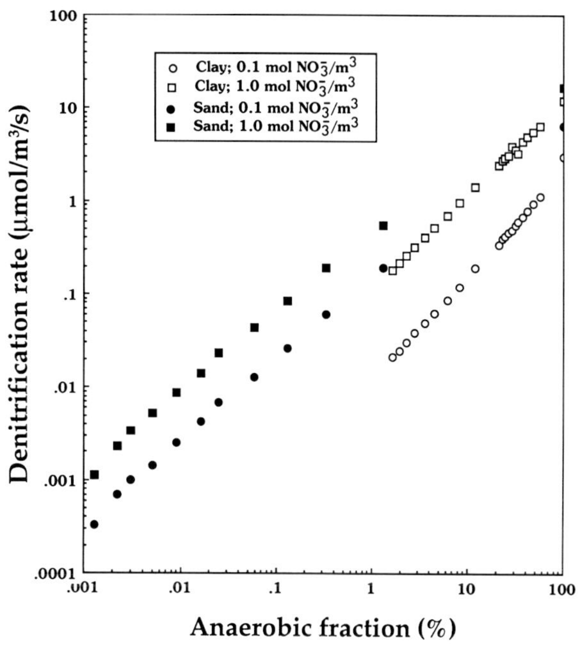

Fig. 10. Relationship between fraction of total soil volume which is anaerobic and denitrification rate, calculated for two model soils of contrasting textures, and for two concentrations of nitrate.

gregate size). Specimen predictions for three model soils are shown in Fig. 9. Intermediate sample sizes, larger than one microsite but too small to be representative of the soil as a whole, are simulated by Monte Carlo methods. The model is thus deterministic at a microsite scale, and stochastic in its macroscopic predictions.

We have examined the response times of microsite denitrification rates and N_2O fractions to a simulated major perturbation such as instantaneous saturation with water (Arah 1990). Results suggest that spherical microsites < c. 1 cm in diameter may usually be treated as if they are in a state of pseudo-equilibrium. Model microsites of similar dimensions with other geometries respond equally rapidly. This considerably simplifies the application of the model. One prediction that emerges is that, in soils where nitrate supply is not limiting, the denitrification rate of macroscopic soil samples should be proportional to the fraction of the total volume that is anaerobic, which may be calculated by a modification of the method of Smith (1980) (Fig. 10). Experiments carried out by Parkin and Tiedje (1984) showed this to be the case, and similar experiments carried out in our laboratory support this finding.

The fraction of the gaseous products of denitrification that is released from a microsite to the atmosphere as N_2O rather than harmless N_2 is determined by the balance between diffusive escape and the rate of the terminal reduction step. Thus variables such as soil water content and aggregate size have a major influence. Chemical factors are also important; for example, reduction of N_2O is inhibited by both O_2 and nitrate

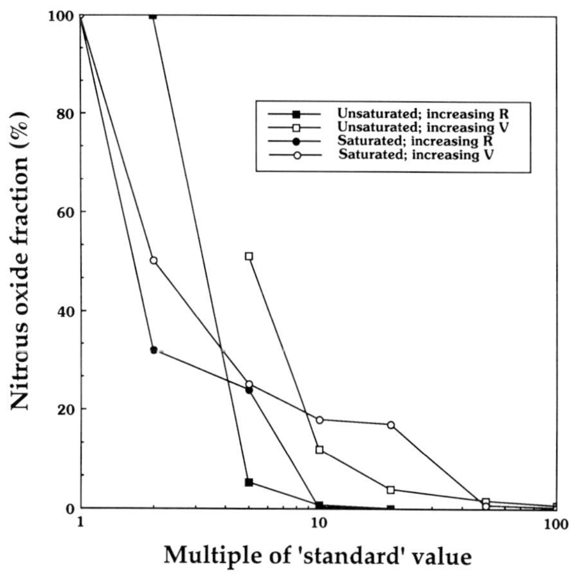

Fig. 11. Calculated variation in N_2O fraction with variation in radius R of spherical microsite and in oxygen demand V; squares indicate unsaturated conditions, circles saturated conditions. Standard value of R = 1 cm; standard value of V = 10^4 mol m^{-3} s^{-1}.

(McConnaughey and Bouldin 1985), and the model has to take account of this. Calculated N_2O fractions for model spherical microsites are extremely sensitive to microsite radius and oxygen demand (Fig. 11), but considerably less so to gaseous diffusivity within the microsite or to external oxygen concentration (Arah and Smith 1990). The model provides a means for rationalising the otherwise extremely confusing interactive effects of the parameters known to affect N_2O emission by denitrification.

The next important stage of development is to make the model applicable to much larger areas, initially to fields (1–100 ha), and subsequently to larger regions. For this to be achievable, the parameters in the model must be simplified to fit in with the information available in soil mapping and GIS databases. Here, there is a need to establish the relationships between the relatively precise parameters of the model (e.g. diffusivities, size distributions, oxygen demand) and the traditionally (and more readily) measured environmental and soil properties such as temperature, rainfall, texture, bulk density, water content, organic matter content, and drainage status. What is required is a probabilistic functional mapping procedure which will relate, say, soil series and cultivation regime to a range of probable aggregate size distributions and organic matter contents. Probable diffusion constants could be calculated from known moisture characteristic curves (in turn derivable, where necessary, from texture data?) and precipitation-evapotranspiration models. Database compilers, in their turn, must be prepared to take account of internal inhomogeneity within their mapping units (Burrough 1989), because of the extreme spatial variability of N_2O emissions.

Process modelling – nitrification

In forests and natural grasslands, most of the available nitrogen is in the ammonium form. Nitrate substrate for denitrification is lacking, and the predominant mechanism for N_2O formation appears to be nitrification (Blackmer and Bremner 1977, Bremner and Blackmer 1981). In many regions, much arable land and managed grassland is well-drained, and receives inputs of nitrogenous fertilizers predominantly or exclusively in the form of ammonium or ammonium-producing compounds. Here, too, nitrification is likely to be the major source of N_2O.

The inclusion of both nitrification and denitrification into a process model of N_2O emissions is an important task for the future. If it is done at the small scale initially, and then extended to larger areas as suggested above for denitrification, we should eventually be able to predict the total N_2O flux from the land surface of a region. Since nitrification is an aerobic rather than an anaerobic process, and since aerobism is effectively the normal condition of the unsaturated zone, nitrification is likely to be less variable than denitrification. Where nitrification dominates N_2O emissions, scaling up of a process-based model should be relatively straightforward.

References

Arah, J. R. M. 1988. Modelling denitrification in aggregated and structureless soils. – In: Jenkinson, D. S. and Smith, K. A. (eds), Nitrogen efficiency in agricultural soils. Elsevier, London, pp. 433–444.

– 1990. Modelling spatial and temporal variability of denitrification. – Biol. Fert. Soils 9: 71–77.

– and Smith, K. A. 1989. Steady-state denitrification in aggregated soils: a mathematical model. – J. Soil Sci. 40: 139–149.

– and Smith, K. A. 1990. Factors influencing the fraction of the gaseous products of soil denitrification evolved to the atmosphere as nitrous oxide. – In: Bouwman, A. F. (ed.), Soils and the greenhouse effect. Wiley, Chichester, pp. 475–480.

–, Smith, K. A., Crichton, I. J. and Li, H. S. 1991. Nitrous oxide production and denitrification in Scottish arable soils. – J. Soil Sci. 42: 351–367.

Aulakh, M. S., Rennie, D. A. and Paul, E. A. 1984. Gaseous nitrogen losses from soils under zero-till as compared with conventional-till management systems. – J. Env. Qual. 13: 130–136.

Benckiser, G., Gaus, G., Syring, K.-M., Haider, K. and Sauerbeck, D. 1987. Denitrification losses from an Inceptisol field treated with mineral fertilizer or sewage sludge. – Z. Pflanzenernahr. Bodenk. 150: 241–248.

Bouwman, A. F. (ed.) 1990. Soils and the greenhouse effect. Wiley, Chichester.

Breitenbeck, G. A. and Bremner, J. M. 1986. Effects of various nitrogen fertilizers on emission of nitrous oxide from soils. – Biol. Fert. Soils. 2: 195–199.

Bremner, J. M. and Blackmer, A. M. 1981. Terrestrial nitrification as a source of atmospheric nitrous oxide. – In: Delwiche, C. C. (ed.), Denitrification, nitrification, and atmospheric nitrous oxide. Wiley, New York, pp. 151–170.

Blackmer, A. M. and Bremner, J. M. 1977. Nitrification of fertilizer nitrogen in soils as a source of atmospheric N_2O. – Agron. Abstr. p. 146.

Burrough, P. A. 1989. Fuzzy mathematical methods for soil survey and land evaluation. – J. Soil Sci. 40: 477–492.

Currie, J. A. 1961. Gaseous diffusion in the aeration of aggregated soils. – Soil Sci. 92: 40–45.

Denmead, O. T. 1979. Chamber systems for measuring nitrous oxide emissions from soils in the field. – Soil Sci. Soc. Am. J. 43: 716–718.

Eichner, M. J. 1990. Nitrous oxide emissions from fertilized soils: summary of available data. – J. Env. Qual. 19: 272–280.

Egginton, G. M. and Smith, K. A. 1986. Losses of nitrogen by denitrification from a grassland soil fertilized with cattle slurry and calcium nitrate. – J. Soil Sci. 37: 69–80.

Folorunso, O. A. and Rolston, D. E. 1984. Spatial variability of field-measured denitrification gas fluxes. – Soil Sci. Soc. Am. J. 48: 1214–1219.

Greenwood, D. J. and Berry, G. 1962. Aerobic respiration in soil crumbs. – Nature (Lond.) 195: 161–163.

Hutchinson, G. L. and Mosier, A. R. 1981. Improved soil cover method for field measurement of nitrous oxide fluxes. – Soil Sci. Soc. Am. J. 45: 311–316.

Matthias, A. D., Blackmer, A. M. and Bremner, J. M. 1980. A simple chamber technique for field measurement of emissions of nitrous oxide from soils. – J. Env. Qual. 9: 251–256.

McConnaughey, P. K. and Bouldin, D. R. 1985. Transient microsite models of denitrification: I. Model development. – Soil Sci. Soc. Am. J. 49: 886–891.
Mosier, A. R. and Parton, W. J. 1985. Denitrification in a shortgrass prairie: a modelling approach. – In: Caldwell, D. E., Brierly, J. A. and Brierly, C. L. (eds), Planetary Ecology. Selected papers from the sixth international symposium on environmental biogeochemistry. Van Nostrand Rheinhold, New York, pp. 441–452.
–, Parton, W. J. and Hutchinson, G. L. 1983. Modeling nitrous oxide evolution from cropped and native soils. – Environmental Biogeochemistry, Ecol. Bull. (Stockholm) 35: 229–241.
Parkin, T. B. 1987. Soil microsites as a source of denitrification variability. – Soil Sci. Soc. Am. J. 51: 1194–1199.
– and Tiedje, J. M. 1984. Application of a soil core method to investigate the effect of oxygen concentration on denitrification. – Soil Biol. Biochem. 16: 331–334.
Parton, W. J., Mosier, A. R. and Schimel, D. S. 1988. Rates and pathways of nitrous oxide production in a shortgrass steppe. – Biogeochem. 6: 45–58.
Robertson, G. P. and Tiedje, J. M. 1987. Nitrous oxide sources in aerobic soils: nitrification, denitrification and other biological processes. – Soil Biol. Biochem. 19: 187–193.
Ryden, J. C. and Dawson, K. P. 1982. Evaluation of the acetylene-inhibition technique for the measurement of denitrification in grassland soils. – J. Sci. Food Agric. 33: 1197–1206.
–, Lund, L. J. and Focht, D. D. 1978. Direct in-field measurement of nitrous oxide flux from soils. – Soil Sci. Soc. Am. J. 42: 731–737.
–, Lund, L. J., Letey, J. and Focht, D. D. 1979. Direct measurement of denitrification loss from soils: II. Development and application of field methods. – Soil Sci. Soc. Am. J. 43: 110–118.
–, Skinner, J. H. and Nixon, D. J. 1987. Soil core incubation system for the field measurement of denitrification using acetylene-inhibition. – Soil Biol. Biochem. 19: 753–757.
Smith, K. A. 1980. A model of the extent of anaerobic zones in aggregated soils, and its potential application to estimates of denitrification. – J. Soil Sci. 31: 263–277.

Ecological Bulletins 42: 124–132. Copenhagen 1992

Impacts of trace gas fluxes in mid-latitude ecosystems

J. Schimel, G. P. Robertson, D. Baldocchi, J. Bogner, E. Davidson, J. Duyzer, D. Ehhalt, D. Fowler, P. Groffman, K. Haider, V. Isadorov, L. Klemedtsson, J. Melillo, K. Smith, W. H. Su, and W. Wieprecht

Schimel, J., Robertson, G. P., Baldocci, D., Bogner, J., Davidson, E., Duyzer, J., Ehhalt, D., Fowler, D., Groffman, P., Haider, K., Isidorov, V., Klemedtsson, L., Melillo, J., Smith, K., Su, W. H. and Wieprecht, W. 1992. Impacts of trace gas fluxes in mid-latitude ecosystems. – Ecol. Bull. (Copenhagen) 42: 124–132.

Corr. author: J. Schimel, Inst. of Arctic Biology, Univ. of Alaska, Fairbanks, AK 99775, USA.

Introduction

The temperate region is highly populated, with most ecosystems in the region subject to at least some human influence. Large areas of forest and grassland, for example, have been converted to agriculture, and many systems are receiving increased inputs of sulfur and nitrogen compounds associated with acid rain and industrial activity. These changes have the potential for having major effects on trace gas production and consumption within the region.

Although there have been many trace gas studies in the midlatitudes, only a few flux estimates, however, have been based on frequent measurements made over an entire year: most studies have been mechanistic in nature and have focused on small areas (e.g., individual fields). Consequently, the effects on trace gas fluxes of converting native to agricultural lands, and of loading ecosystems with nutrients (N and S) and potential toxins are just beginning to be understood on a regional scale.

Because mechanistic controls on trace gas fluxes are best understood for temperate ecosystems, these systems provide an opportunity for modeling fluxes at a regional level that is not available for most other regions. Moreover, the geographic information base necessary for developing reliable large-scale models is available. Temperate regions thus offer the opportunity to rapidly develop the approaches necessary for spatial and temporal extrapolations in all regions worldwide.

We have identified five critical questions that need to be addressed in order to improve substantially our understanding of mid-latitude trace gas fluxes and their impacts on global processes. Four questions address specific scientific issues and the fifth is a methodological problem:

1. How important are mid-latitude ecosystems as net CO_2 sinks?
2. How important are mid-latitude ecosystems as net sinks for CH_4?
3. How important are mid-latitude agronomic systems as sources for the atmospheric N_2O increase?
4. How important are mid-latitude ecosystems as sinks for atmospheric oxidants and sources of oxidant precursors?
5. What further techniques are needed for modeling, extrapolation, and validation of trace gas fluxes in heterogeneous landscapes to regional and global levels.

Each of these questions will require a combination of measurement, manipulation, and modeling approaches. Developing regional extrapolations and models will require extensive international cooperation to be successful.

We suggest a single unified approach for addressing each of these questions. At its center is a network consisting of core and satellite sites that will encompass an appropriate variety of community composition, land-use, chemical deposition, climate, and soil-type categories. Where possible these sites will be part of existing national research networks. At core sites appropriate trace gas fluxes should be measured and the mechanistic studies necessary to develop process-based models undertaken. Process-based models will be integrated into large-scale models that can then be validated and used to estimate regional gas fluxes and how they may respond to changes in climate, chemical deposition, and land-use.

Project 1: How important are mid-latitude ecosystems as a net CO_2 sink?

Atmospheric CO_2 concentrations continue to rise as a result of human activity. The CO_2 increase reflects only a fraction (40–60%) of the CO_2 that is released via anthropogenic sources (Berner and Lasaga 1989, Tans et al. 1990). Tracer and modeling studies (linked to the global CO_2 monitoring network) indicate that the oceans cannot account for the imbalance in the global carbon budget (Tans et al. 1990).

Some estimates suggest that terrestrial ecosystems are a source of CO_2, yet new data and analyses suggest that increased mid-latitude carbon storage may account for the "missing" carbon (Tans et al. 1990). Several events support this idea. Large portions of the mid-latitudes have undergone substantial land-use changes since 1900, including extensive reforestation with the resulting C-storage in woody biomass and soil organic matter. With much of the present forest area in the region now in a midsuccessional, aggrading stage, it is possible that mid-latitude forests are a significant net C-sink. Moreover, increased levels of precipitation-N and atmospheric CO_2 may have boosted net primary production, amplifying this potential sink. In addition, new agricultural practices (e.g. reduced tillage) focus on increasing soil organic matter pools, thereby acting as a C-sink. On the other hand, the potential mid-latitude C-sink may be reduced or reversed if acid deposition, accumulation of toxic ozone levels, or nitrogen saturation (Aber et al. 1989, Loucks 1989) result in significant forest decline and production losses in agricultural ecosystems.

We believe the potential for significant carbon storage in mid-latitude forest is sufficiently strong to warrant immediate, intensive investigation. If such storage is occurring, and if changes in CO_2 concentration and chemical deposition affect the potential trend, it may help to balance estimates of the global CO_2 budget. This could also suggest new strategies for ameliorating the global atmospheric CO_2 increase.

Objectives

1) To determine the consequences of changing land use (reforestation, succession, deforestation, and urbanization) on C-storage in mid-latitude ecosystems.
2) To evaluate the effect of changing physical climate (temperature, moisture) on CO_2 exchange.
3) To evaluate the effects of chemical deposition (N, S, and acid) on CO_2 exchange.
4) To determine the effects of elevated CO_2 on the gross and net CO_2 exchange of representative mid-latitude biomes.

Research plan

Measurement

We propose to establish site networks in temperate forests and agricultural ecosystems, which will include forests of varying successional age and both forest and agricultural sites on different soil types and management regimes. Where possible, these networks will be the same as those used for the CH_4 and N_2O studies (Projects 2 and 3). We will measure net primary production (NPP), soil respiration, decomposition, C-fixation, ecosystem CO_2 exchange (by micrometeorological methods), and the climatic and biotic variables that drive these processes. These will be measured intensively over at least 2 years to determine seasonal patterns and key environmental controls of CO_2 exchange. After this intensive sampling period is over, measurements will be continued at a reduced level to determine interannual variation and long-term changes in site C-dynamics.

Manipulations

Plots in each core site of the networks will be manipulated to improve our understanding of the long-term effects of various climatic drivers (physical and chemical) on ecosystem function and trace gas exchange. These manipulations will include: increased air and soil temperature, altered moisture, elevated CO_2, and fertilization (N and S). Field and laboratory studies should examine the interactions of different perturbations (e.g., temperature and moisture).

Manipulative studies need to evaluate whole plant physiological responses, as well as soil microbial responses (e.g., root respiration, decomposition and mineralization rates). These studies must operate continuously for several years (at least 3 and preferably 5 to 10 years) to study several growth cycles, which will affect the cumulative N-availability and carbon storage of the system.

For studies in agricultural systems, manipulations should be based in field-greenhouse systems, in which all of the parameters of interest (temperature, moisture, nutrients) can be controlled. In forest studies, the size of the vegetation makes air temperature and CO_2 experiments difficult. Most such experiments have been done on seedlings and saplings, but the physiological responses of immature plants differ from their mature counterparts (Kramer and Kozlowski 1979). In forest systems, therefore, work should start with soil manipulations on plots several tens of meters per side. This should include combinations of soil temperature (several degrees above ambient), moisture (increased and decreased), and nutrient addition.

Parallel experiments should start at a more limited

scale, evaluating whole tree responses to increased CO_2 and temperature. These studies may be done in either open-top chambers or in field-greenhouses. Both systems have their limitations: in open chambers it is more difficult to control CO_2 and temperature, but chambers leave plants exposed to actual environmental conditions (e.g. light and precipitation), for which greenhouses require extensive engineering. Whichever approach is used, it must address the role of CO_2 enrichment on all of the components of the ecosystem: photosynthesis, respiration, allocation, leaf nitrogen content, and their consequent effects on decomposition and mineralization. Moreover, the experiments must be run through a sufficient number of growth cycles to identify potential system-level feedbacks.

Resources
For the field experiments, there are a number of national research networks that will be valuable, including the Long Term Ecological Research (LTER) sites in the USA, site networks in western Europe, and in the Peoples Republic of China. There are a number of teams worldwide capable of measuring CO_2 fluxes by micrometeorological methods, including groups in the USA, Canada, Europe, and Asia. Facilities for doing whole tree studies are somewhat limited and new facilities may need to be established.

Timetable

1991–	Establish site networks. Begin experimental work. Begin model development.
1992–93–	Continue field and laboratory experiments.
1994–	Integration and validation.

Project 2: Are mid-latitude ecosystems important net sinks for CH_4?

Especially in the northern hemisphere, the temperate region is highly populated, and humans may be significantly altering the regional CH_4 budget. Industrial activities have dramatically changed local environments and altered regional precipitation chemistry. Landfills in industrial countries and waste-water lagoons in developing countries may represent intense but highly localized sites of CH_4 production: landfills may be collectively responsible for as much as 7% of the global CH_4 budget (Cicerone and Oremland 1988). Drainage may have shifted many wetlands from being net to neutral CH_4 sources or even to net CH_4 sinks.

Nitrogen inputs in precipitation have increased substantially over the past 40 years in much of the temperate northern hemisphere . Today, some areas receive as much as 50 kg N ha^{-1} y^{-1} in atmospheric deposition (wet and dry fall). This can represent as much as 15 Tg N added annually to temperate region forests, such an increase in N-deposition may significantly reduce CH_4 oxidation in forest soils (Melillo et al. 1989, Steudler et al. 1989).

We need to evaluate the effects of N-deposition, landfills, and wetland drainage on CH_4 fluxes in mid-latitude ecosystems. Two subprojects are outlined; the primary project addresses the effects of N-availability and fertilization on CH_4 consumption, and the second addresses CH_4 emissions resulting from landfills and drained wetlands.

Project 2a: The effects of N-availability (added-N, native-N) on CH_4 oxidation

The linkages between N-availability and CH_4 flux are poorly understood but perhaps significant; Steudler et al. (1989) reported that two years of low-level N-additions to both coniferous and deciduous forest soils in central Massachusetts, USA reduced CH_4 uptake rates by 45 and 34%, respectively. Moreover, both forests and agricultural ecosystems with high rates of nitrogen cycling and nitrification appear to exhibit low CH_4 oxidation rates (Melillo et al. 1989). If such relationships hold in general, major changes in land-use (e.g., native to agricultural) and widespread chemical deposition may be having major impacts on regional CH_4 fluxes. This project is designed to test the generality of the linkage between N-availability and CH_4 fluxes.

Objectives

1) To quantify the role of mid-latitude ecosystems as sinks for atmospheric CH_4.
2) To quantify the relationship between N-availability and CH_4 oxidation in grassland and forest systems of the temperate zone.
3) To formulate a model for quantitative estimates of CH_4 oxidation in forest and grassland that accounts for the effects of three major controls on oxidation-temperature, moisture, and N-cycling rate (includes both N-deposition and native-N).

Research plan

This project should have both intensive and extensive components.

Intensive research plan
1. Locate 10 grassland sites and 10 forest sites around the world that are well characterized in terms of climate, vegetation, soils, and N-cycling rates.
2. Establish fertilization plots at each site. Two levels of fertilization would be established, 50 and 100% increases over annual net N-mineralization rates, which will be measured prior to study.

3. Perform monthly measurements of CH_4 uptake; these will be made throughout the diurnal cycle. At the same times, measure temperature, moisture, and various N-stocks.
4. Perform month-long integrative measurements of soil N-dynamics to establish mineralization and nitrification rates.
5. Develop a model to estimate CH_4 production and consumption as a function of temperature, moisture, and N-cycling.

Extensive research plan
Extensive research will be focussed on testing the models developed in the intensive component of the project. To do this, additional well-characterized sites (forest and grassland) will be identified. Methane fluxes will be estimated from the models and tested by field measurements of CH_4 uptake.

Project 2b: Net CH_4 emissions from landfills and wetlands

Landfills, feedlots, and wetlands may be major temperate-zone biogenic CH_4 sources. Maximum CH_4 fluxes from landfills may be up to 600 g m^{-2} yr^{-1} (Bogner et al. 1989), while wetland fluxes may be up to 100 g m^{-2} yr^{-1} (Harriss 1989). Temperate wetlands have decreased in area over the last century while landfills have increased. The changing role of these areas is a poorly understood component of the global CH_4 cycle. We need to determine how wetland drainage and changing landfill designs have affected and will continue to affect the global CH_4 cycle.

Objectives

1) To estimate long-term changes in mid-latitude CH_4 production from wetlands, landfills and feedlots.
2) To elucidate CH_4 production/consumption relationships in landfills and how these are controlled by changing landfill management, with a view to reducing CH_4 emissions.

Research plan

Basic field measurements
1. Establish field sites for CH_4 flux measurement in landfills and feedlots. These should cover humid, semi-arid, and arid sites. Establish sites in landfills with various management regimes.
2. Establish field sites in temperate wetlands and drained wetlands.
3. Measurements at each sites should include: CH_4 flux by chamber techniques, soil gas profiles (CH_4, CO_2, O_2), stable-C-isotope profiles, and soil moisture/temperature/soil gas profiles.
4. Measure both diurnal and seasonal CH_4 fluxes in wetlands, landfills, and feedlots over a two year period.

Manipulation experiments
Manipulations of soil water content, gas pressure, and landfill type will be combined with $^{14}CH_4$ field studies to evaluate CH_4 and CO_2 production, consumption, and transport.

Extrapolation and modeling
Gas flux modeling for landfills will be based on two dimensional diffusional and convective gas flux models incorporating production/consumption reactions. This will be combined with areal estimates of different landfill types.

Reduction in CH_4 source due to wetland drainage will be estimated from the difference in net CH_4 production by drained and undrained wetlands and by the reduction in wetland area, as determined from geographic data bases.

Resources
There are several valuable research site networks in North America, Europe and Asia for determining CH_4 fluxes from natural systems. There are established programs on landfill trace gas emissions in both the USA and Europe.

Timetable

1991–	Establishment of site networks.
	Begin monitoring studies for both natural and landfill projects.
	Begin model development.
1992–1993–	Continue studies.
1994–	Begin validation exercises.

Project 3: How important are mid-latitude agronomic ecosystems as sources for the increase in atmospheric N_2O concentration?

The N_2O concentration of the atmosphere has increased from 280–285 ppb 100 years ago to c. 310 ppb today. The present rate of increase is estimated at c. 0.25% yr^{-1}, corresponding to a net global emission of 3.5 Tg yr^{-1} (Robertson et al. 1989, Stewart et al. 1989) and c. 4–5% of global warming effects can be attributed to N_2O. In spite of its relatively low concentration, the long atmospheric lifetime of this gas (100–200 yr) and its strong IR absorption make it significant in the earths radiation balance. Nitrous oxide is also the primary non-synthetic compound involved in the depletion of stratospheric ozone.

Bouwmann (1990) estimated that c. 90% of N_2O emissions are from soils, but the specific sources are not all well characterized. World-wide agricultural inputs of N from biological and industrial N_2 fixation are c. 60 and 90 Tg yr^{-1} respectively (Hauk 1988), and the loss of only 2–3% of this N as N_2O could account for the entire observed annual increase in atmospheric concentration. Few studies, however, have found more than 1% of fertilizer being lost as N_2O, but data are far from complete (Stewart et al. 1989).

Almost 45% of the world fertilizer consumption currently takes place in North America and western Europe (most of the remainder being used in Asia). The need to increase food production in temperate Asia and eastern Europe has and will continue to stimulate increased fertilizer use in these areas (Crosson and Rosenberg 1989). The extent to which fertilizer use may exacerbate the current N_2O increase – either directly by N_2O production in the farm field or indirectly via N_2O production in riparian, coastal marine, and other downstream systems – needs to be understood.

We also need to know the system-level mechanisms responsible for N_2O fluxes. Nitrous oxide production in soils results predominantly from two microbial processes: nitrification and denitrification. Nitrification may be the most important source of N_2O in well-drained agricultural soils, whereas denitrification may be more important in poorly drained soils (and wetlands receiving NO_3^--rich drainage water).

Key questions regarding N_2O emissions from mid-latitude agricultural systems include:

1) What is the relative importance of nitrification vs. denitrification vs. other potential biological sources?
2) To what extent can these processes be ameliorated by acceptable changes in agricultural practices, e.g.: improved water management (irrigation and drainage), improved fertilizer management, inhibition of nitrification/denitrification, alterations to tillage practices and livestock management.
3) How much to the NO_3^--N that leaves agricultural fields as surface runoff and groundwater is ultimately lost as N_2O following transfer in drainage water to wetlands and estuaries?

Objectives

1) To improve regional estimates of N_2O emission from mid latitude agricultural ecosystems by the use of quantitative modeling.
2) To determine the principal factors governing N_2O emissions from agricultural ecosystems. This is critical both for modeling and developing management practices that will reduce N_2O emissions without jeopardizing food production.
3) To determine the extent of N_2O emissions from natural systems receiving N-inputs from agricultural lands as a result of N-transport in drainage water and the atmosphere.

Research plan

Regional extrapolations need to be based on precise and representative measurements of N_2O fluxes from agricultural ecosystems on the scale of the individual farm field (i.e., of the order of 1 to 100 ha, depending on the region). This is necessary because management systems, including N-fertilization, vary widely between adjacent fields. Scaling from the individual fields to the regional level will rely on geographically-based models that incorporate key management practices. For many countries good data bases already exist, which contain information on N-inputs, tillage, N-losses, and other management factors expected to affect N_2O fluxes.

We propose the establishment of core networks in major agricultural regions, each with sites covering a range of crops, soil types, and management systems typical of the region. At each site, N_2O fluxes should be determined and related to the controlling factors, some of which (N-input, irrigation, inhibitors, etc.) may be manipulated. N_2O fluxes from these sites should be periodically monitored over 24 h periods for two years. The sampling frequency should be centered around conditions of particular significance (e.g., following rainfall). Parallel measurements will be made on those factors which control N_2O fluxes, including N-pools, N-cycling rates, temperature, moisture, etc. Fluxes of N_2O should also be determined in selected sites (terrestrial and aquatic) receiving N-inputs as a result of agricultural operations.

We also suggest that the development and application of promising N_2O detection technologies be accelerated. At present, N_2O flux data are primarily dependant on the use of small (i.e. <1 m^2) chambers and gas chromatographic analysis. Low N_2O flux rates relative to background concentrations have made N_2O fluxes difficult to measure by micrometeorological methods (Desjardins and MacPherson 1989). Even though tunable diode laser sensors (TDLAS) for N_2O are still at an experimental stage, they appear to be sufficiently sensitive to measure N_2O fluxes using the eddy correlation technique. In order to arrive at large area flux estimates, sensors also need to be developed for aircraft-based systems. The use of new measurement techniques may also be used with more traditional chamber methods. Tunable diode lasers or Fourier transform infrared (FTIR) may be used in conjunction with extended chambers (up to 10 m long) to take advantage of the sensitivity gain achieved by containment, while at the same time overcoming much of the spatial variability associated with the traditional chamber design by enclosing a long strip of land. Cross calibration of chambers against tower-based measurements using TDLAS will make it possible to use data already gathered by chambers and to continue their use where logistical/financial constraints preclude the use of other techniques.

Along with these measurements, process-based models of N_2O fluxes will be developed and integrated into ecosystem and regional models (e.g. Parton et al. 1988). After development, the models will be tested by extensive flux survey measurements.

Resources

Many groups have been studying N_2O dynamics in agricultural systems around the world, and a number of well characterized longterm agricultural experiment sites have been established, including such sites as the Broadbalk plots in England, which have been studied for over a century.

Timetable

1991– Workshop to bring together workers who have been examining nitrification, denitrification, and N_2O fluxes, to coordinate this program with their efforts.

1992– Establish site networks.
Begin measurements.
Begin model development.

1993– Continue experimental work and modeling.

1994– Begin validation efforts.

Project 4: How important are mid-latitude ecosystems as sinks for atmospheric oxidants and sources of oxidant precursors?

The mid-latitudes contain a substantial fraction of the terrestrial surface of the globe, and have been assumed to provide a substantial fraction of the natural CO and NOx emissions to the atmosphere (Johansson 1989, Khalil and Rasmussen 1990). The magnitude of the emissions, and the processes controlling them are poorly understood, yet the supply of CO and NOx to the atmosphere constitutes a primary control of oxidant production (O_3, H_2O_2; Bouwman 1990). Terrestrial ecosystems also represent a major sink by dry deposition for tropospheric O_3, a process that while better understood than NOx emissions, is still subject to uncertainties. Fluxes have been measured in very few mid-latitude ecosystems, and the presence of large areas of enhanced O_3 concentrations within and downwind of industrial regions presents a phytotoxic stress to many plant communities (Loucks 1989). Effects of this stress on physiological processes and on the exchange of heat, water vapor, CO_2 and a range of trace gases have been observed but cannot be quantified on regional scales with our current understanding.

Volatile organic compounds (VOCs) emitted by plant canopies provide a further control on the production of photochemical oxidants. Mid-latitude ecosystems are thought to be important source areas of volatile organic compounds (VOC) including isoprene, terpenes, and possibly alkenes, yet the magnitude of the fluxes and the controlling physical, chemical, and biological processes remain poorly understood.

Lastly, mid-latitude ecosystems contain large areas in which regional modification of chemical deposition, notably but not exclusively in Europe and North America, has altered the cycling of major plant nutrients (e.g., nitrogen and sulfur). The effects of N, S and acid deposition on nutrient cycling and trace gas exchange (including N_2O, NOx, O_3, and CH_4) remain unclear.

Objectives

1) To quantify the strength of mid latitude ecosystems as sources for atmospheric oxidant precursors (NOx, VOC) and sinks of oxidants (O_3 and NOx).
2) To evaluate the response of mid-latitude biogenic sources of NOx and VOC to changes in the climate and chemical deposition.

Research plan

Three areas of scientific activity are required to meet the above objectives:

1. Long-term field measurements of O_3, NOx, and VOC fluxes in representative ecosystems, and of the associated physical, chemical, and biotic controlling variables.
2. Process directed field and laboratory studies to determine the fundamental mechanisms and controls on the production, transport and dispersion of oxidants and their precursors.
3. Modeling of O_3, NOx, and VOC gas exchange processes, extrapolation to regional scales and field validation.

We propose establish networks of sites in areas subject to intensive ozone deposition (eastern North America and western Europe) and in control regions. Stations should be established within these regions to measure concentrations of O_3, NOx, and VOCs. Sets of measurements should be carried out diurnally through the year to determine seasonal patterns of exchange and to evaluate the primary environmental controlling factors. Experimental work should be carried out to determine the specific process controls necessary to model gas exchanges. These mechanistic models will be used to develop regional and temporal extrapolations.

Resources

There are several established national projects evaluating biotic interactions with the oxidant cycle, including groups in North America (LTER), western Europe (European experiment on transport and transformation of environmentally relevant trace constituents in the troposphere over Europe [EUROTRAC]), the former Soviet Union, and the Peoples Republic of China.

Timetable

1991–	Workshop to coordinate efforts of the different national and international programs and to plan the activities needed to scale these projects to the global perspective called for in this program.
1992–4–	Coordinated experimental work. Work on model development.
1995–	Begin model validation studies.

Objectives

1) To develop models of CO_2 fluxes capable of depicting fluxes over landscapes and producing regional-scale estimates of CO_2 fluxes.
2) To develop a georeferenced data base necessary for extrapolation of process-level models of CO_2 fluxes to a regional estimate.
3) To develop and apply methods for validating these models.

Project 5: Modeling, extrapolation and validation of trace gas fluxes in mid latitude ecosystems: A CO_2 pilot study

A need in most IGBP research programs is to extrapolate flux estimates from plot-based measurements to regional scales. This extrapolation requires the development of large-geographic-scale, process-based models and their subsequent validation. Several programs have initiated research to this end for areas with relatively homogeneous vegetation (e.g. First International Satellite Land Surface Climatology Project Field Experiment [FIFE], Atmospheric Boundary Layer Experiment [ABLE], Northern Wetland Study [NOWES], etc.). The next step is to develop scaling techniques for complex landscapes. We propose a pilot study to develop such techniques, and have targeted CO_2 because the necessary components are all available: conceptual understanding of the processes, flux measurement techniques (chamber, micrometeorological, and aircraft based) and powerful process-based models. This study will relate directly to Project 1, and will also help develop techniques needed for the other projects and for other regions as well.

A key tenet of scaling efforts is to integrate data on trace gas fluxes into overall ecosystem nutrient cycling models. Development of large-area-estimates of trace gas fluxes requires linking process-level models with georeferenced data sets containing variables relevant to both process and region levels. Geographic information system (GIS) development has been vigorous for the mid-latitudes and a wide range of variables at different scales are available for extrapolation efforts. Ecosystem models have already been adapted to incorporate N_2O and CO_2 fluxes into regional models (e.g. CENTURY for grassland; Parton et al. 1988, Burke et al. 1989) and ongoing efforts will expand this application to other gases (e.g., CH_4) and ecosystem types (forest, agriculture). The availability of the data sets and ecological understanding described above, in combination with the technical resources and institutions in the mid-latitudes, make it logical to conduct this pilot study in the mid-latitude regions.

Research plan

We propose a specific case study where each step in the extrapolation from process studies to a regional flux estimate is coordinated. Strong interaction and cooperation will be necessary between investigators performing detailed process studies, mesoscale modeling, and meteorological measurements using either aircraft or towers. This coordination is crucial since the object of the study is to cross-validate measurements made at differing spatial and temporal scales. CO_2 is the best candidate gas for this study due to its intermediate half-life, multiple source-sink characteristics in the soil, vegetation, and atmosphere, and heterogeneous spatial and temporal fluxes from various target areas within a region.

During the first stage of the pilot study, a measurement area of c. 100 km on a side should be selected. Preferably it will consist of a limited number of ecosystem types (e.g., a forest-agricultural mix would be ideal). At several sites in this area, measurements will be made of trace gas fluxes (including diurnal and seasonal patterns) and of other parameters needed for driving flux models. At a smaller number of sites, micrometeorological methods will be used to produce estimates of fluxes at scales of up to several kilometers. Preferably these will be automated methods so that a large number of environmental conditions will be covered and a large data set will be obtained.

After two years of process level measurements, a large boundary layer experiment will be performed. In this experiment, aircraft will be used to estimate fluxes over a large area. The fluxes will be derived from aircraft data and supporting measurements of boundary layer conditions using a budget approach so that flux estimates will be obtained for an area of up to several tens of km^2 (Schuepp et al. 1990). These estimates will be compared with those derived from models capable of depicting fluxes over the same spatial and temporal scales. The results of these comparisons will be used to evaluate both the modeling and aircraft approaches to estimating large-area fluxes in heterogeneous landscapes. We can then use high resolution data bases to identify areas responsible for the divergence of results. This divergence analysis will form the basis for changes in the models to more accurately reflect flux patterns across the region.

Specific needs

A. Data bases

The development and use of GIS for regional and global scale extrapolation of trace gas fluxes has increased in recent years. Global coverage of soils, vegetation, climate, and land-use parameters are available in a variety of formats at different locations. The IGBP global environment data directory (GEDD), which will be produced by the IGBP working group on Data and Information Systems, will provide a centralized summary of available data sets. A major challenge in the extrapolation area will be developing and evaluating geographic data that are relevant to both small-scale process models and large-area validation techniques. Balancing detail and resolution versus size and cost in global data bases will be a major concern.

B. Micrometeorological methods

Micrometeorological methods using towers and aircraft have been used for years but have mainly been applied on special selected sites with relatively homogeneous vegetation cover. In order to be useful for our purposes research is necessary to extend the applicability of these methods to heterogeneous or patchy areas of scales up to 100 km^2.

At this stage several aircraft campaigns using measurement strategies and flight patterns based upon budget or eddy correlation techniques have been successful for gases such as H_2O, CO_2 and even O_3 (Desjardins and MacPherson 1989). The nature of aircraft studies however requires that flux estimates derived from the process studies are available on timescales smaller than days. This problem could be overcome if micrometeorological methods are applied using high towers (up to 200 m) to measure fluxes over longer periods. Unfortunately this method is only applicable over spatial scales of a maximum of several kilometers. Even to this scale there have not been many attempts to compare fluxes derived from process studies with micrometeorological measurements

C. Modeling and extrapolation

It will be important to evaluate extrapolation models relative to validation approaches that will be taken. The extrapolation techniques must produce flux estimates consistent with the scale of the aircraft and/or micrometeorological approaches that produce area-averaged flux estimates. Moreover, the extrapolation techniques must be able to produce estimates at higher resolution than the validation efforts to determine where the modeling and/or validation methods fail. This information will be important to evaluation and refinement of both modeling and validation techniques.

Resources

There are a many groups worldwide with some expertise in measuring or modeling CO_2 exchanges that relates to this project. There are few groups, however, that have the expertise to handle all the components of the program. The availability of personnel trained in ecological applications of geographic data bases and modeling may be a particular problem; IGBP should consider sponsoring educational and training initiatives in these areas.

Timetable

1991–	Workshop for developing conceptual models and planning the program. Personnel from involved groups meet to identify data base needs, models, and validation approaches that the program needs. The site for the CO_2 pilot study will be chosen and the organizational structure will be planned. Other efforts (e.g. FIFE, ABLE) will provide an organizational model.
1991–1995–	Experimental work, modeling and integration studies proceed. Yearly meetings will be held to exchange information on model and database development and evaluation.

References

Aber, J. D., Nadelhoffer, K. J., Steudler, P. A. and Melillo, J. M. 1989. Nitrogen saturation in northern forest ecosystems – hypotheses and implications. – Bioscience 39: 378–386.

Berner, R. A. and Lasaga, A. C. 1989. Simulation des geochemischen Kohlenstoffkrieslaufs. – Spektrum d. Wissenschaft 5: 54–63.

Bogner, J., Vogt, M. and Piokowski, R. 1989. Landfill gas generation and migration: review of current research II. – In: Proc. anaerobic digestion rev. meeting, Jan 25–26, 1989. Solar Energy Res. Inst., Golden, CO.

Bouwman, A. 1990. The role of soils and land use in the Greenhouse effect. – In: Bouwman, A. (ed.), Soils and the greenhouse effect. Wiley, New York.

Burke, I. C., Schimel, D. S., Yonker, C. M., Parton, W. J. and Joyce, L. A. 1989. Regional modeling of grassland biogeochemistry using GIS. – Lanscape Ecology 4: 45–54.

Cicerone, R. J. and Oremland, R. S. 1988. Biogeochemical aspects of atmospheric methane. – Glob. Biogeochem. Cyc. 2: 229–237.

Crosson, P. R. and Rosenberg, N. J. 1989. Strategien für die Landwirtschaft. – Spektrum d. Wissenschaft 11: 108–115.

Desjardins, R. L. and MacPherson, J. I. 1989. Aircraft-based measurements of trace gas fluxes. – In: Andreae, M. O. and Schimel, D. S. (eds), Exchange of trace gases between terrestrial ecosystems and the atmosphere. Wiley, New York, pp. 135–152.

Harriss, R. C. 1989. Experimental design for studying atmosphere biosphere interactions. – In: Andreae, M. O. and Schimel, D. S. (eds), Exchange of trace gases between terrestrial ecosystems and the atmosphere. Wiley, New York, pp. 291–301.

Hauk, R. D. 1988. A human ecosphere perspective of agricultural nitrogen cycling. – In: Wilson, J. R. (ed.), Advances in nitrogen cycling in agricultural ecosystems. CAB Int., Wallingford, U.K., pp. 2–19.

Johansson, C. 1989. Fluxes of NOx above soil and vegetation. – In: Andreae, M. O., and Schimel, D. S. (eds), Exchange of trace gases between terrestrial ecosystems and the atmosphere. Wiley, New York, pp. 229–246.

Khalil, M. A. K. and Rasmussen, R. A. 1990. The global cycle of carbon monoxide: trends and mass balance. – Chemosphere. 20: 227–242.

Kramer, P. J. and Kozlowski, T. T. 1979. Physiology of Woody Plants. – Academic Press, New York.

Loucks, O. L. 1989. Large-scale alteration of biological productivity due to transported pollutants. – In: Botkin, D. B., Caswell, M. F., Estes, J. E. and Orio, A. A. (eds), Changing the global environment. – Academic Press, New York, pp. 99–116.

Melillo, J. M. Steudler, P. A., Aber, J. D. and Bowden, R. D. 1989. Atmospheric deposition and nutrient cycling. – In: Andreae, M. O. and Schimel, D. S. (eds), Exchange of trace gases between terrestrial ecosystems and the atmosphere. Wiley, New York. pp. 263–280.

Parton, W. J. Mosier, A. R. and Schimel, D. S. 1988. Rates and pathways of nitrous oxide production in a shortgrass steppe. – Biogeochem. 6: 45–58.

Robertson, G. P., Andreae, M. O., Bingemer, H. G., Crutzen, P. J., Delmas, R. A., Duyzer, J. H., Fung, I., Harriss, R. C., Kanakidou, M., Keller, M., Melillo, J. M. and Zavarzin, G. A. 1989. Group report: Trace gas exchange and the chemical and physical climate: critical interactions. – In: Andreae, M. O., and Schimel, D. S. (eds), Exchange of trace gases between terrestrial ecosystems and the atmosphere. Wiley, New York, pp. 303–320.

Schuepp, P. H., Leclerc, M. Y., MacPherson, J. I. and Desjardins, R. L. 1990. Footprint prediction of scalar fluxes from analytical solutions of the diffusion equation. – Boundary Layer Meteorol, (in press).

Steudler, P. A., Bowden, R. D., Melillo, J. M. and Aber, J. D. 1989. Influence of nitrogen fertilization on methane uptake in temperate forest soils. – Nature (Lond.) 341: 314–316.

Stewart, J. W. B., Asselmann, I., Bouwmann, A. F., Desjardins, R. L., Hicks, B. B., Matson, P. A., Rodhe, H., Schimel, D. S., Svensson, B. H., Wassmann, R., Whiticar, M. J.,and Yang, W. X. 1989. Group report: Extrapolation of flux measurements to regional and global scales. – In: Andreae, M. O. and Schimel, D. S. (eds), Exchange of trace gases between terrestrial ecosystems and the atmosphere. Wiley, New York, pp. 155–174.

Tans, P. P., Fung, I. Y. and Takahashi, T. 1990. Observational constraints on the global atmospheric CO_2 budget. – Science 247: 1431–1438.

Ecological Bulletins 42: 133–148. Copenhagen 1992

Biomass burning studies: the use of remote sensing

Pam Kennedy

Kennedy, P. 1992. Biomass burning studies: the use of remote sensing. – Ecol. Bull. (Copenhagen) 42: 133–148.

Events over the last ten to fifteen years have shown that both natural fires and those associated with human activities constitute a primary disturbance of the vegetation cover, and act as major determinants of atmospheric chemistry. Despite some limitations, satellite remote sensing can provide much information about the parameters required to analyse the complex mechanisms and interactions linking vegetation cover, fire, atmospheric chemistry and climate. This paper reviews the current status of research utilizing remotely sensed data for studying fire and its effects on ecosystems and the role of resulting gas and particulate emissions in biogeochemical cycles.

P. Kennedy, Inst. of Remote Sensing Appl., CEC Joint Res. Centre, Ispra Establish., TP 440, I-21020, Ispra, (Varese), Italy.

Introduction

One phenomenon that has recently received attention in the context of monitoring global change is that of fires in vegetation (i.e. biomass burning of forest, grass and cropland). Fires occur naturally in most ecosystems due to lightning, but many of the fires frequently observed, especially in the tropical belt, are of human origin. In grasslands, for example, fire is often used by pastoralists to improve access to ranges and to stimulate the regrowth of young, more palatable plants. As an agricultural practice, burning may be used as a method of pest control or of accelerating the recycling of nutrients. In tropical rain forests, fire is a primary agent of deforestation and can be used as an indicator of environmental degradation (Malingreau et al. 1985). The impact of fire on a given ecosystem depends on its extent, intensity, temperature and frequency of occurrence. These, in turn, depend on the amount and state of the fuel, which are themselves cumulative functions of the weather conditions during one or more preceding growing seasons and of the past history of burning. Extensive burning, even as small localized fires can have major impacts on the local ecosystem e.g. the reduction in the plant canopy can lead to an increase in surface runoff and accelerate soil erosion, or, the destruction of the natural habitat may result in a loss or a reduction of species diversity.

Globally, biomass burning has now been recognized as a major source of CO_2 (carbon dioxide), important trace gases and aerosol particles to the atmosphere (Hao et al. 1990). Regionally, the emphasis has been put on biomass burning associated with deforestation in the Amazon basin (e.g. Tucker et al. 1984, Malingreau and Tucker 1988, Kaufman et al. 1989, 1990), but, recent findings suggest that the amounts of CO_2 and other trace gases emitted by savanna fires are about three times larger than those estimated to result from deforestation activities (Hao et al. 1990). The release of large quantities of chemically and radiatively active gases e.g. CO (carbon monoxide), CO_2, CH_4 (methane) and NO_x (oxides of nitrogen) can, through photochemical reactions, cause substantial production of tropospheric ozone and can lead to damage of biological systems. These gases can also affect the oxidizing capacity of the air and thereby the concentrations and life-time of many other atmospheric constituents, including pollutants (Brasseur and Verstraete 1989). Biomass burning also cause considerable emissions of smoke aerosol particles which may partly serve as cloud condensation nuclei and thus have important implications in terms of modifying local weather conditions or global climatic regimes.

Measurements of gases emitted from biomass burning, are, by their very nature, area specific (e.g. Greenberg et al. 1984, Crutzen et al. 1985, Andreae et al. 1988) and cannot help resolve the global budgets of CO_2 and other chemically and radiatively active trace gases. Such emission estimates have to be weighted by fire frequency and spatial extent. In order to achieve global estimates of the emitted gases it is necessary to study fire in a broader environmental context and on a larger spatial scale. There is a need to gain a better understanding of the spatial and temporal dynamics of fire, to estimate the spatial extent and temperature of individual fires and to estimate fuel amounts and burning efficiency. Research to date has suggested that satellite data can be used to detect the occurrence of fire (e.g. Matson and Holben 1987, Malingreau 1990), though this can be better assessed at some spatial resolutions than at others (Belward 1990, Belward and Lambin 1990). Other studies have demonstrated techniques to derive the temperature and fractional area of sub-pixel resolution fires (e.g. Matson and Dozier 1981), whilst Paltridge and Barber (1988) demonstrated the use of satellite data for estimating fuel loadings.

Even though the reliability of utilizing satellite observations for estimating the above parameters remains to be confirmed, there is no doubt that fire detection from satellites poses many interesting challenges. For example, there is the problem of scale: fires occur at a variety of spatial scales from small localized events to long burning fronts, and there is the issue of sampling: the ground resolution cell of the sensor may be different from the actual size of the fire, or the satellite overpass time may not coincide with the diurnal peak of fire events. This paper discusses these challenges and points to future directions of research at the interface between satellite remote sensing and biomass burning.

Current sensor systems

Earth resource technology satellites: LANDSAT series

With the development of the LANDSAT programme at the beginning of the 1970's, several projects were conducted to test the reliability of MSS (Multi-Spectral Scanner) imagery for forest fire mapping and inventory. The absence of thermal infrared sensors (on LANDSAT–1 and –2) meant that most of the research for fire detection was indirect i.e. fuel supply mapping or an evaluation of burned areas (e.g. Benson and Briggs 1978, Hall et al. 1980, Tanaka et al. 1983). More recently, LANDSAT–MSS channel–4 (near-infrared) data have been employed to study bushfire activity throughout Burkina Faso (Parnot 1988). The two near-infrared channels (1.55–1.75 μm and 2.08–2.35 μm) of the Thematic Mapper (TM) sensor on-board LANDSAT–5 have been used for direct temperature determination of volcanic activity (Francis and Rothery 1987, Rothery et al. 1988). However, the high cost, the 16-day repeat observational interval and the likelihood that even small fires would saturate the 30 m resolution pixels, are all features which render these data inappropriate for extension to fire monitoring studies. Furthermore, the temporal coverage provided by these satellites reduces the probability of obtaining cloud-free data, especially over areas near the equator which are currently undergoing rapid and large-scale transformations.

NOAA series of satellites

The current sequence of NOAA (National Oceanic and Atmospheric Administration) satellites has been in continuous operation since Oct 1978. Recent studies utilizing data from the AVHRR (Advanced Very High Resolution Radiometer) sensor on-board this series of satellites have demonstrated their potential both for fire detection (e.g. Matson et al. 1984, Muirhead and Cracknell 1985, Flannigan and Vonder Haar 1986, Stephens and Matson 1987, Malingreau and Tucker 1988) and for fuel mapping (McKinley et al. 1985, Werth et al. 1985). Four of the primary attributes of this sensor for fire detection studies are:

1) Spectral resolution. The AVHRR acquires data in five spectral channels. The spectral location of channel–3 (3.55–3.93 μm), coincides with the radiation of maximum intensity from a blackbody with a temperature c. 780 to 800 K, and is therefore well suited to the detection of elevated heat sources, which indicate the presence of active fires. In comparison, channel–4 (10.3–11.3 μm) coincides with the radiation of maximum intensity from a blackbody with a temperature of 275 K. Data from channels–1 and –2 (visible and near-infrared respectively), have been combined in vegetation indices such as the NDVI (Normalized Difference Vegetation Index=(NIR−RED)/(NIR+RED), where NIR is the near-infrared response and RED is the red response), and found to be of considerable value for a wide range of vegetation studies (e.g. Tucker et al. 1985, Justice 1986).

2) Spatial resolution. Full resolution data (1.1 km) can be recorded for selected areas of the world in the LAC (Local Area Coverage) mode, or transmitted directly from the satellite in the High Resolution Picture Transmission (HRPT) mode for areas within a radius of 2500 km of a receiving station. Lower resolution (4 km) GAC (Global Area Coverage) data are recorded on-board, after resampling of the full resolution data. Although there are inevitable limitations associated with the use of such low spatial resolution data, which can be coarser than the dimensions of many fires, the GAC format offers the only currently available data source for obtaining wall-to-wall coverage of the tropical belt at appropriate intervals of time.

3) Frequent temporal coverage. When in full configuration, there are two NOAA satellites in complementary near-polar orbits, one crossing the equator at local solar times of approximately 07:30 and 19:30, and the other at 02:30 and 14:30. At present, these are NOAA–10, launched in 1986 and NOAA–11, launched in 1988 respectively. The orbital characteristics are such

Table 1. Case studies using thermal data for fire detection.

Author	Satellite	Channels	Geographical area	Fire paratmeters
Matson et al. (1984)	AVHRR (LAC)	Channel-3	Western USA	Forest fire detection.
Tucker et al. (1984)	AVHRR	All channels + NDVI	Rondonia, Brazil	Deforestation study.
Malingreau et al. (1985)	AVHRR (GAC)	Channel-3 + NDVI	Kalimantan and north Borneo	Forest fire detection and impact analysis.
Muirhead and Cracknell (1985)	AVHRR (LAC)	Channel-3	United Kingdom	Evaluation of straw burning.
Flannigan and Vonder Haar (1986)	AVHRR	Channels-3 and -4	Canada	Forest fire monitoring.
Nelson and Holben (1986)	MSS, AVHRR, GOES	All channels	Rondonia, Brazil	Forest, non-forest discrimination for assessing extent of deforestation.
Matson et al. (1987)	AVHRR (GAC)	Channels-1 and -3	Mexico, Brazil, Russia, Mozambique	Forest fire detection.
Malingreau and Tucker (1987)	AVHRR	Channels-1, -2 and -3	Amazon basin	Deforestation study by identification of disturbed areas.
Matson and Holben (1987)	AVHRR	Channels-3, -4 + NDVI	Brazil	Fire detection and areal assessment of forest damage.
Setzer (1988)	AVHRR	Channels-3 and -4	Legal Amazon	Large-scale deforestation study.
Malingreau and Tucker (1988)	AVHRR (LAC)	Channel-3	Amazon basin, Brazil	Fire count/burning calendar.
Citeau et al. (1988)	METEOSAT	Thermal channel	Senegal	Monitoring occurrence and development of bush fires.
Malingreau et al. (1989)	AVHRR (GVI, GAC, LAC)	Channel-3 + NDVI	Kalimantan, Borneo, Amazon basin, West Africa	Dynamics of forest transformations.
Grégoire (1990a, b)	AVHRR (LAC)	Channel-3	West Africa	Detection of bush fires in the Sudanian savanna.
Malingreau (1990)	AVHRR (LAC)	Channel-3	Tropics/sub-tropics	Global monitoring of fires in tropical and sub-tropical ecosystems.
Kaufman et al. (1990)	AVHRR (LAC)	Channels-1, -2, -3 and -4	Brazilian Amazonia	Forest fire detection and estimation of gaseous emissions.
Lee and Tag (1990)	AVHRR (LAC)	Channels-3 and -4	Yellowstone National Park, USA	Automated fire detection algorithm.
Setzer and Pereia (1991)	AVHRR, LANDSAT-TM	Channel-3	Brazilian Amazonia	Forest fire detection and estimation of gaseous emissions.

that with two satellites in operation, the possibility exists for twice daily and twice nightly coverage at any point on the equator. This coverage increases with latitude.

4) Overpass time. The current afternoon, approximately 14:30 overpass time of NOAA–11 is the best available in terms of fire detection and monitoring in the tropical belt, where, for example, in the Sudanian zone of West Africa, agricultural fires are at their maximum in early afternoon.

Geostationary satellites

The GOES (Geostationary Operational Environmental Satellite) is part of a network of geosynchronous orbiters designed to provide meteorological observations for almost the entire world.

Aside from its usefulness in monitoring and measuring cloudiness, GOES visible and near-infrared data have been used to detect fires indirectly by the identification of smoke plumes (Segal et al. 1989). Based on the assumption that the smoke particulate concentration is directly proportional to smoke reflectance it is possible to monitor the relative changes in smoke reflectance and concentration. Biomass burning may then be quantified indirectly by the quantity of smoke that is generated (e.g. Fishman et al. 1986). Limitations associated with the use of these data include the low spatial resolution (8 km) of the infrared channel, (which restricts detection to large-scale i.e. >30 km smoke plumes), and the broad spectral bandwidths of both the visible (0.55–0.75 μm) and infrared (10.5–12.5 μm) channels, which renders them inappropriate for any vegetation assessment studies, or for multi-spectral approaches to fire detection and monitoring. Data from METEOSAT (primarily a meteorological satellite) have also been used for a range of terrestrial applications, including regional soil and vegetation studies and biomass and crop production estimates. Citeau et al. (1988) used data from the thermal channel (channel–3) to observe the occurrence and development of bush fires in Senegal. Although the frequency of data acquisition (i.e. every 30 min) offers excellent monitoring capabilities, the spatial resolution (5×5 km pixels) restricts studies to the detection of large-scale fire events. In addition, global monitoring is prevented by the latitudinal limits of useable data, i.e. between approximately 65° N/S for qualitative studies and 55° N/S for quantitative work.

Thermal data for fire studies

Some of the most commonly cited case studies using thermal data from different satellites, are listed in Table 1. The research undertaken in these and other studies is described in detail in the following sections.

Introduction: fundamental concepts for fire detection

Matson and Dozier (1981), Matson et al. (1984) and Muirhead and Cracknell (1984, 1985) have demonstrated that the 3.8 μm thermal channel (channel–3) of the AVHRR sensor provides the capability to detect high temperature fields or hot spots such as steel plants, waste gas flares from oil fields and straw burning. Recent research has illustrated the utility of these data for fire monitoring in deforestation studies (e.g. Malingreau et al. 1985, Matson and Holben 1987) and in detecting and monitoring continental-scale features such as the more or less continuous fire belt, 100 to 200 km wide, which has been identified throughout the savanna zone of West and Central Africa (Malingreau 1990). Channel–3 is sensitive to a combination of reflected and emitted radiation. It is located near the spectral maximum for radiative emissions for objects radiating at temperatures c. 800 K, i.e. the near-normal temperature for burning grass (Langaas and Muirhead 1988). Owing to Wien's displacement law the mid-infrared flux, or blackbody radiance becomes stronger as the temperature increases. This means that for high temperature targets the maximum blackbody radiance occurs at shorter wavelengths, such that the thermal channels of the AVHRR sensor, i.e. channel–4 (10.3–11.3 μm) and channel–5 (11.5–12.5 μm) receive less blackbody radiation. Matson et al. (1987) found that the satellite-derived temperatures for a controlled forest burn and a phosphorous plant were, respectively, 16.2 K and 33.9 K higher in channel–3 than in channel–4. Detailed explanations of this phenomenon in relation to the AVHRR sensor can be found in Kerber and Schutt (1986) and Schutt et al. (1986).

Fire detection: the techniques, limitations and potential solutions

The hot spots or high temperature sources which give rise to a response in channel–3 can be identified relatively easily if it is assumed that fire of some temperature and dimension (minima unknown) will create a response, or even saturate the corresponding pixel. Saturation temperatures given in the literature are somewhat inconsistent. Robinson (1987) quotes that the upper threshold for channel–3 saturation for NOAA–9 occurs at 312 K (brightness temperature), whilst Matson and Holben (1987) state that channel–3 saturates at c. 320 K. However, because saturation will occur at c. 315 to 320 K, which is higher than the temperatures for most non-fire features on the ground, it can be assumed for most observations, that a saturated pixel represents a response from a fire. This assumption loses validity when, on some occasions the combined emitted and reflected radiance from hot sparsely vegetated surfaces (e.g. in the Sahelian zone of Africa during the dry season) can also cause a pixel to saturate. In Senegal, for example, non-burnt, sparsely vegetated arid areas in the north, can exhibit exactly the same thermal characteristics as recently burnt areas in the savanna zones further south (Frederiksen et al. 1990). Such difficulties can sometimes be overcome by utilizing data from the night-time passes of the NOAA satellites (when in situ soil surface temperatures are more likely to be well below the sensor saturation threshold), by monitoring a pixel response over time, or by comparing the temperature differential between the response in channel–3 and channels–4 and –5 (Matson and Dozier 1981, Matson et al. 1984). The latter technique can also be used to distinguish hot spots from pixel error or highly reflective cloud (Muirhead and Cracknell 1984), though such procedures may be less effective in the tropics because the atmospheric transmissivity in channels–4 and –5 is greatly affected by the water column.

Using a theoretical bi-spectral model developed by Dozier (1981) it is possible to utilize the temperature difference in the 3.8 and 11 μm channels, to estimate subresolution areal fractions and temperatures of high temperature regions in a cool environment (Matson and

Dozier 1981). The ability to determine fire temperatures is of particular interest for estimating the release of CO_2, trace gases and aerosol particles from vegetation fires. Smouldering combustion at temperatures <800 K results in release of large amounts of trace gases and particles, whilst in the case of flaming combustion at temperatures >800 K, burning efficiency is higher, so that the levels of gases and particulates released is correspondingly lower (Ward 1990).

Despite the fact that the Dozier model has a sound physical foundation, there are a number of assumptions associated with the ground scene (composition of the temperature fields within a pixel), the atmosphere and the sensor, which restrict its application for meaningful quantitative thermal analysis. These assumptions include; 1) that any scene (pixel) consists of only two uniform classes; the high temperature target (fire), and the low temperature background, 2) that all fires within a pixel are considered to be grouped together, 3) that both fire and its background have blackbody emissivity properties, 4) that the background is spatially homogeneous in terms of thermal properties. These, and other limitations associated with the application of this model are discussed in detail in Langaas and Muirhead (1988).

The Dozier model has been largely restricted to single-date, small scale case studies, (Matson and Dozier 1981, Flannigan and Vonder Haar 1986, Matson and Holben 1987) due largely to the difficulties of implementing an automated procedure to derive local environment background temperature information, by sampling neighbouring fire pixels. Recently, attempts have been made to develop techniques to parameterize the background temperature in a way which allows the model to be used in a more automated manner, so that fires can be studied over long time series and at regional scales (Lee and Tag 1990).

The ability to discriminate fire/non-fire pixels, whether for input to the Dozier model or for operational fire detection and monitoring, is possible only if the readings from the pixel in which they are found correspond to temperatures between the environment background temperature and the temperature distribution that would saturate the pixel measurement. Since both channel–3 and channel–4 saturate at c. 320 K these data can be used to count the number of fires, and to provide estimates of the magnitude of the fire, but cannot be used to accurately determine the fire size (Kaufman et al. 1990).

The saturation threshold is fixed by sensor design. In general terms, a subpixel hotspot will cause saturation if hot spot radiance plus background radiance exceeds the energy flux for sensor saturation. Although the exact level of sensitivity or the lower thresholds of detectability remain undefined, it clearly does not take a 1.1 km target of a full resolution LAC (Local Area Coverage) pixel to stimulate a response in channel–3. The temperature contrast between the hotspot and the environment background can be more important in determining how well a fire or hotspot can be detected, than the actual energy flux from the hotspot. Kaufman et al. (1990), suggested that fires with temperatures exceeding 500 K occupying 0.01–0.1% of a pixel would be enough to saturate the 3.7 μm channel, and have a small effect in the 11 μm channel. Channel saturation suppresses the ability to derive meaningful effective subresolution blackbody temperatures of a fire, though fire identification may still be performed. One way to avoid saturation problems would be to raise the saturation threshold and stretch the dynamic range of the sensor so that each quantum of flux density represents a greater temperature range. This is an important issue in terms of determining the optimum saturation temperatures for the fire detection sensors to be carried on-board future satellite platforms.

The fire detection thresholds, on the other hand, will obviously vary depending for example on the characteristics and homogeneity of the environment background temperature and the extent to which data are contaminated by cloud-top reflectance. It is evident that a single threshold will be inappropriate for universal application, because not only will it vary in a spatial dimension (i.e. with different types of ecosystems), but it has become increasingly clear from research in West Africa that the date of occurrence is also of critical importance (Grégoire 1990a). Fires occurring late in the dry season may be difficult to discriminate from dry sparsely vegetated land surfaces which become increasingly hotter throughout the duration of the dry season. Detection of these late season fires will be dependent upon determining different thresholds from those used to detect early season fires when the problem of high background temperatures is less pronounced.

The methods which have been adopted for operational fire detection and monitoring have relied either upon a visual comparison of enhanced channel–3 and channel–4 data (Matson et al. 1984), or on some empirically defined threshold values of DN (digital number) or temperature, to discriminate between pixels containing hot spots and those of the non-burning surroundings and cloud-contaminated pixels (Kaufman et al. 1990, Pereira and Setzer 1991, Setzer and Pereira 1991). An automated method for fire detection in the Amazon basin is currently being used in an operational way. Originally all pixels saturated in channel–3 were considered to be pixels containing fires (Malingreau and Tucker 1988). This approach has been enhanced, and fire pixels are identified on the basis of fulfilling a set of decision rules (Kaufman et al. 1988, 1990). The three rules are: 1) channel–3 temperature is ≥316 K, 2) channel–3 temperature is equal to or greater than channel–4 temperatures plus 10 K, and 3) channel–4 temperature is >250 K. The first criterion requires that in order for a pixel to be considered to contain a fire it has to be hot. The second criterion imposes that the radiative temperature has to be much higher in channel–3 than in channel–4 if the pixel contains a fire rather than a warm

Table 2. Case studies using remote sensing for vegetation analysis.

Author	Satellite	Channels	Geographical area	Fire paratmeters
Pre-fire information				
Shasby et al. (1981)	LANDSAT	MSS	Nevada	Forest fuel mapping.
Barber and Paltridge (1986)	AVHRR	Channels-1, -2 and modified NDVI	Grasslands, Victoria, Australia	Fuel moisture content mapping.
Miller et al. (1986)	AVHRR	Channels-1, -2 and NDVI	Eastern Oregon, USA	Fuel type mapping.
Malingreau and Tucker (1987)	AVHRR	Channel-3, NDVI and GVI	Amazon basin	Seasonal fuel potential monitoring using time series analysis of a vegetation index.
Paltridge and Barber (1988)	AVHRR	Channels-1, -2 and NDVI	Grasslands, Victoria, Australia	Fuel moisture content/fire potential mapping.
Post-fire information				
Hall et al. (1980)	LANDSAT	MSS	Alaska	Analysis of initial recovery of burned tundra.
Tanaka et al. (1983)	LANDSAT	MSS	Etajima Island, Japan	Classification of burnt areas according to degree of carbonization.
Malingreau et al. (1985)	AVHRR	Channel-3, NDVI and GVI	Kalimantan and N. Borneo	Use of the NDVI temporal profiles to monitor the impact of drought and fire on forest vegetation.
Milne (1986)	LANDSAT	MSS	Eastern Australia	Mapping and monitoring vegetation change associated with bushfire events.
Hick et al. (1986)	AVHRR + LANDSAT	Visible and near-infrared	Western Australia	Evaluating extent of bushfire damage.
Parnot (1988)	LANDSAT	MSS (Channel-4)	Burkino Faso	Cartographic inventory of burnt areas.
Chuvieco and Congalton (1988)	THEMATIC MAPPER	All bands	Eastern Spain	Discrimination and mapping of burnt areas.
Langaas (1989)	AVHRR and hand-held radiometer	Channels-1, -2, -3 and -4	Senegal	Study of spectral characteristics of burnt areas.
Frederiksen et al. (1990)	AVHRR	All bands	Senegal	Multitemporal mapping of burnt areas of savanna.
Grégoire (1990a, b)	AVHRR	Channel-3, NDVI	Savanna in W. Africa	Use of NDVI temporal profiles to monitor vegetative regeneration after bush fires.
Pereira et al. (1991)	AVHRR + LANDSAT-TM	AVHRR (Chs.-1, -2 and -3) TM (Bands 3, 4 and 5)	Central Brazil	Active fire detection and burn scar identification.

surface. The third criterion requires that the pixel does not include strongly reflective clouds which may saturate channel–3. This approach was used to undertake a fire count for the 1987 dry season in Brazil. The identified pixels were additionally examined in the visible channel (0.58–0.68 μm) to verify that there was an associated smoke plume (Kaufman et al. 1990, Setzer and Pereira 1991).

In conclusion to this section it should be emphasized that fire information obtained from analysing AVHRR data can only be considered to represent a sample of the fires occurring on the ground. The low spatial resolution of LAC data and the time of the satellite over-pass, both set limitations on detecting fire events and thus hinder the compilation of complete statistical data sets. At present, the afternoon pass of the NOAA–AVHRR (14:30 equatorial crossing of NOAA–11), is near-optimal in terms of ensuring that for many ecosystems throughout the tropical belt observations are made at the peak of burning. However, this does not preclude the possibility of missing fire events e.g. due to cloud cover, which can be at a maximum at this time over tropical regions. It is also possible that some fire episodes will be short and start at any time of day. Geostationary satellite data could be used to provide information at more frequent time intervals throughout a day, but the existing satellites (METEOSAT, GOES) lack the appropriate ground resolution in the thermal channels.

Visible and near-infrared data for fire studies: techniques for studying vegetation and atmospheric characteristics

Analysis of vegetation

Several sub-continental areas around the world are currently undergoing rapid, wide-ranging changes in land cover. Many of these large-scale disturbances are centred in the tropical countries such as Brazil, Columbia, Indonesia, Ivory Coast and Thailand and are associated with the various activities of deforestation. Despite the recognized importance of tropical deforestation and the widely accepted view that its impacts have to be seriously considered by the international community, it still remains extremely difficult to produce an overall assessment of the on-going process. Recent research has investigated the potential of using satellite data to improve the capability of surveying vast regions of the globe and for conducting regular assessments of the dynamics of the world's ecosystems. The underlying common denominator in these continental and global-scale studies is that the majority utilize data from the AVHRR sensor. The multi-temporal characteristics, manageability and cost effectiveness of these data have enabled researchers to address global, continental and regional vegetation studies (e.g. Justice et al. 1985, Prince and Tucker 1986, Townshend and Justice 1986).

The methodological procedures adopted for using the visible and near-infrared data of channel–1 and –2 respectively for vegetation studies have been the subject of much research (e.g. Gatlin et al. 1981, Townshend and Tucker 1981, Tucker et al. 1981, 1984). The fundamental principle behind the successful use of these data is that the two channels can resolve the differential reflectance properties of green vegetation. Channel–1 corresponds closely to the chlorophyll absorption band, whilst channel–2 is sensitive to the highly reflective plateau in the near-infrared which is associated with healthy green vegetation. Ratios of reflectance in these two regions of the spectrum have been found to be related to various vegetation characteristics such as green biomass, percentage vegetation cover and photosynthetic activity (Tucker 1979, Tucker et al. 1981, Tucker and Sellers 1986). The majority of studies utilizing AVHRR data for vegetation analysis have employed the NDVI. This yields a measure of photosynthetic capacity, such that the higher the value of the ratio, the more photosynthetically active the cover type (Sellers 1985).

In terms of utilizing visible and near-infrared data in fire studies, there are two main areas of application. These are in obtaining pre-fire information (e.g. fuel-potential mapping), and in gaining information after a fire event, by detecting and monitoring large-scale ecosystem transformations. The former is dependent upon obtaining spatial and temporal information about vegetation status, such as biomass density and moisture content (e.g. Malingreau and Tucker 1987). The latter relies largely on the ability to identify disturbed areas, i.e. burnt areas and/or areas undergoing stages of regrowth after a burning event (e.g. Malingreau et al. 1985). Studies which have utilized visible and near-infrared satellite data for obtaining pre- and post-fire information are listed in Table 2.

Fuel potential mapping

The amount of available fuel is considered to be one of the main variables in any fire-danger rating system (Deeming et al. 1978). As a consequence, several projects have attempted to obtain detailed fuel loading maps from a digital classification of LANDSAT imagery. Many such maps have been combined with additional layers of information such as elevation, soil moisture, insolation and fire history to derive fire-risk information (e.g. Cosentino et al. 1981, Root et al. 1986). One of the major limitations of this approach was found to be the inability of satellite data to discriminate the under-storey component of forest vegetation (Salazar 1982).

One approach adopted by Barber and Paltridge (1986) for developing an operational system for monitoring grassland fire potential in Victoria, Australia was to derive an empirical relationship between information collected in the field and channel–1 and –2 data from the AVHRR. They combined ground-derived information on fractional vegetation cover with a slightly modified NDVI, and then correlated the results with field measured fuel moisture content. This information was then used to map the relative flammability of the grassland during the dry season. By taking advantage of the derived relationships, and the high temporal frequency of AVHRR data, it was possible to reproduce the information regularly so that the relevant authorities could utilize the maps as an up-to-date information source for determining fire danger periods, and for declaring 'total fire ban days'.

Miller et al. (1986), adopted a more complicated approach utilizing AVHRR/NDVI data to produce a fuel-type map for an area of 170,000 km^2 in eastern Oregon. Principal components analysis were used to reduce a time series of NDVI images to four scenes which contained approximately 96% of the variation found in the original data set. An unsupervised clustering algorithm was used to identify fortyfive spectral classes, which were then aggregated into eleven fuel type classes using field knowledge, aerial photographs and LANDSAT false colour composites.

In addition to these approaches, it is also possible to utilize AVHRR/NDVI data as a tool for studying the phytodynamics of plant communities. For example, analyses of NDVI profiles over time can provide information about when the vegetation conditions are ripe for burning, and when the vegetation no longer represents optimal fuel loading conditions (Malingreau

Table 3. Case studies using remote sensing for atmospheric analysis. [1] TOMS = Total Ozone Mapping Spectrometer, [2] MAPS = Measurement of Air Pollution from satellites, [3] SMMR = Scanning Multichannel Microwave Radiometer.

Author	Satellite	Channels	Geographical area	Fire parameters
Chung and Le (1984)	AVHRR	Channels 1, 2 and 4	N.W. Canada	Detection of forest fire smoke plumes.
Fraser et al. (1984)	GOES	Visible		Aerosol mass and transport.
Tucker et al. (1986)	AVHRR	Channels 1, 2 + NDVI		Relationship between atmospheric CO_2 variations and satellite derived NDVI.
Reichle et al. (1986)	MAPS[1] on-board the space shuttle			Measurement of middle and upper tropospheric carbon monoxide mixing ratios.
Fishman et al. (1986)	TOMS[2], GOES, MAPS	Visible	Amazon basin	Study of tropospheric ozone in the tropics.
Kaufman (1987)	AVHRR	Channels 1, 2		Remote sensing of aerosol absorption.
Fung et al. (1987)	AVHRR	Channels 1, 2 + NDVI		Establish relationship between NDVI and field measurements to determine geographic distribution of monthly exchange of CO_2 between atmosphere and biosphere.
Choudhury (1988)	SMMR[3]	37 GHz frequency		Correlation between polarization difference, NDVI and atmospheric CO_2 concentrations.
Kaufman and Sendra (1988)	LANDSAT	MSS		Derivation of aerosol optical thickness.
Kaufmann et al. (1988)	AVHRR	Channels 1, 2		Smoke loading characteristics.
Segal et al. (1989)	GOES	Visible and infrared	Yellowstone National Park, USA	Identification and monitoring of smoke plume density and transportation path.
Kaufman et al. (1990)	AVHRR	Channels 1, 2	Brazilian Amazonia	Emission rates and characteristics of aerosol part of smoke from biomass burning.

and Tucker 1987, Malingreau 1990). More information on the behaviour of the NDVI for different ecosystems for so called normal years is required, before such thresholds can be determined and tested for estimating fire potential and fire risk.

Burnt area identification and post-fire recovery

Studies conducted to remotely map and inventory burned surfaces have demonstrated the utility of satellite data with respect to more traditional methods i.e. aerial surveys and field work (Tanaka et al. 1983, Milne 1986, Abriol et al. 1987), Table 2. There have, however, been very few studies specifically addressing the problem of understanding the spectral definition of burned land. Several authors have reported an increase in response in the visible wavelengths, along with a decrease in the near-infrared (Tanaka et al. 1983, Abriol et al. 1987).

Langaas (1989) found that for fire scars in southern Senegal, reflectances measured by a hand-held radiometer increased relatively rapidly after a fire event. After eleven days the channel–1 (red) response had increased to c. 75% of pre-fire levels, channel–2 (near-infrared) to 55% and the derived NDVI to 45% of pre-fire levels. This rapid recovery suggested that AVHRR imagery would have to be acquired on a high temporal basis in order to optimize the use of the large brightness differential between burnt and non-burnt areas for burn scar identification.

Subsequent research by Frederiksen et al. (1990), has shown that significant differences in the digital values recorded in both the visible and near-infrared channels (1 and 2 of the AVHRR sensor), and in the satellite-derived NDVI values, occur at the border of fire-affected and non-fire affected areas. In general, the change from bright vegetation to dark ash-covered areas results in a corresponding reduction in recorded reflectance. In practice the magnitude and identification of this contrast depends upon such factors as: 1) the age of the burnt area, in terms of whether there has been

Table 4. Sensor characteristics and data requirements which should be considered when designing future satellite systems for fire detection and monitoring.

Issues/criteria	Ecosystem component	Thermal sensing	Atmospheric component
Channel location	Visible: red and near-infrared channels required for studying the ecological components of fire i.e. fuel loadings, vegetation stress etc. Thermal: channels can be used to elucidate the boundaries between non-burnt : burnt areas.	Mid-infrared channel is essential: multiple infrared data can be useful for threshold determination and for making atmospheric corrections.	Visible channel data required for gaining information on the chemistry and transport of species in smoke plumes.
Radiometric resolution and dynamic range	Current status: adequate	Mid-infrared dynamic range required, which is sufficient to accomodate fire temperatures, and which eliminates current problems of saturation.	Current status: adequate.
Spatial resolution	Compromise between mid and low-resolution for individual fire detection and for global fire monitoring respectively.		High, to detect the spectral characteristics and effective size of particulate emissions, mid to low for global surveys.
Spatial coverage	Consideration needs to be given to: (a) Increment in the current number and distribution of receiving stations for HRPT-AVHRR imagery. (b) Identification of regions not currently imaged. (c) Degree of distortion experienced at track edges (polar orbiter).		
Temporal resolution	Very high temporal resolution imagery is initially required to understand the temporal and spatial distribution of fire (Geostationary platforms). More appropriate temporal sampling strategies can then be determined.		
System design and other considerations	(a) Time synchronous data collection. (b) Cost of data. (c) Manageable data volumes. (d) Easily available to user. (e) Accuracy of navigational data for registration to time series.		

removal of ash and thus changes in the colour and/or reflective properties of the surface, and/or regrowth of the vegetation, and 2) the intensity of the fire, in terms of influencing the fractional cover burnt within areas experiencing a fire event. In general, the NDVI was found to be the best individual spectral discriminator during the beginning of the dry season (Frederiksen et al. 1990).

Chuvieco and Congalton (1988) used TM data to study forest fires on the mediterranean coast of Spain. They employed the NDVI as a measure of vegetation status in order to monitor the different levels of damage caused by the fire. As with other sources of stress, e.g. drought, fire causes a sensible change in the near-infrared/red ratio, such that the corresponding NDVI response can provide an indication of burning intensity. The impact from fire and drought will affect vegetation physiology in a way which reduces the NDVI response. The basic assumption underlying this method is that the larger the fall, or drop in NDVI values, the more severe has been the impact of the fire or drought (Malingreau et al. 1985). Limitations associated with this approach are; 1) that it is not always easy to discriminate between burnt areas and other land cover categories giving a similar spectral response, and 2) the spectral characteristics of the burnt vegetation may become concealed by increasing values of the NDVI responding to the rapid regrowth or greening-up of vegetation after a fire event.

Chuvieco and Congalton (1988) also explored the possibility of utilizing single channel data and ratios in order to classify burnt land. They concluded that the results obtained from digital processing techniques could not provide sufficient accuracy for discriminating burned and unburned vegetation in areas where the vegetation cover is sparse (Tanaka et al. 1983), and where there is a strong influence from both soil radiance and desiccated unburned vegetation (Benson and Briggs 1978). Spatial masking by visual analysis provided the most effective way of identifying the perimeter of the scorched area, whilst digital processing of thermal and visible data provided additional discrimination within the burnt areas. Variations in the character and appearance of burnt areas result from the variability of antecedent burning conditions, burning intensity and efficiency. Such factors determine the temperature regimes, the size of the fire and the percentage of material burnt within the affected area (Malingreau 1990).

Hick et al. (1986) reported on a comparative study utilizing aerial photographs, LANDSAT–MSS and NOAA–AVHRR imagery for fire scar delineation in a schlerophyll open woodland area near Perth, Australia. They found that channel–2 data of the AVHRR sensor could be used to discriminate between fire scars and the surrounding non-burnt areas, provided the spatial extension of the burnt area was of sufficient magnitude in relation to the geometric resolution of AVHRR data. In a similar comparative study carried out in central Brazil using TM and AVHRR data, it was found that on average, the burned area determined using AVHRR channel–3 data, tended to overestimate by c. 43% the actual area of the burn scars estimated from the higher resolution TM data (Pereira et al. 1991).

The multi-temporal information provided by satellite data can also be of use in evaluating the post fire effects on ecosystems, and in monitoring the occurrence and rate of regrowth of the burned vegetation (e.g. Hall et al. 1980). Malingreau et al. (1985) used a sequence of tri-weekly NDVI values derived from a Global Vegetation Index (GVI) product to observe the temporal pattern of vegetation in east Kalimantan, in relation to the severe drought conditions of 1982/1983 and two fire events between Aug–Oct 1982 and April–May 1983. The damage inflicted during the pre-fire drought and by the actual fire events was identifiable by a clear reduction in the NDVI response. Progressive regeneration was also evident from a gradual increase in NDVI values, indicative of vegetation recovery occurring in response to the onset of the 1983/84 rainy season. The actual severity of a fire's impact on a vegetation community will depend on a number of factors relating to the type and efficiency of the fire, its time of occurrence with respect to climatic conditions, the seasonal development of the vegetation and on its ability to recover after a fire event.

Atmospheric analysis

During controlled and natural fires, large amounts of trace gases (e.g. CO_2, CO, CH_4, NOx), and particulates (organic matter and graphitic carbon) are released to the atmosphere. At present, methods for estimating trace gas emissions are based on measurements of the ratio of the trace gas concentration to the concentration of CO_2, and on a crude global estimate of the rate of emission of CO_2 during biomass burning. The only direct measurements from space of a tropospheric trace gas (CO) have been from the Space Shuttle (Reichle et al. 1986). Estimates of other trace gases e.g. of tropospheric ozone have been derived indirectly by subtracting empirical data from the total ozone content as measured by the Total Ozone Mapping Spectrometer (TOMS), on-board the Nimbus–7 polar orbiting satellite (Fishman et al. 1986). Table 3 lists some of the most commonly cited remote sensing studies of smoke and trace gases.

Smoke plume detection and analysis

Kaufman et al. (1990) proposed two methods to obtain information about the emission products from biomass burning in Amazonia. Both techniques are based on the analysis of NOAA/AVHRR, 1 km resolution imagery and depend on the detection and count of fires using data from channels–3 and –4. One method requires the determination of the average emission of particles per fire in a given area and season, and the other method, estimation of the total burned area and subsequent computation of the total burned biomass and emitted trace gases using models found in the literature. These two approaches are described briefly here.

In the former technique, the estimates of average emission particles are translated into trace gas emissions using established, measured ratios of particle concentrations, and trace gas concentrations in smoke for several fire conditions (Ward and Hardy 1984, Ward 1986). The mass of particles in the smoke and their single scattering albedo (w_0) are determined from the visible and near-infrared channels of the AVHRR (channels–1 and –2 respectively). The w_0 is the ratio of light scattering to light extinction by the smoke, which is related to the graphic carbon content of the smoke (Ackerman and Toon 1981), it is c. 0.97 for smouldering conditions and 0.66 for flaming conditions. Previous research revealed the possibility of employing AVHRR satellite data to derive the single scattering albedo of smoke, which in turn offers the capability to distinguish between smouldering and flaming conditions (Kaufman 1987, Fraser et al. 1984). The total mass of the emitted particulates is computed as a product of the emitted mass per fire and the total number of fires. Inaccuracies in the application of this approach can often arise from the uncertain effects of water vapour on aerosol scattering, and from the assumption that there is a direct relationship between the particulate characteristics measured from prescribed fires (Ward 1986), and the particulate characteristics derived from space observations.

The second technique is based on the assumption that both the mass of emitted particulates and the mass of emitted trace gases are proportional to the mass of the burned biomass, or the mass of the released CO_2. In order to obtain an approximate estimate of the total area burned, the number of pixels identified to contain fires were multiplied by 1.2, (the area of a pixel in km^2 at nadir). This information was then used to determine an estimate of the percentage of forest and grassland which was actually burnt. Using models found in the literature (Seiler and Crutzen 1980) it is then possible to estimate the amount of dry matter and, in turn, to estimate associated emissions of carbon dioxide (Seiler and Crutzen 1980) and emissions of trace gases (Andreae et al. 1988).

The results from this research indicated that 350,000 independent fires, corresponding to c. 20 mill ha of different types of vegetation burned during the dry season of 1987 in Brazil's Amazon basin. It was estimated that 1700 mill tons of CO_2 were emitted from the fires (Kaufman et al. 1990, Setzer and Pereira 1991).

Research programme for the future

Development of a global fire monitoring system rationale

Despite the recognized importance of several regional-scale case studies such as those reported from the Amazon basin and Borneo (Malingreau and Tucker 1987, Matson and Holben 1987, Malingreau 1990), Central Africa and entire regions throughout South East Asia, remain virtually unknown in terms of fire activity. A number of qualitative observations suggest that these areas also represent regions undergoing extensive transformations (Malingreau 1986, 1990). The scale, and possible consequences of these events have aroused much interest in the scientific community. The growing international concern regarding the increase in atmospheric concentrations of several trace gases, for example, stimulated the formulation of the IGAC (International Global Atmospheric Chemistry) project, which has subsequently been recognized as one of the core projects of the much broader inter-disciplinary programme of the IGBP (International Geosphere-Biosphere Programme).

The main goal of the IGAC project is to increase our quantitative understanding of the present state of tropical biosphere-atmosphere interactions and effects of change on trace gas exchange and atmospheric chemistry. One of the many activities identified within the IGAC research plan is a proposal to study the impact of biomass burning on the atmosphere. Four main tasks have been identified which address the major gaps in our knowledge of understanding trace gas emissions resulting from biomass burning. These are: 1) spatial and temporal characterization of biomass burning, 2) chemical characterization of biomass burning emissions and plume chemistry, 3) short and long-term effects of fires in terrestrial ecosystems and 4) ecological effects of the deposition of biomass burning products (IGBP Report No. 13 1990).

It is being proposed that satellite remote sensing could contribute to the objectives of this IGAC activity, and to other similar projects, by providing a potential solution to the problem of efficiently and economically detecting and monitoring fire at regional and global scales. Remote sensing contributions could be visualized in two ways:

1) Direct approach: the construction of biomass burning source inventories at regional and global scales. This would include the provision of quantitative information on the spatial and temporal characteristics of fires (at least for the time of the satellite overpass), and estimates of the rate of deforestation.

2) Indirect approach: determination of the quantity and characteristics of emission products for a full range of land cover types. This would include either estimating the emission of particulates per fire (e.g. Kaufman et al. 1989, 1990), or measuring gaseous emissions in the troposphere and stratosphere using such instruments as the SCIAMACHY (Scanning Imaging Absorption SpectroMeter for Atmospheric CHemistrY).

Current status and research priorities

At the present time, approaches to the development of a global fire monitoring programme are limited by several factors. The most important shortcomings are: 1) data availability: the widespread use of fire and biomass burning dictates that the satellite coverage must be exhaustive. There are still portions of the world which are not recurrently covered by satellite observations at the desired resolution, 2) temporal sampling: the afternoon overpass time of the AVHRR sensor provides only a single time frame in the diurnal pattern of fire dynamics. A more continuous picture can be constructed from the coarser resolution METEOSAT data, 3) spatial resolution: at the present time global coverage can only be achieved using the 4 km AVHRR–GAC data. It is of paramount importance therefore to evaluate the effect of the LAC–GAC sampling transform (four out of five contiguous LAC pixels in every third scan line are averaged to compute one GAC pixel) upon fire detection capabilities, 4) spectral calibration: channel–3 of the AVHRR sensor saturates at c. 320 K (Matson and Holben 1987). Bare soil surfaces (e.g. in the dry season of the tropics) can have a radiative temperature higher than the saturation level, and cause confusion between hot soil pixels and pixels containing active fires.

There are several other aspects related to global scale studies which require intensive research before development problems can be resolved. These include new image processing algorithms to improve data quality, and advances in computer hardware, and data archiving facilities to deal with the large volumes inherent to global data sets (Verstraete et al. 1990).

In spite of these limitations, there are important issues which have to be addressed in terms of evaluating more precisely the utility of existing satellite data sets for studying biomass burning at regional and continental scales. Since the AVHRR sensor offers the only currently available system for obtaining large area coverage at appropriate temporal and spatial resolutions, it becomes a central tool in the research for developing a global fire monitoring system.

The most important questions which have to be addressed are:

1) To what extent can satellite data be used to identify and characterize active fires at regional, continental and global scales? In addition to research investigating appropriate thresholding techniques for fire detection, it will be necessary to carry out comparative spatio-temporal analyses of 4 km and 1 km AVHRR data to evaluate the effect of sampling and resolution. Preliminary investigations suggest that the under-sampling of the higher resolution LAC data has two possible effects:

Table 5. Planned sensor launches with near and mid-infrared channels. [1]) EOS, MODIS-N = Earth Observing System, MODerate Resolution Imaging Spectrometer-Nadir, [2]) TOMS = Thermal Infrared Multispectral Scanner, [3]) MRIR = Medium Resolution Imaging Radiometer, [4]) ERS-1, ATSR = European Remote Sensing Satellite, Along-Track Scanning Radiometer, [5]) ATMOS-SCIAMACHY = Atmospheric and environmental research satellite SCanning Imaging Absorption spectroMeter for Atmospheric CHemistrY, [6]) ROSIS = Reflective Optics System Imaging Spectrometer.

Platforms/ sensor	Launch date	Infrared channels (μm)	Pixel size (km)	Observation hour	Comments
NOAA-KLM AVHRR-3	1992 to 1995	1.58 to 1.64 day 3.63 to 3.9 night	1.1 to 4 km	2:30, 8:30, 13:30, 20:30 Daily	Limited on-board data storage.
GOES-I-M IMAGER	1989 to 2000	3.8 to 4.0	4 km	Every 30 min.	Geostationary, over equator.
GOES-RESEARCH	1995+	To be decided	To be decided	Every 3 or 30 min.	Equatorial, variable longitude, dedicated to research, still in design phase.
EOS[1]) MODIS-N	1995+	Eleven channels between 3.75 and 9.73	1 km	13:30 (2 d)	36 channels in total. Two at 250 m, five at 500 m and twenty-nine at 1 km resolution.
EOS TOMS[2])	1995+	Four channels between 8.2 and 10.2	0.03 km	13:00 (2 d)	Six channels in total between 8.2 and 11.7 μn. Saturation problems inevitable.
NOAA-MRIR[3])	1994	Ten channels (0.45 to 12.5)	0.25 km	9:00, 13:00, 21:00, 1:30	Operational polar orbiting platform, still in design phase.
ERS[4]) ATSR	1991 (launched)	Four channels 1.6, 3.7, 11 and 12	1×1 km	Local mean solar time at the equator : 10:30 am	Sun-synchronous circular orbit. In initial commissioning phase the satellite will have a 3 d repeat cycle.
ATMOS[5]) SCIAMACHY	1995	Series of channels 0.24 to 2.38	32×70 km^2 (at nadir)	Day-time observations	Primary objective is the measurement of trace gases in the troposphere and stratosphere.
ATMOS ROSIS	1995	Fifteen channels 0.4 to 1.05	250 m (at nadir)		Primary objectives: vegetation cover, type and stress, burning of biomass.

1) fires can either be under-represented, i.e. their occurrence can be totally missed, depending on the location and configuration of the fire in relation to the pixels sampled for GAC data, or, 2) fires can be over-represented in terms of areal coverage i.e. the importance of small fires can be exaggerated (Malingreau and Belward 1990)

2) How accurately can time series data depict the true return period or frequency of fires? Although it should be possible to determine the frequency of burn for each pixel on the basis of a simple fire/non-fire criteria, research is required to determine the representativeness of the temporal sample dictated by the satellite overpass time.

3) To what extent can satellite data be used to determine the areal extent of burnt land? Automated techniques need to be developed for burn scar identification. Results obtained from AVHRR data should be compared with the findings from higher spatial resolution LANDSAT or SPOT data, and with higher temporal resolution data such as from GOES, or METEOSAT, in order to evaluate the limitations of the AVHRR's low spatial and temporal resolution.

4) To what extent can satellite data be used to provide estimates of burning efficiency? Investigations need to be carried out to determine whether satellite data can be used to indicate the percentage of consumed biomass to unburnt material. Currently, such estimates are derived from very few field measurements, e.g. 30% for closed forest and 97% for savanna (Delmas 1982). It could be postulated that the near infrared albedo will decrease with the amount of black char left on the ground, and that this is linked to the burning efficiency. Comparative analyses between the observations obtained from AVHRR data, and data from sensors with higher spatial and temporal resolution would be essential.

5) To what extent can time series data be used to identify relative changes in the spatial and temporal distribution of total available fuel? This would require analysis of time series vegetation indices, or single channel (near-infrared) data, to assess the possibility of monitoring changes in vegetation status.

6) To what extent can satellites be used to measure the gaseous and particulate emissions from biomass burning? This requires investigations to establish more

accurately the relationships between the particulate characteristics measured from prescribed fires (Ward 1986), and the particulate characteristics derived from space. Direct measurements of emission products will be possible using future instruments such as the SCIAMACHY.

Sensor systems planned for the future

With reference to the issues discussed in the above section it is possible to identify the aspects which should be considered when planning future satellite systems. Some of the most important considerations are listed in Table 4. These are described in detail in the following sections, in relation to the proposals made for the satellites to be launched in the next decade.

Thermal data availability

Many of the sensors planned for future American space-based observing programmes e.g. EOS (Earth Observing System), or for the next generation of meteorological satellites, include mid-infrared channels for direct fire detection (Table 5). This is also true of the European Remote Sensing Satellite (ERS–1), which was launched in July 1991 (Table 5). Feasibility studies have also been conducted for placing a mid-infrared channel on the next DMSP (Defense Meteorological Satellite Program) Block launch. Although no commitment has been made, it is likely to be time-shared with a near-infrared channel, with the latter operating during the day. This capability will not, however, fill the gap left by the AVHRR omissions in mid-infrared coverage, since it is also proposed that channel–3 of the NOAA AVHRR–3 instrument (Table 5), should be time-shared between daytime cloud and ice observations at 1.58–1.64 μm, and nocturnal thermal sensing. The latter would obviously provide very little information for biomass burning studies.

The 30-min resolution of GOES–Next would provide excellent facilities to study diurnal fire patterns, and to identify and calibrate the biases inherent in fixed image acquisition times. Furthermore, if the proposal for GOES–Next to operate as two platforms in the western hemisphere is realized, then simultaneous operation of GOES–East and West will permit stereo viewing, which, among other things may enable smoke plume height determinations. Unfortunately, these improved qualities are counter-balanced by three factors which may restrict GOES–Next's utility for fire evaluations: 1) boreal forest fires could not be studied because image quality deteriorates above 50° north and south, 2) the 4 km resolution with 8-bit quantization may put small fires below the sensor's detection threshold – especially at the periphery of the image, and 3) data volumes will be enormous!

The experimental platform, GOES–Research (Table 5) could offer remedies to some of the deficiencies of GOES–Next if the proposed 3–min observation interval, plus multiple channel infrared coverage at resolutions of 2 to 0.5 km are all realized. Although the platform would still be confined to an equatorial stationary orbit, the design specifications would be ideal for obtaining detailed statistical information on the distribution of fire, both in time and space throughout the entire tropical belt.

Visible and near-infrared data availability

Vegetation analysis

Many of the future hopes to improve evaluation of the vegetation and atmospheric components of fire rely on the EOS and ERS–1 satellites. The Moderate Resolution Imaging Spectrometer (MODIS), which is proposed as part of the EOS programme, will provide complete and comprehensive 0.5 to 1 km resolution data for ocean, land and atmosphere remote sensing. It will have 500 m resolution in the visible channels, and 1 km resolution for other channels, over a 1500 km swath width for global coverage every two days. The instrument would consist of two optical units called MODIS–T (T=Tilt) and MODIS–N (N=Nadir). The former would comprise of 64 channels between 4 and 10.4 μm, the latter would contain 36 channels between 0.4 and 14.54 μm (NASA 1990). It has also been proposed that temperature saturation thresholds in the mid-infrared channels of MODIS–N should be increased to levels of 700 K (NASA EOS 1986).

The ROSIS (Reflective Optics System Imaging Spectrometer), a nadir looking instrument with fifteen spectral channels in the visible and near-infrared region, is planned as part of the ATMOS (Atmospheric Trace Molecules Observed by Spectroscopy) payload (DLR 1991), with a proposed launch date of 1995 (Table 5). High spatial resolution data (250 m at nadir), obtained from the ROSIS instrument is expected to provide information on trends in foliage cover, on vegetation stress, primary production and the cycle of carbon, water and trace gases (DLR 1991).

Atmospheric analysis

The only comprehensive project proposal which directly addresses the question of measuring global stratospheric and tropospheric trace gases from space is the SCIAMACHY (SCanning Imaging Absorption spectroMeter for Atmospheric CHemistrY) project (Burrows et al. 1990). SCIAMACHY is part of the German initiative ATMOS, which is currently undergoing Phase A studies for possible flight on the European Space

Agency's European Orbiting Polar Platform (EOPP). The main objective of SCIAMACHY is the measurement of the global abundances of such gases as O_3, NO_2 and CH_4. Column amounts of the gases will be determined by the use of differential absorption spectroscopy. Solar radiation from the majority of the wavelength range (between 2.40 and 24.00 μm) covered by SCIAMACHY penetrates to the Earth's surface. As a result, simultaneous measurements in limb and nadir (of column amounts), will enable the tropospheric and stratospheric abundances of atmospheric constituents to be retrieved.

Concluding comment

The ability to meet the challenges and desired research objectives outlined in this paper, will be largely dependent upon establishing an international commitment to co-ordinate the necessary laboratory experiments, field surveys, aircraft studies and analysis of satellite data (e.g. the IGAC project). An increased understanding of the spatial and temporal characteristics of fire, is an essential prerequisite, before it will be possible to utilize general circulation models to understand and predict, on a global scale, the transportation and fate of trace gases emitted from biomass burning.

Acknowledgements – I would like to thank J.-P. Malingreau, M. Verstraete, J.-M. Grégoire and A. Belward from the Joint Res. Centre of the EEC, for their valuable comments and suggestions in the preparation of this document.

References

Ackerman, T. P. and Toon, O. B. 1981. Absorption of visible radiation in atmosphere containing mixtures of absorbing and non-absorbing particles. – Appl. Optics 20: 3661–3668.

Abriol, R., Romeu, J. and Vinas, O. 1987. Deteccio i evaluacio de les superficies forestales cremades durant l'any 1984 a Catalunya, mitjan, çant tecniques de Teledeteccio. – Revista Catalana de Geografia 2: 29–46.

Andreae, M. O., Browell, E. V., Garstang, M., Gregory, G. L., Harris, R. C., Hill, G. F., Jacob, D. J., Pereira, M. C., Sachse, G. W., Setwer, A. W., Silver Dias, P. L., Talbot, R. W., Torres, A. L. and Wofsy, S. C. 1988. – Biomass burning emissions and associated haze layers over Amazonia. J. Geophys. Res. 93: 1509–1527.

Barber, J. and Paltridge, G. W. 1986. Fuel moisture content of vegetation from AVHRR – an operational fire potential monitoring system. – In: Proc. 1st Australian AVHRR conference, perth, 22–24 Oct 1986, pp. 184–193.

Belward, A. S. 1992. Spatial attributes of AVHRR imagery for environmental monitoring. – Int. J. Remote Sensing 13: 192–208.

– and Lambin, E. 1990. Limitations to the identification of spatial structures from AVHRR data. – Int. J. Remote Sensing 11: 921–927.

Benson, M. L. and Briggs, I. 1978. Mapping the extent and intensity of major forest fires in Australia using digital analysis of Landsat imagery. – In: Proc. of the int. symp. on remote sensing for observ. and inventory of earth resources, pp. 1965–1980.

Brasseur, G. and Verstraete, M. M. 1989. Atmospheric chemistry-climate interactions. – In: Berger, A., Schneider, S. and Duplessy, J. Cl. (eds), Climate and geo-sciences: a challenge for science and society in the 21st century. Kluwer Academic Publ., Dordrecht, pp. 279–302.

Burrows, J. P. Schneider, W., Perner, D. Crutzen, P. J., Harris, G. W., Moortgat, G. K., Chance, K. V., Geary, J. C., Goede, A. P. H., Aarts, H., Vreis de J., Smorenburg, C., Visser, H., Hoelze, E. and Pitz, W. 1990. – The SCIAMACHY project – global stratospheric and tropospheric trace gas sensing from space.

Chung, Y. S. and Le, H. V. 1984. Detection of forest-fire smoke plumes by satellite imagery. Atmos. Environ. 18: 2143–2151.

Choudhury, B. J. 1988. Relating Nimbus-7 37 GHz data to global land-surface evaporation, primary productivity and the atmospheric CO_2 concentration. – Int. J. Remote Sensing 9: 169–176.

Chuvieco, E. and Congalton, R. G. 1988. Using cluster analysis to improve the selection of training statistics in classifying remotely sensed data. – Photogram. Eng. Remote Sensing 54: 1275–1281.

Citeau, J., Demarcq, H., Mahé, G. and Franc, J. 1988. Une Nouvelle Station est neé. – Veille Climatique Satellitaire, pp. 23–29.

Cosentino, M. J., Woodcock, C. E. and Franklin, J. 1981. Scene analysis for wildland fire-fuel characteristics in a Mediterranean climate. – Proc. 15th Int. Symp. on Remote Sensing of Environ., pp. 635–646.

Crutzen, P. J., Delany, A. C., Greenberg, J., Haagenson, P., Heidt, L., Lueb, R., Pollock, W., Seiler, W., Wartburg, A. and Zimmerman, P. 1985. Tropospheric chemical composition measurements in Brazil during the dry season. – J. Atmos. Chem. 2: 233–256.

Deeming, J. E., Burgan, R. E. and Cohen, J. D. 1978. The National Fire Danger System. – U.S. Dept of Agric., Forest Service, Ogden.

Delmas, R. 1982. On the emission of carbon, nitrogen and sulfur in the atmosphere during bushfires in inter-tropical savanna zones. – Geophys. Res. Lett. 9: 761–764.

DLR – Oberpfaffenhofen. 1991. Atmospheric and environmental research satellite ATMOS. – Document Number DLR-ATMOS-MR/91/1–02, Jan 1991.

Dozier, J. 1981. A method for the satellite identification of surface temperature fields of sub-pixel resolution. – Remote Sensing of Environ. 11: 221–229.

Fishman, J., Minnis, O. P. and Reichle H. G. Jr. 1986. Use of satellite data to study ozone in the tropics. – J. Geoph. Res. 91: 14451–4465.

Flannigan, M. D. and Vonder Haar, T. H. 1986. Forest fire monitoring using NOAA satellite AVHRR. – Can. J. For. Res. 16: 975–982.

Francis, P. W. and Rothery, D. A. 1987. Using the Landsat Thematic Mapper to detect and monitor active volcanoes: an example from Lascar volcano, northern Chile. – Geology 15: 614–617.

Fraser, R. S., Kaufman, Y. J. and Mahoney, R. L. 1984. Satellite measurements of aerosol mass and transport. – J. Atmos. Environ. 18: 2577–2584.

Frederiksen, P. S., Langaas, S. and Mbaye, M. 1990. NOAA AVHRR and GIS-based monitoring of fire activity in Senegal – a provisional methodology and potential applications. – In: Goldammer J. G. (ed.), Fires in tropical biota. Ecol. Studies 82, Springer, Berlin, pp. 400–417.

Fung, I. Y., Tucker, C. J. and Prentice, K. C. 1987. Application of advanced very high resolution radiometer vegetation index to study atmosphere–biosphere exchange of CO_2. – J. Geoph. Res. 92: 2999–3051.

Gatlin, J. A. Tucker, C. J. and Schneider, S. R. 1981. Use of NOAA–6 AVHRR channels 1 and 2 for monitoring vegetation. – In: Proc. of the 1981 int. geoscience and remote sensing symp., Inst. of Elect. and Electronics Eng., New York, p. 1154.

Greenberg, J. P., Zimmerman, P. R., Heidt, L. and Pollock, W. 1984. Hydrocarbon and carbon monoxide emissions from biomass burning in Brazil. – J. Geoph. Res. 89: 1350–1354.

Grégoire, J. M. 1990a. Effects of the dry season on the vegetation canopy of some river basins of West Africa as deduced from NOAA–AVHRR data. – Hydrol. Sci. J. 35: 323–338.

– 1990b. Watershed degradation – use of thermal data and vegetation indices as indicators of environmental changes – hydrological implications of changes in land surface cover. – In: Belward, A. and Valenzuela, C.R. (eds), Euro course on 'Remote sensing and geographical information systems for resource management in developing countries'. Kluwer Academic Publ., pp. 235–252.

Hall, D. K., Ormsby, J. P., Johnson, L. and Brown, J. 1980. Landsat digital analysis of the initial recovery of burned tundra at Kokolik river, Alaska. – Remote Sensing of the Environ. 10: 263–272.

Hao, W. M., Liu, M.-H. and Crutzen, P. J. 1990. Estimates of annual and regional fire releases of CO_2 and other trace gases to the atmosphere from fires in the tropics, based on the FAO statistics for the period 1975–1980. – In: Goldammer, J. G. (ed.), Fires in tropical biota. Ecol. Studies 82, Springer, Berlin, pp. 440–462.

Hick, P. T., Prata, A. J., Spencer, G. and Cambell, N. 1986. NOAA–AVHRR satelite data evaluations for areal extent of bushfire damage in western Australia. – In: Proc. 1st Australian AVHRR conf. Perth, 22–24 Oct 1986.

IGBP Report No. 13. 1990. Terrestrial biosphere exchange with global atmospheric chemistry. Terrestrial biosphere perspective of the IGAC project: companion to the Dookie report. – IGBP of the Int. Council of Scientific Unions (ICSU), Stockholm.

Justice, C. O. (ed.), 1986. Monitoring the grasslands of semi-arid Africa using NOAA–AVHRR data. – Int. J. Remote Sensing 7: 1–239.

– , Townshend, J. R. G., Holben, B. N. and Tucker, C. J. 1985. Analysis of the phenology of global vegetation using meteorological satellite data. – Int. J. Remote Sensing 6: 1271–1318.

Kaufman, Y. J. 1987. Satellite sensing of aerosol absorption. – J. Geoph. Res. 92: 4307–4317.

– and Sendra, C. 1988. Algorithm for atmospheric corrections. – Int. J. Remote Sensing 9: 1357–1381.

– , Tucker, C. J. and Fung, I. 1988. Remote sensing of biomass burning – Method. – In: Proc. of int. radiation symp. 18–24 Aug 1988, Lille, France.

– , Tucker, C. J. and Fung, I. 1989. Remote sensing of biomass burning in the tropics. – Advanced Space Res. 9: 265–268.

– , Setzer, A., Justice, C., Tucker, C. J. and Fung, I. 1990. Remote sensing of biomass burning in the tropics. In: Goldammer, J. G. (ed.), Fires in tropical biota. Ecol. Studies 82, Springer, Berlin, pp. 371–399.

Kerber, A. G. and Schutt, J. B. 1986. Utility of AVHRR channels 3 and 4 in land-cover mapping. – Photogram. Eng. Remote Sensing 52: 1877–1883.

Langaas, S. 1989. A Study of spectral characteristics of burnt areas in Senegal, with recommendations for an AVHRR based bushfire monitoring methodology. – Dept of Surveying, Agric. Univ. of Norway, Aas.

– and Muirhead, K. 1988. Monitoring bush fires in West Africa by weather satellites. – In: Proc. of 22nd int. symp. on remote sensing of environ. 20–26 Oct 1988, Abidjan, Ivory Coast, 2: 563–574.

Lee, T. F. and Tag. P. M. 1990. Improved detection of hotspots using the AVHRR 3.7 μm channel. – Bull. Am. Meteorol. Soc. 71: 1722–1730.

Malingreau, J. P., Stephens, G. and Fellows, L. 1985. Remote sensing of forest fires: Kalimantan and North Borneo in 1982–83. – Ambio. 6: 314–321.

– 1986. Global vegetation dynamics: satellite observations over Asia. – Int. J. Remote Sensing 7: 1121–1146.

– 1990. The contribution of remote sensing to the global monitoring of fires in tropical and subtropical ecosystems. – In: Goldammer, J. G. (ed.), Fires in tropical biota. Ecol. Studies 82, Springer, Berlin, pp. 337–370.

– and Tucker, C. J. 1987. The contribution of AVHRR data for measuring and understanding global processes: large-scale deforestation in the Amazon basin. – In: Proc. of IGARSS 1987 Symp. Ann Arbor 18–21 May 1987, pp. 443–448.

– and Tucker, C. J. 1988. Large scale deforestation in the southeastern Amazon basin of Brazil. – Ambio. 17: 49–55.

– , Tucker, C. J. and Laporte, N. 1989. AVHRR for monitoring global tropical deforestation. – Int. J. Remote Sensing 10: 855–867.

– and Belward, A. S. 1990. Vegetation monitoring using AVHRR data at different resolutions. – In: Proc. of 4th AVHRR data users' conf., Rothenburg ob der Tauber, Federal Republic of Germany, 5–8 Sept 1989, pp. 141–144.

Matson, M. and Dozier, J. 1981. Identification of subresolution high temperature sources using a thermal IR sensor. – Photogram. Eng. Remote Sensing 47: 1311–1318.

– , Schneider, S. R., Aldridge, B. and Satchwell, B. 1984. Fire detection using the NOAA-series satellites. – NOAA Techn. Rep. NESDIS 7, Dept of Commerce, Washington, D.C.

– and Holben, B. 1987. Satellite detection of tropical burning in Brazil. – Int. J. Remote Sensing 8: 509–516.

– , Stephens, G. and Robinson, J. 1987. Fire detection using data from the NOAA–N satellites. – Int. J. Remote Sensing 8: 961–970.

McKinley, R. A., Chine, E. P. and Werth, L. F. 1985. Operational fire fuels mapping with NOAA–AVHRR data. – Pecora X Symp., pp. 295–304.

Miller, W. A., Howard, S. M. and Moore, D. G. 1986. Use of AVHRR data in an information system for fire management in the western United States. – In: Proc. 20th int. symp. on remote sensing of environ. 4–10 Dec 1986, Nairobi, Kenya, pp. 67–79.

Milne, A. K. 1986. The use of remote sensing in mapping and monitoring vegetational change associated with bushfire events in eastern Australia. – Geocarto Int. 1: 25–34.

Muirhead, K. and Cracknell, A. P. 1984. Identification of gas flares in the North Sea using satellite data. – Int. J. Remote Sensing 5: 199.

– and Cracknell, A. P. 1985. Straw burning over Great Britain detected by AVHRR. – Int. J. Remote Sensing 6: 827–833.

NASA EOS. 1986. Earth observing system science and mission requirements working group report, Vol. 1. Tech. Mem. 86129, NASA Goddard Space Flight Centre, Greenbelt, MD.

NASA. 1990. EOS – Earth observing system reference handbook, NASA, Goddard Space Flight Center, Greenbelt, MD.

Nelson, R. and Holben, B. 1986. Identifying deforestation in Brazil using multiresolution satellite data. – Int. J. Remote Sensing 7: 429–448.

Paltridge, G. W. and Barber, J. 1988. Monitoring grassland dryness and fire potential in Australia with NOAA/AVHRR data. – Remote Sensing of Environ. 25: 381–394.

Parnot, J. 1988. Inventaire des feux de brousse au Burkina Faso saison seche 1986–1987. – In: Proc. of the 22nd symp. on remote sensing of environ. 20–26 Oct 1988. Abidjan, Ivory Coast, Vol. 2, pp. 563–574.

Pereira, M. C. and Setzer, A. W. 1991. Spectral characteristics of deforestation fires in NOAA/AVHRR images. – Int. J. Remote Sensing, (in press).

– , Setzer, A. W. and dos Santos, J. R. 1991. Fire estimates in savannas of central Brazil with thermal AVHRR/NOAA calibrated by TM/LANDSAT. – Proc. 24th int. symp. on remote sensing of environ., Rio de Janeiro, Brazil, 27–31 May 1991.

Prince, S. D. and Tucker, C. J. 1986. Satellite remote sensing of rangelands in Botswana, II; NOAA–AVHRR and herbaceous vegetation. – Int. J. Remote Sensing 7: 1555–1570.

Reichle, H. G. Jr., Conners, V. S., Holland, J. A., Hypes, W. D., Wallio, H. A., Casas, J. C., Gormsen, B. B., Saylor, M. S. and Hasketh, W. D. 1986. Middle and upper tropospheric carbon monoxide mixing ratios as measured by a satellite-borne remote sensor during November, 1981. – J. Geophys. Res. 91: 10865–10887.

Robinson, J. M. 1987. The role of fire on earth: a review of the state of knowledge and a systems framework for satellite and ground-based observations. – NCAR: CT-112. Coop. Thesis No. 112. NCAR, Boulder, CO.

Root, R. R., Stitt, S. C. F., Nyquist, M. O., Waggoner, G. S. and Agee, J. K. 1986. Vegetation and fire fuel models mapping of North Cascades National Park. – ACSM-ASPRS Annual Convention. Tech. Papers 3: 78–85.

Rothery, D. A., Francis, P. W. and Wood, C. A. 1988. Volcanic monitoring using short wavelength infrared data from satellites. – J. Geoph. Res. 93: 7993–8008.

Salazar, L. A. 1982. Remote sensing techniques aid in preattack planning for fire management, Pacific southwest forest and range experiment station, Berkeley.

Schutt, J. B., Holben, B. N. and Kerber, A. G. 1986. Separating emitted and reflected components in AVHRR channel–3 for vegetation–soil discrimination. – Photogram. Eng. Remote Sensing 52: 1877–1883.

Segal, M., Weaver, J. and Purdom, J. F. W. 1989. Some effects of the Yellowstone fire smoke plume on northeast Colorado at the end of summer 1988. – Mon. Weather Rev. 117: 2278–2284.

Seiler, W. and Crutzen, P. J. 1980. Estimates of gross and net fluxes of carbon between the biosphere and the atmosphere from biomass burning. – Climatic Change 2: 207–247.

Sellers, P. J. 1985. Canopy reflectance, photosynthesis and transpiration. – Int. J. Remote Sensing 6: 1335–1372.

Setzer, A. W. 1988. Relatorio de atividades do projeto IBDF-INPE "SEQE" ANO 1987. – INPE publ., INPE-4537 RPE/565. Sao Jose dos Campos, Brazil.

– and Periera, M. C. 1991. Amazonia biomass burnings in 1987 and an estimate of their tropospheric emissions. – Ambio. 20: 19–22.

Shasby, M. B., Burgan, R. R. and Johnson, R. R. 1981. Broad area forest fuels and topography mapping using digitial Landsat and terrain data. – Proc. 7th int. symp. on machine processing of remotely sensed data, pp. 529–537.

Stephens, G. and Matson, M. 1987. Regional and global fire detection using AVHRR data. – In: Proc. of 21st int. symp. on remote sensing of environ., Ann Arbor Michigan, Oct 26–30 1987.

Tanaka, S., Kimura, H. and Suga, Y. 1983. Preparation of a 1:25 000 Landsat map for assessment of burnt area on Etajima Island. – Int. J. Remote Sensing 4: 17–31.

Townshend, J. R. G. and Justice, C. O. 1986. Analysis of the dynamics of African vegetation using the normalized difference vegetation index. – Int. J. Remote Sensing 7: 1435–1445.

– and Tucker, C. J. 1981. Utility of AVHRR data of NOAA–6 and –7 for vegetation mapping. – In: Allan, J. A. (ed.), Matching remote sensing technologies and their applications. Remote Sensing Soc., Reading, pp. 97–109.

Tucker, C. J. 1979. Red and photographic infrared linear combinations for monitoring vegetation. – Remote Sensing of Environ. 8: 127–150.

– , Holben, B. N., Elgin, J. H. Jr. and McMurtrey, J. E. III. 1981. Remote sensing of total dry-matter accumulation in winter wheat. – Remote Sensing of Environ. 11: 171–189.

– , Gatlin, J. A. and Schneider, S. R. 1984. Monitoring vegetation in the Nile delta with NOAA–6 and NOAA–7, AVHRR. – Photogram. Eng. Remote Sensing 50: 53–61.

– , Townshend, J. R. G. and Goff, T. E. 1985. African land-cover classification using satellite data. – Science. 227: 369–375.

– and Sellers, P. J. 1986. Satellite remote sensing of primary production. – Int. J. Remote Sensing 7: 1395–1416.

Verstraete, M. M., Belward, A. S. and Kennedy, P. J. 1990. The institute for remote sensing applications contribution to the global change research programme. – Proc. of 16th annual conf. of the remote sensing soc. Swansea, Sept 1990, Remote Sensing Soc., Nottingham, U.K. pp. vi-xxii.

Ward, D. E. 1986. Field scale measurements of emission from open fires. – Technical paper presented at the defense nuclear agency global effects review, Defense Nuclear Agency, Washington.

– 1990. Factors influencing the emissions of gases and particulate matter from biomass burning. – In: Goldammer, J. G. (ed.), Fires in tropical biota. Ecol. Studies 82, Springer, Berlin, pp. 418–436.

– and Hardy, C. C. 1984. Advances in the characterization and control of emissions from prescribed fires. – In: Proc of 77th annual meeting of the air poll. cont. ass. San Francisco.

Werth, L. F., McKinley, R. A. and Chine, E. P. 1985. The use of wildland fire fuel maps produced with NOAA–AVHRR scanner data. – Pecora X Symp., pp. 326–331.

Ecological Bulletins 42: 149–155. Copenhagen 1992

Pulses of nitric oxide and nitrous oxide flux following wetting of dry soil: an assessment of probable sources and importance relative to annual fluxes

Eric A. Davidson

Davidson, E. A. 1992. Pulses of nitric oxide and nitrous oxide flux following wetting of dry soil: an assessment of probable sources and importance relative to annual fluxes. – Ecol. Bull. (Copenhagen) 42: 149–155.

Large pulses of emissions of NO and N_2O have been observed following wetting of dry soil. Ecosystems that experience prolonged dry seasons cover over 28 mill. km^2, suggesting that pulses of soil emissions early in the wet season at these sites warrant consideration in regional and global budgets of NO and N_2O. Most studies have shown that the magnitude of pulses diminishes with successive wetting events. Soil inorganic-N was also readily available at the time of first wetting in each of the ecosystems studied. Pulses of NO usually exceeded pulses of N_2O. Nitrifier denitrification appears to be the dominant source of N_2O when soil moisture does not exceed field capacity. Nitrification or a combination of NO_2^- production via nitrification and abiological reactions involving NO_2^- is the dominant source of NO. Death of microbial biomass (with subsequent release of readily available N and C following wetting) and accumulation of soil NO_2^- during droughts probably contribute to pulses of NO and N_2O production following wetting. Where post-wetting pulses have been compared to annual flux budgets, pulses of N_2O appear insignificant, whereas pulses of NO appear important.

E. A. Davidson, Woods Hole Res. Center, P.O. Box 296, Woods Hole, MA 02543, USA.

Introduction

Several biomes experience alternating wet and dry seasons and extreme fluctuations in soil moisture, including chaparral and grasslands of Mediterranean climates, seasonally dry tropical forests, and tropical savannas. The combined area of ecosystems with prolonged dry seasons (not counting deserts and semi-deserts) is over 28 mill. km^2 (Bolin et al. 1979). Large fluxes of NO and N_2O immediately following wetting of dry soil have become a well observed phenomenon in these ecosystems (Table 1). Hence, early wet season rains across this large collective area could result in soil emissions of NO that affect regional tropospheric ozone chemistry (Crutzen 1979) and soil emissions of N_2O that contribute to global warming (Lashof and Ahuja 1990) and to destruction of stratospheric ozone (Cicerone 1987). Unfortunately, the biological and abiological causes of these pulse emissions are poorly understood, and the importance of observed pulses following wetting events relative to annual budgets of NO and N_2O emissions are generally unknown. However, several consistent patterns emerge from the literature which shed light on sources of NO and N_2O and the relative magnitudes of post-wetting production. The goals of the present review are to explore the probable controlling mechanisms and to assess the importance of pulses of NO and N_2O following wetting of dry soil.

Common features of observations

Not surprisingly, most studies of wetting events have been conducted in climates where dry periods are common. The Mediterranean climates of Spain (Slemr and Seiler 1984) and California USA (Anderson et al. 1988, Davidson 1992), the savanna and woodlands of Venezuela (Hao et al. 1988, Johansson et al. 1988, Johansson and Sanhueza 1988), and seasonally dry tropical forests of Mexico (Davidson et al. 1991) experience prolonged

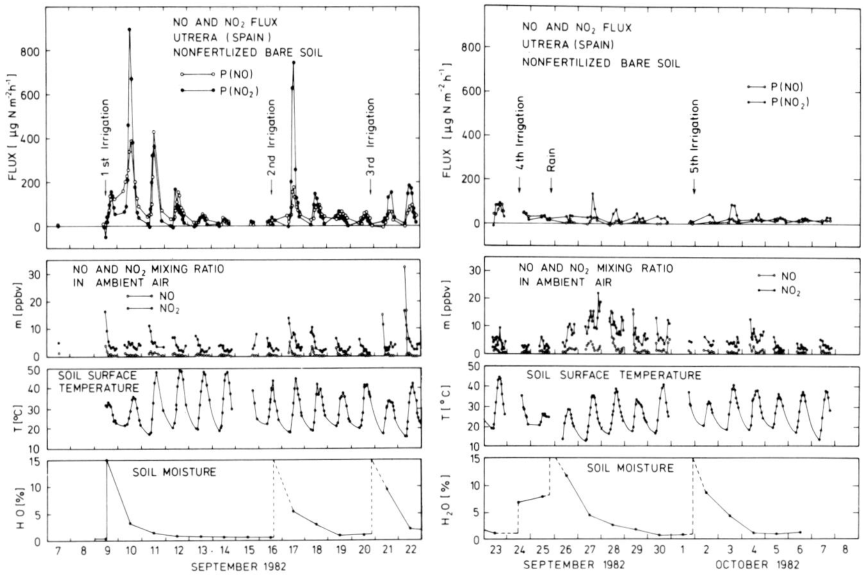

Fig. 1. Variation of the NO and NO_2 flux rates measured in Utrera on non-fertilized soils without vegetation. The lower part of the figure shows the temporal variation of NO and NO_2 mixing ratios in ambient air as well as the variation of soil surface temperature and soil moisture. (From Slemr and Seiler 1984. Reproduced with kind permission of Kluwer Academic Publ. © Kluwer Academic Publ.).

dry seasons when the soil dries to below 3% gravimetric moisture. Temperate grasslands (Williams et al. 1987), forests (Davidson and Swank 1986) and agricultural fields (Mosier and Hutchinson 1981, Anderson and Levine 1987) experience more brief summer droughts that also can result in significant soil drying and wetting cycles.

A common feature of many of these studies is that the largest pulse of NO and/or N_2O flux was observed immediately after the first wetting event following a dry period, and that subsequent wetting events produced smaller pulses (Mosier and Hutchinson 1981, Slemr and Seiler 1984, Williams et al. 1987, Johansson et al. 1988, Davidson et al. 1991). An example of the effects of repeated irrigation and rainfall at a bare soil site in Spain is given in Fig. 1. Soil moisture alone is not a good

Table 1. Comparison of wetting effects on NO and N_2O fluxes in several ecosystems.

Ecosystem/location	Treatment	Nitric oxide (NO) dry	Nitric oxide (NO) wet	Nitrous oxide (N_2O) dry	Nitrous oxide (N_2O) wet
		ng N cm^{-2} h^{-1}			
[a]Chaparral/California USA	unburned	4	8	ND	ND
	burned	5	22	<2	19
[b]Grassland/California USA	dry	0.4		0.0	
	wetted < field capacity		18		9
	wetted > field capacity		5		42
[c,d]Savanna/Venezuela		2	35	0.4	1.6
[e,f]Seasonal forest/Mexico		0.4	57	0.2	12
[g]Grassland/Colorado USA		1	10	ND	ND
[h]Cropland/Spain	unfertilized	10	40	ND	ND
	fertilized	5	140	ND	ND
[i]Cropland/Virginia USA		25	10	120	250
[j]Cropland/Colorado USA		ND	ND	2	400

[a] Anderson et al. 1988; continuous drip was used for N_2O fluxes; a single wetting of 1.3–2.5 cm was used for NO fluxes
[b] Davidson unpubl. data; intact soil cores
[c] Johansson and Sanhueza 1988; data for NO
[d] Hao et al. 1988; data for N_2O
[e] Davidson et al. 1991; data for NO
[f] Vitousek et al. 1989; data for N_2O
[g] Williams et al. 1987
[h] Slemr and Seiler 1984
[i] Anderson and Levine 1987
[j] Mosier and Hutchinson 1981
ND: Not Determined

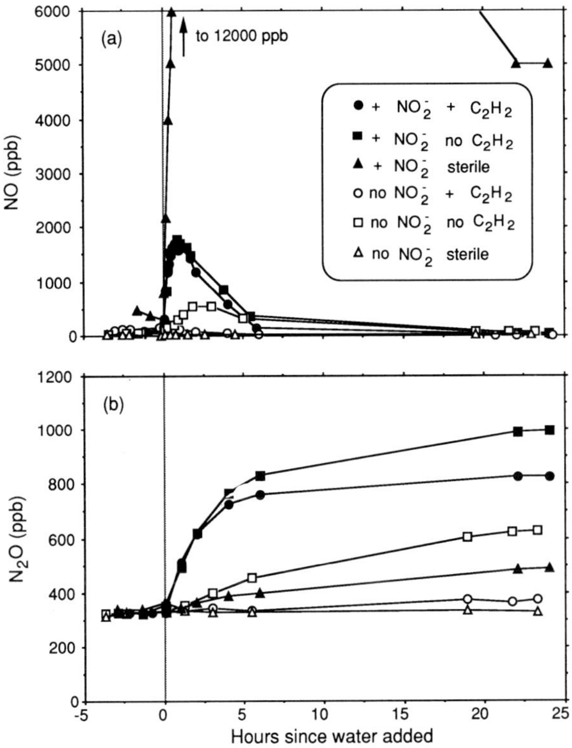

Fig. 2. Profiles of NO (a) and N_2O (b) concentrations in the headspace gas (0.5 l) over sieved soil (30 g) from a California grassland wetted to 33% gravimetric moisture (below field capacity), with sterilization (autoclaving twice), previous exposure to C_2H_2 (10 kPa C_2H_2), and KNO_2 addition (2 µg NO_2-N g^{-1} dry soil) imposed in a factorial design. Data shown are for samples from a single plot. Analysis of variance of data of samples from 3 plots indicate that the effects of sterilization, C_2H_2, NO_2^-, and NO_2^--by-treatment interactions were significant at p = 0.01 (Davidson 1992).

predictor of N-gas fluxes; the duration of the antecedent period of dryness also appears to be a critical controlling factor.

Another common feature of most of these studies is that inorganic-N was readily available at the time of wetting. In the case of agricultural soils, fertilizer applications occurred prior to or concurrent with wetting experiments (Mosier and Hutchinson 1981, Slemr and Seiler 1984). In a temperate hardwood forest dominated by an N-fixing species, NO_3^- exported from the system via stream water was elevated relative to control watersheds (Davidson and Swank 1986). Extractable (KCl) inorganic-N (µg N g^{-1} dry soil) ranged from 8 to 130 for NH_4^+, 2 to 14 for NO_3^-, and up to 3.6 for NO_2^- in a Venezuelan savanna (Johansson and Sanhueza 1988), a Mexican seasonally dry forest (Davidson et al. 1991), a Californian chaparral (Anderson et al. 1988), and a Californian grassland (Davidson 1992). Although the causes of large inorganic-N pools may differ among ecosystems (i.e., fertilization, N-fixing species, burning and lack of N-limitation in the ecosystem), inorganic-N did not appear to be limiting to microbial processes at the time of wetting.

Possible sources of NO and N_2O

Several biological and abiological sources of NO and N_2O have been identified (Firestone and Davidson 1989, and references therein), but the most probable sources of large fluxes are: 1) classical denitrification: a sequence of nitrogen oxide reduction ($NO_3^- \rightarrow NO_2^- \rightarrow NO \rightarrow N_2O \rightarrow N_2$) usually linked to respiratory electron transport and ATP generation: 2) nitrifier denitrification: reduction of NO_2^- to N_2O and perhaps to NO by NH_4^+ oxidizing bacteria: 3) nitrifier oxidation: intermediates or by-products of intermediates in the pathway of NH_4^+ oxidation to NO_2^- may include NO or perhaps N_2O: 4) abiological reactions of NO_2^- with soil organic matter: probably a more important source of NO than N_2O (Blackmer and Cerrato 1986): and 5) chemodenitrification: abiological self decomposition of HNO_2 to NO + NO_2 (Nelson 1982).

If biological processes require time to recover from drought stress, the immediate pulse of NO and N_2O flux commonly observed following wetting of extremely dry soil might suggest an abiological source. However, this interpretation is based more on a teleological assumption that soil microorganisms need time to "wake up" after a long, stressful drought, rather than any direct evidence. Persistence of denitrifying enzymes in air-dried soil has been demonstrated (Smith and Parsons 1985), and de novo synthesis of denitrifying enzymes has been measured within 4 to 8 hours post-wetting (Rudaz et al. 1991). Survival of nitrifying bacteria in very dry soil has also been demonstrated, and production of NO_2^- and NO_3^- has been detected within 2 hours of wetting (Davidson et al. 1990). Finally, production of NO and N_2O that can be inhibited by sterilization or exposure to C_2H_2 and therefore is clearly biological, has been observed within minutes of wetting (Fig. 2).

In soils from regions with a Mediterranean climate, a cycle of severe drying (≤ -2.8 MPa matric potential) and wetting caused release of between 17 and 70% of the microbial biomass carbon (Bottner 1985, Kieft et al. 1987). This release of carbon demonstrates the importance of microbial turnover as a source of available C and N, but it also demonstrates that at least one-third of the microbial biomass survives extreme desiccation stress. A viable population appears capable of becoming metabolically active and producing N trace gases immediately upon wetting.

Evidence for biological sources does not rule out abiological sources or interactions of biological and abiological processes that might lead to NO and N_2O production. Indeed, addition of NO_2^- to sterile soil caused a small increase in N_2O production and increased NO production by a factor of 100 or more (Fig. 2). These results must be interpreted with caution, because auto-

Table 2. Effects of C_2H_2 treatment and identification of sources of NO and N_2O in laboratory experiments using intact soil cores sampled during the dry season from a seasonally dry tropical forest of Mexico and wetted to 3 moisture contents. Rates of net production were calculated from linear regression of changes in headspace gas concentration during the first 3 hours post-wetting. Means (and standard errors) of 3 replicates are given (Davidson et al. unpubl. data).

Soil moisture	Nitric oxide (NO) A. no C_2H_2 (all sources)	Nitric oxide (NO) B. 10 Pa C_2H_2 (denitrification)	Nitric oxide (NO) A-B[a] (nitrification)	Nitrous oxide (N_2O) A. no C_2H_2 (all sources)	Nitrous oxide (N_2O) B. 10 Pa C_2H_2 (denitrification)	Nitrous oxide (N_2O) A-B[a] (nitrification)
H_2O g^{-1} dry soil	ng N g^{-1} dry soil h^{-1}					
0.02 ambient	0.002 (0.001)	0.000 (0.001)	0.002 (0.001)	0	0	0
0.14 < f.c.[b]	0.663 (0.148)	<0[c]	0.663 (0.148)	0.366 (0.086)	0.016 (0.014)	0.350 (0.097)
0.27 < f.c.	0.528 (0.102)	<0[c]	0.528 (0.102)	0.494 (0.184)	0.089 (0.020)	0.405 (0.204)
0.45 > f.c.	0.290 (0.062)	0.055 (0.056)	0.235 (0.118)	15.697 (3.636)	31.678 (2.988)[d]	?[e]

[a] means are subtracted and error terms are added
[b] field capacity (f.c.) is defined as the soil moisture content at which micropores are water-filled, but macropores are air-filled
[c] net consumption of NO was observed
[d] some inhibition of N_2O reduction may have occurred
[e] given the large means and standard errors for denitrification-N_2O, a modest source of N_2O from nitrification could have existed but not have been detected by this method of C_2H_2 inhibition.

claving is known to eliminate ambient soil NO_2^- and to alter structures of soil organic matter. Nevertheless, the potential for significant abiotic NO production is clear. In nonsterile soil, production of NO_2^- and H^+ by NH_4^+ oxidizing bacteria could create microsites where $[NO_2^-]$ and acidity favor self decomposition of HNO_2 to NO + NO_2 (Davidson 1992). In dry soil, ions are concentrated in thin water films, possibly further enhancing concentration-dependent reactions involving NO_2^-. Adsorption of NO, NO^+, and NO_2^- to soil surfaces has been noted (Mortland 1965), but the significance of this adsorption and of potential desorption of during wetting events is unknown.

Identification of sources of NO and N_2O

Sources can be inferred from responses to different types of fertilizer amendments. For example, Johansson (1984) found enhanced NO fluxes from forest soil following addition of NO_3^-, suggesting a denitrifying source. In contrast, Slemr and Seiler (1984) found that NO fluxes increased in the order of the following amendments: none, $NaNO_3$, NH_4NO_3, NH_4Cl, and urea. These data suggest that NO_3^- could be a substrate for NO production, but that oxidation of NH_4^+ was the dominant process related to NO production. Unfortunately, interpretation of these results is unclear because a number of possible N transformations confuse the sources. Several groups of heterotrophic microorganisms are capable of reducing NO_3^- to NO_2^-, and NO_2^- could then be the substrate for NO and N_2O production via the pathways listed above. Similarly, oxidation of NH_4^+ may produce NO directly, or it may be indirectly related to NO production by first producing NO_2^- and NO_3^-.

More direct evidence is provided by using C_2H_2 or nitrapyrin to selectively inhibit the first step of NH_4^+ oxidation by nitrifying bacteria. (Tortoso and Hutchinson 1990, Davidson 1992). In a California grassland and a seasonally dry tropical forest of Mexico, exposure of sieved soils and intact soil cores to C_2H_2 in the laboratory resulted in inhibition of net NO and N_2O production (Fig. 2, Table 2). A source of NO was observed during the first 30 to 60 min after wetting that was not sensitive to C_2H_2, but thereafter, net NO production was strongly dependent upon sustained nitrification in the absence of C_2H_2. No depletion of ambient soil NO_3^- was observed, so inhibition of NO production was directly related to inhibition of NH_4^+ oxidation and/or NO_2^- production, and not due to indirect inhibition of denitrification via NO_3^- depletion. Nitrifying bacteria could be the direct source of NO either via a by-product of the oxidative pathway or via nitrifier denitrification. An equally plausible source is a combination of NO_2^- production by nitrifying bacteria followed by NO-producing abiological reactions of NO_2^- with soil organic matter or self decomposition of HNO_2.

Production of N_2O was also sensitive to C_2H_2 (Table 2), and responses to NO_2^- addition were greatest in nonsterile soils (Fig. 2b). Furthermore, N_2O production via nitrification is usually positively correlated with soil moisture (Table 2, Goodroad and Keeney 1984), suggesting a reductive pathway at microsites of low pO_2. Hence, nitrification is the probable source of N_2O (Table 2), and reduction of NO_2^- (nitrifier denitrification) is the probable pathway. Denitrification was a major source of N_2O and became a very modest source of NO only when the wetting treatment raised soil moisture above field capacity (i.e., some of the macropores were water-filled). Production of N_2O was positively related to soil moisture, while net NO production was negatively related to soil moisture (Table 2).

These laboratory results are consistent with field data

where both NO and N_2O fluxes were measured during wetting experiments in the chaparral of California, USA, the savanna of Venezuela, a seasonally dry forest of Mexico, and an agricultural site in Virginia, USA. In most cases, NO fluxes exceeded N_2O by a factor of 2 to 20 in moderately moist soil (Table 1). Fluxes of N_2O exceeded NO fluxes only in two cases where wetting increased soil moisture to near or above field capacity: a California chaparral site that was continuously wetted with drip irrigation and a Virginia site that had 18% gravimetric moisture before the precipitation event.

Combining the results of laboratory C_2H_2 treatments with observations from field studies, the following generalizations can be inferred: 1) after moderate soil wetting (i.e., soil moisture below field capacity), NO fluxes typically exceed N_2O fluxes and both result from or are dependent upon nitrification. Nitrifier denitrification is the dominant source of N_2O. The dominant source of NO is nitrification or a combination of NO_2^- production via nitrification and abiological reactions of NO_2^-. 2) Wetting events that elevate soil moisture above field capacity enhance N_2O fluxes more than NO fluxes. Some NO flux related to nitrification remains and perhaps a small contribution from denitrification occurs. Denitrification is the dominant source of N_2O.

Although NO production by denitrifying bacteria has been clearly demonstrated in the laboratory (Johansson and Galbally 1984, McKenney et al. 1982, Remde et al. 1989), identification of a significant denitrification source for NO in field settings is less certain. In laboratory studies where NO is produced under anaerobic conditions, the NO is purged from the soil by rapid mass flow of a carrier gas. As the flow rate is decreased, thus allowing more time for NO consumption, net NO production also decreases. In field settings, movement of gases through soil occurs primarily via diffusion and not via mass flow. Denitrification occurs when soil moisture partially impedes O_2 diffusion through pore spaces, thus creating microsites of anaerobiosis within the soil. Soil moisture contents that restrict inward diffusion of O_2 would also restrict outward diffusion of NO and increase the probability of NO consumption.

Unfortunately, a more specific mathematical function or model relating soil moisture to fluxes of NO and N_2O cannot be derived from the literature. Soil moisture is most commonly expressed gravimetrically, which does not afford comparisons of aeration across sites with widely varying soil texture. At present, we can only make generalizations about "moderately moist" and "very wet" soils or "above and below field capacity." It would be interesting to see if data on soil matric potential and/or water-filled pore space would permit prediction of ratios of NO and N_2O flux as a function of soil moisture and aeration in a variety of ecosystem types.

Causes of pulses of NO and N_2O fluxes following wetting

The duration of an antecedent dry period appears to be a key clue to the cause of trace gas pulses following wetting. Death and mineralization of between 17 and 70% of the microbial biomass (Bottner 1985, Kieft et al. 1987) releases readily available C and NH_4^+. Extractable NH_4^+ increased from 12 to 17 µg N g^{-1} dry soil during the first two hours following soil wetting during the dry season at a seasonally dry forest of Mexico (Davidson unpubl. data). If nitrification is the dominant source of NO and N_2O, then increased rates of nitrification in response to increased NH_4^+ availability is an obvious mechanism.

Increased available-C could also play an important role. Where denitrification is a dominant source of N_2O in very wet soil, and even where it is a minor source of N_2O in moderately moist soils, readily available organic-C is needed as a substrate for denitrifying bacteria and for all heterotrophic bacteria that contribute to O_2 consumption. Consumption of O_2 by heterotrophs caused by increased available-C would also affect nitrifier denitrification. The ratio of $N_2O:NO_2^-$ produced by NH_4^+ oxidizing bacteria increases as pO_2 decreases (Goreau et al. 1980, Goodroad and Keeney 1984). A similar effect of pO_2 on the ratio of $NO:NO_2^-$ by a pure culture of *Nitrosomonas europaea* has also been observed (Lipschultz et al. 1981), but this pathway of NO production is less certain than for N_2O.

Pulses of N_2O and especially NO following wetting may also be related to accumulation of NO_2^- in dry soil. Nitrite is seldom observed in moist soils, presumably because it readily diffuses in soil solution from sites of production to sites of consumption. In dry soil, however, all diffusion of ions becomes restricted to thin water films, and diffusion rates drop precipitously with decreases in soil moisture (Papendick and Campbell 1981). Sinks for NO_2^- may also decline in dry soil, although a population of NO_2^- oxidizers appears to remain viable. At 2 or 3% gravimetric soil moisture, diffusional constraints in very thin water films could account for accumulation of NO_2^- near sites of production. Upon wetting, the NO_2^- would readily diffuse throughout the soil and some fraction could be reduced to NO and N_2O. Assuming a soil bulk density of 1 g cm^{-3} and an active soil depth of only 1 cm as an N-gas source, a soil NO_2^- pool of only 1 µg N g^{-1} dry soil could account for an average gaseous flux of 100 ng N g^{-1} dry soil h^{-1} over a 10 h period. Given that NO_2^- pools have been observed in excess of 3 µg N g^{-1} dry soil, that the source of gas production may be deeper than 1 cm (especially for N_2O), and that the pulse usually peaks about 4 to 6 h after wetting, reduction of only a small fraction of the ambient NO_2^- pool could account for a large fraction of the reported pulses of NO and N_2O.

Both mechanisms suggested here – 1) increased NH_4^+ and C availability due to death and mineralization of microbial biomass, and 2) accumulation of soil NO_2^- during the dry season – are consistent with observations of smaller N-gas pulses following subsequent wettings. The magnitude of microbial biomass C and N turnover and the amount of NO_2^- that accumulates in dry soil should be positively related to the duration of a drought.

Significance of pulses relative to annual flux budgets

Assessing the significance of observed pulses is difficult because it requires not only careful measurement of the pulses, but also numerous measurements throughout the rest of the year to obtain annual budgets. Tentative budgets have been developed for the Mexican seasonally dry forest (Davidson et al. 1991) and the Venezuelan savanna (Johansson and Sanhueza 1988).

At the Venezuelan savanna, soil moisture declined as low as 2% gravimetric moisture during a drought in the midst of the rainy season. The proposed annual NO emission budget is dominated by the effects of wetting events, because the observed NO pulse following wetting was extrapolated across 60 precipitation events of >5 mm throughout the rainy season.

Most of the precipitation at the Mexican seasonally dry forest falls during c. 6 major storms in the wet season. The soil dries between precipitation events, but generally not to the degree reported for the Venezuelan savannas. Simulated rain in the midst of the wet season in the Mexican forest had little effect on NO and N_2O emissions. At the end of the dry season, NO and N_2O fluxes increased by factors of 140 and 60, respectively, following a 2 cm simulated rain (Table 1), but the observed N_2O pulse was only c. 2% of the estimated annual budget. In contrast, the observed NO pulses equalled between 10 and 22% of the estimated annual budget of NO emissions. The accuracy of the annual estimates is uncertain, but the large difference between the significance of N_2O and NO pulse emissions following the first rain is clear. Furthermore, NO emissions from soils do not always result in export of NO_x from the boundary layer, because NO is readily oxidized to NO_2 and then is adsorbed onto leaf surfaces if a plant canopy is present (Jacob and Wofsy 1990). However, because NO pulses resulting from soil wetting usually occur at the very beginning of the wet season, when plant canopies in these ecosystems are often absent or reduced, the NO emissions probably result in significant NO_x export to the atmosphere.

Acknowledgements – I thank P. Matson, NASA Ames Res. Cent., and M. Firestone, Univ. of California at Berkeley, for support and for helpful discussions of ideas. I also thank the Nat. Res. Council Ass. Program for support at the NASA Ames Res. Cent. while this manuscript was written, and I thank Kluwer Academic Publ. for permission to reproduce Fig. 1.

References

Anderson, I. C., Levine, J. S., Poth, M. A. and Riggan P. J. 1988. Enhanced biogenic emissions of nitric oxide and nitrous oxide following surface biomass burning. – J. Geophys. Res. 93: 3893–3898.

– and Levine, J. S. 1987. Simultaneous field measurements of biogenic emissions of nitric oxide and nitrous oxide. – J. Geophys. Res. 92: 965–976.

Blackmer, A. M. and Cerrato, M. E. 1986. Soil properties affecting formation of nitric oxide by chemical reactions of nitrite. – Soil Sci. Soc. Am. J. 50: 1215–1218.

Bolin, B., Degens, E. T., Dumigneaud, P. and Kempe, S. 1979. The global biogeochemical carbon cycle. – In: Bolin, B., Degens, E. T., Kempe, S. and Ketner, P. (eds), The global carbon cycle. Wiley, New York, pp. 1–53.

Bottner, P. 1985. Response of microbial biomass to alternate moist and dry conditions in a soil incubated with ^{14}C- and ^{15}N-labelled plant material. – Soil Biol. Biochem. 17: 329–337.

Cicerone, R. J. 1987. Changes in stratospheric ozone. – Science 237: 35–42.

Crutzen, P. J. 1979. The role of NO and NO_2 in the chemistry of the troposphere and stratosphere. – Ann. Rev. Earth Planet. Sci. 7: 443–472.

Davidson, E. A., Vitousek, P. M., Matson, P. A., Riley, R., Garcia-Mendez, G. and Maass, J. M. 1991. Soil emissions of nitric oxide in a seasonally dry tropical forest of Mexico. – J. Geophys. Res. 96: 15439–15445.

– 1992. Comparison of sources and sinks of nitric oxide and nitrous oxide following wetting of dry soil. – Soil Sci. Soc. Am. J. 56: 95–102.

– , Stark, J. M. and Firestone, M. K. 1990. Microbial production and consumption of nitrate in an annual grassland. – Ecology 71: 1968–1975.

– and Swank, W. T. 1986. Environmental parameters regulating gaseous nitrogen losses from two forested ecosystems via nitrification and denitrification. – Appl. Environ. Microbiol. 52: 1287–1292.

Firestone, M. K. and Davidson, E. A. 1989. Microbiological basis of NO and N_2O production and consumption in soil. – In: Andreae, M. O. and Schimel, D. S. (eds), Exchange of trace gases between terrestrial ecosystems and the atmosphere. Wiley, New York, pp. 7–21.

Goodroad, L. L. and Keeney, D. R. 1984. Nitrous oxide production in aerobic soils under varying pH, temperature, and water content. – Soil Biol. Biochem. 16: 39–43.

Goreau, T. J., Kaplan, W. A., Wofsy, S. C., McElroy, M. B., Valios, F. W. and Watson, S. W. 1980. Production of NO_3^- and N_2O by nitrifying bacteria at reduced concentrations of oxygen. – Appl. Environ. Microbiol. 40: 526–532.

Hao, W. M., Scharffe, D., Crutzen, P. J. and Sanhueza, E. 1988. Production of N_2O, CH_4, and CO_2 from soils in the tropical savanna during the dry season. – J. Atmos. Chem. 7: 93–105.

Jacob, D. J. and Wofsy, S. C. 1990. Budgets of reactive nitrogen, hydrocarbons, and ozone over the amazon forest during the wet season. – J. Geophys. Res. 95: 16737–16754.

Johansson, C. 1984. Field measurements of emission of nitric oxide in fertilized and unfertilized forest soils in Sweden. – J. Atmos. Chem 1: 429–442.

– and Galbally, I. E. 1984. Production of nitric oxide in loam under aerobic and anaerobic conditions. – Appl. Environ. Microbiol. 47: 1284–1289.

– , Rodhe, H. and Sanhueza, E. 1988. Emission of NO in a tropical savanna and a cloud forest during the dry season. – J. Geophys. Res. 93: 7180–7192.

– and Sanhueza, E. 1988. Emission of NO from Savanna soils during rainy season. – J. Geophys. Res. 93: 14193–14198.

Kieft, T. L., Soroker, E. and Firestone, M. K. 1987. Microbial biomass response to a rapid increase in water potential when dry soil is wetted. – Soil Biol. Biochem. 19: 119–126.

Lashof, D. A. and Ahuja, D. R. 1990. Relative contributions of greenhouse gas emissions to global warming. – Nature (Lond.) 344: 529–531.

Lipschultz, F., Zafiriou, O. C., Wofsy, S. C., McElroy, M. B., Valois, F. W. and Watson, S. W. 1981. Production of NO and N_2O by soil nitrifying bacteria. – Nature (Lond.) 294: 641–643.

McKenney, D. J., Shuttleworth, K. F., Vriesacker, J. R. and Findlay, W. I. 1982. Production and loss of nitric oxide from denitrification in anaerobic Brookston clay. – Appl. Environ. Microbiol. 43: 534–541.

Mortland, M. M. 1965. Nitric oxide adsorption by clay minerals. – Soil Sci. Soc. Am. Proc. 29: 514–519.

Mosier, A. R. and Hutchinson, G. L. 1981. Nitrous oxide emissions from cropped fields. – J. Environ. Qual. 10: 169–173.

Nelson, D. W. 1982. Gaseous losses of nitrogen other than through denitrification. – In: Stevenson, F. J. (ed.), Nitrogen in agricultural soils. Am. Soc. of Agron., Madison, WI, USA, pp. 327–364.

Papendick, R. I. and Campbell, G. S. 1981. Theory and measurement of water potential. – In: Parr, J. F., Gardner, W. R. and Elliot, L. F. (eds), Water potential relations in soil microbiology. Soil Sci. Soc. Am. Spec. Publ. No. 9, Madison, WI, USA, pp. 1–22.

Remde, A., Slemr, F. and Conrad, R. 1989. Microbial production and uptake of nitric oxide in soil. – FEMS Microb. Ecol. 62: 221–230.

Rudaz, A., Davidson, E. A. and Firestone, M. K. 1991. Sources of nitrous oxide production following wetting of dry soil. – FEMS Microb. Ecol. 85: 117–124.

Slemr, R. and Seiler, W. 1984. Field measurements of NO and NO_2 emissions from fertilized and unfertilized soils. – J. Atmos. Chem. 2: 1–24.

Smith, M. S. and Parsons, L. L. 1985. Persistence of denitrifying enzyme activity in dried soils. – Appl. Environ. Microbiol. 49: 316–320.

Tortoso, A. C. and Hutchinson, G. L. 1990. Contributions of autotrophic and heterotrophic nitrifiers to soil NO and N_2O emissions. – Appl. Environ. Microbiol. 56: 1799–1805.

Vitousek, P. M., Matson, P. A., Volkmann, C., Maass, J. M. and Garcia, G. 1989. Nitrous oxide flux from dry tropical forests. – Global Biogeochem. Cycl. 3: 375–382.

Williams, E. J., Parrish, D. D. and Fehsenfeld, F. C. 1987. Determination of nitrogen oxide emissions from soils: Results from a grassland site in Colorado, United States. – J. Geophys. Res. 92: 2183–2179.

Ecological Bulletins 42: 156–163. Copenhagen 1992

Tropical land use change and trace gas emissions

M. Keller, I. Galbally, M. Baer, E. Davidson, D. Fitzjarrald, G. Harris, C. Johansson, P. Matson, C. Nobre, E. Sanhueza and J. Stewart

Keller, M., Galbally, I., Baer, M., Davidson, E., Fitzjarrald, D., Harris, G., Johansson, C., Matson, P., Nobre, C., Sanhueza, E. and Stewart, J. 1992. Tropical land use change and trace gas emissions. – Ecol. Bull. (Copenhagen) 42: 156–163.

Corr. author: M. Keller, Inst. of Tropical Forestry, Call Box 25000, Rio Piedras, Puerto Rico 00928, USA.

Introduction

Important changes in regional and global atmospheric chemistry may be associated with agricultural development and land use changes in tropical forest. Tropical soils and vegetation represent globally significant sources of reactive hydrocarbons and CO (Crutzen and Gidel 1983, Logan 1985), NO, N_2O, (Keller et al. 1983, 1986, Seiler and Conrad 1987, Kaplan et al. 1988, Goreau and de Mello 1988, Johansson et al. 1988) and CH_4 (Bartlett et al. 1988). Disturbance of vegetation and soils can alter emission rates for these important gases.

Recent studies of soil emissions and atmospheric chemistry in tropical forest areas suggest that deforestation and land conversion to pasture may have global implications. Luizao et al. (1989) found that pasture soils near Manaus produced 5 times more N_2O annually than adjacent forests. A review of available measurements and a modelling exercise by Keller et al. (1991) concluded that tropical deforestation in humid areas would lead to increased regional levels of ozone and mineral acidity in rainfall. Observations show that this is already occurring in some humid tropical forest areas (Cros et al. 1988). Changes in land use in the seasonally dry forest, savanna, and arid zones may also result in changes in trace gas emission and atmospheric chemistry. Such changes have barely been investigated. Investigations of undisturbed savanna areas are underway as are limited studies in dry forests. In the tropical arid zone little if any information is available regarding biologically mediated trace gas exchange. We suggest the implementation of exploratory studies of trace gas exchange in seasonally dry forest and arid tropical regions. Additionally we recognize that planned reservoirs for hydroelectric developments in tropical regions represent a potentially major source of reduced trace gases. No specific project is suggested for the study of this land use change but we recommend further consideration at a later date. The emphasis of our project is currently on humid forest and savannas.

Objectives

1) Examine the impact of tropical land conversion on trace gas fluxes and regional atmospheric composition and chemistry.
2) Determine the factors that regulate and control trace gas exchange resulting from tropical land conversion.
3) Develop extrapolation approaches and models that estimate and predict effects of current and future land use changes on trace gas emissions on local, regional, and global scales.

Background

Forest conversion

The conversion of tropical forests to agriculture and pastures can be achieved by a variety of methods ranging from slash and burn to mechanical methods of wood harvesting and land clearing with bulldozers and burning. The net results of these operations is a drastic change in soil physical properties, nutrient dynamics, species composition, hydrology, and microclimate.

Soil characteristics change markedly following conversion of tropical forests to pasture. For example, loss of aggregate stability in pastures results in changes to hydraulic conductivity, soil moisture, aeration, and microbial production of trace gases. Surface soils in pastures and especially in agricultural fields are warmer than under forests. Soils in clearings experience much greater extremes and more frequent cycles of wetting and drying than under forests. Under some management systems overall soil moisture may increase compared to forest resulting from reduced evapotranspiration.

Inputs of carbon quantity and distribution will vary with vegetation. Agricultural soils receive low aboveground inputs except in areas with green manure. Conversion of forest to agriculture generally results in the depletion of soil carbon and nitrogen. For pasture, changes in carbon and nitrogen stocks may not be dramatic. Carbon and nitrogen inputs to the soil of a tropical forest primarily derive from above ground litter and root turnover. In pastures, a larger C and N input occurs below ground, as the relatively large grass roots, such as those derived from bamboos, turns over. Differences in rooting densities and mychorrizal proliferation are unknown, but could affect the outcome of competition for NH_4^+ and NO_3^- among plants, heterotrophic microorganisms and autotrophic nitrifiers.

Conversion of forest to agriculture and pasture results in dramatic changes in floral and faunal composition. These changes may be translated into modified trace gas emissions from the altered ecosystem. The role of species in the flux of trace gases is not trivial. The emission of NMHC by vegetation is highly species specific. Insects, particularly termites, emit substantial amounts of trace gases, especially methane. Insects and other soil fauna modify soil structure and nutrient distribution by burrowing and nesting activities. It is not clear how the changes from forest to agriculture or pasture systems affect soil fauna, but we recognize that such changes may affect soil gas exchange.

The conversion of forest to pasture or agriculture changes the height and distribution of vegetation. This results in changes in turbulent transport. The forest top is a boundary which is porous to heat, moisture and trace gas movements. Transport properties are different over aerodynamically flatter and smoother pastures and agricultural fields.

Savanna conversion

The types of changes which occur when forest vegetation is converted to agriculture or pasture uses are similar for conversion of savanna. For instance, vegetation changes will result in different emissions of NMHC's. Soils subjected to intensive agriculture and fertilization undergo radical changes in carbon and nutrient cycling. Some aspects of the savanna systems are radically different than the humid forest and will be emphasized here.

The savannah zone is c. 15 mill km^2, it includes a range of climatic and vegetation types from semi-deciduous forest and woodlands to grasslands and shrubland. The rate of savanna conversion to cultivated land is now proceeding at a very rapid rate, because these zones represent the main areas where food production can be expanded with expanding population. In the past, acidic and nutrient-depleted soils have restricted the use of this area for cultivation despite the benign climate conditions. Recent developments of Al-tolerant grain and other important agricultural crops that require high input management practices (e.g. Brazilian cerrado, Venezuelan llanos) have altered these restrictions. At the low input end of the spectrum, shrubland is managed in a cultivation-bush fallow rotation where burning of brush provides some nutrient inputs to the soil. Across this spectrum, increased burning of plant material is used either to manage agriculture or to clear land for cultivation.

Under intensive management the inputs include; 1) lime addition to neutralize acidity, 2) fertilizer inputs of phosphorous, nitrogen and potassium, and 3) Al–tolerant crop varieties. Net effects of the intensive management are more accumulation of C and N in above and below ground vegetation, altered rates of C, N and P cycling, and changes in soil biology. These changes may be expected to result in changes in soil gas emissions and in soil-plant interactions (with potential effects on vegetation gas emissions).

Soil compaction also changes with time under intensive management and intensive grazing of managed pastures. Such changes affect the soil moisture and gas exchange.

The regional scale meteorology in the savanna zone is sharply different in wet and dry seasons. During the rainy season, the convective regime prevails, but the convective regime is modulated by mid-latitude disturbances propagating into the subtropics. Because of internal dynamics of mid-latitude atmospheric flows, there can be dry spells in those savanna regions lasting from one to three weeks during the rainy season. The dry season is marked by strong subsidence. Low level transports are primarily horizontal and confined to the large scale, subsidence-induced inversion layer ($Z<2$ km). There may occur, during the dry season, episodes of anomalously intense subsidence (RH in central Brazil may reach values as low as 15%). This enhanced downward transport of upper tropospheric air will greatly alter near surface chemistry.

Biomass burning occurs during the dry season. Smoke transport follows low level flow ($Z<2$ km), confined below the inversion, and travels long distances before smoke mixes. Deep vertical mixing also may occur when smoke reaches convective areas to the south and north of burning areas [the so-called "convergence zones", ITCZ (Inter-Tropical Convergence Zone) and SACZ (South Atlantic Convergence Zone)]. Whether smoke mixes immediately into the upper troposphere or "ages" at low levels will affect the chemical products (Chatfield and Delaney 1990).

Table 1. A sampling of current and anticipated work on biosphere–atmosphere trace gas exchange in the tropics.

Institute/Study period	Area	Species
IVIC, Venezuela and Univ. of Stockholm, Sweden/ 1990–92	Savanna, Venezuela	NO_y
IVIC Venezuela, MPI, Germany/ongoing	Savanna, Venezuela	NO_y, CO, CH_4, N_2O, Organic acids, H_2S
MPI, Germany 1995/++	Global (remote sensing)	CH_4, N_2O, CO, O_3, NO_2, H_2CO, H_2O
DECAFE Group/ ongoing	Forest, Congo, Ivory Coast	NO_x, CH_4, N_2O, CO, NMHC, Aerosol
CSIRO, Australia/ ongoing	Agriculture	N_2O, NH_3
UNAM, Mexico, NASA Ames, USA, Stanford Univ., USA/ongoing	Dry forest, Mexico	NO, N_2O
Univ. of Wyoming, USA, Nat. Center for Atmos. Res., USA/ Univ. of Costa Rica, Costa Rica, 1990–93	Wet forest and agricultural land, Costa Rica	CH_4, N_2O, NO, O_3
INPA, Brazil, NASA Ames, USA, Stanford Univ., USA/ongoing	Pastures	NO, N_2O
NASA Ames, USA, Stanford Univ., USA/ ongoing	Various habitats	NO, N_2O
INPE, Brazil, INPA, Brazil, CENA, Brazil, Inst. of Hydrol., U.K.	Forests, clearings	H_2O

Current work

A sampling of current and anticipated work on trace gas exchange in tropical forest and savanna ecosystems is presented in Table 1. Several of the studies are explicitly designed to study effects of land use change. ABLE sponsored by NASA (USA) and INPE (Brazil) and the European collaboration on DECAFE are examples of successful multi-national studies of atmospheric chemistry in the humid tropical forest. These studies can serve as examples for us as we plan detailed studies of trace gas emissions which focus on land use change in the tropics.

Research plan

It is clear that land use change in the tropics is as varied as it is widespread. The group decided to limit our discussion of research for the near future to certain areas. We identified four cases representative of large changes in tropical regions. These cases should not be studied exclusive of other changes. They were chosen because of their widespread occurrence, their areal importance, and their relation to previous work. The conversion cases are; 1) humid forest to pasture; 2) humid forest to intensive agriculture; 3) savanna to pasture; and 4) savanna to intensive agriculture.

The choice of parallel conversions of savanna and forest was purposeful. We feel that studying contrasts between these environments will be useful both because of their differences in climate and because of their broad biogeochemical differences. Fire is a tool used in land conversion and management in many of the cases considered. Studies of direct emissions from fires are considered by Group D. Studies of the after-effects of fires used for land conversion and management on trace gas emissions are implicit to studies we consider here.

Key questions

Research on the effects of land conversion on trace gas emission should address changes on several levels including; a) soil processes, b) soil biology, c) vegetation and canopy structure, d) transport on local and regional scales, and e) photochemistry. The group identified several key questions for guiding research within these levels which apply equally to the four conversion cases identified above.

1) What is the time scale of change in soil characteristics and trace gas fluxes from undisturbed environments to pasture versus agriculture? We wish to examine changes in recently-burned sites through managed use to degradation or forest regeneration.

2) What is the relative importance of soil physical versus biological characteristics in controlling trace gas fluxes following conversion? Changes in microbial activities, roots and root-symbiont distributions, and soil fauna must be considered as well as changes in soil physical characteristics, soil temperature and soil moisture.

3) What are the effects of plant species on trace gas exchange (emission and deposition)? Are leaf area and distribution an important control on canopy exchange and can we accurately determine leaf area?

4) What are the effects of variations in canopy structure on gas storage and exchange? What effects do patchy heterogeneous environments have on regional scale atmospheric dynamics? What constraints does patchiness place on wider scale measurements and modelling?

5) What are the effects of incomplete mixing on chemistry? Is present knowledge of reaction kinetics sufficient? Can present simplified chemical reactions schemes adequately simulate the complex chemistry of natural emissions?

Criteria for design of experiments

We hope that the questions presented above will provide direction to individual investigators and to designers of large "expeditions." We now discuss general considerations for experimental plans for large multinational, interdisciplinary collaborative projects which we consider necessary to tackle many of the problems of atmospheric chemical changes brought about as a result of land use change.

Site selection

Localities for large scale observation and experimentation should be chosen in areas of active land use change where natural ecosystems are being converted to pastures and to intensive agriculture. These localities should encompass several sites that allow contrasts of various soil types. Knowledge of the land use history and a history of supporting environmental measurements in the chosen area are desirable. Remote sensing coverage should be available for those localities. As a secondary priority, sites must be accessible and provide adequate infrastructure for investigators. Building infrastructure may be necessary.

Studies of trace gas emissions at sites of active land use change cannot be accomplished in one or even a series of brief intensive investigations. Care must be taken in site selection and characterization. Sites should be adequately mapped including topography, soils, and vegetation cover. Knowledge of soil characteristics such as bulk density, mineralogic composition, nutrient contents, porosity and diffusity and vegetation characteristics (in various compartments – roots, stems, leaves), leaf area index, stem density, and species composition (where possible) should be gleaned from available sources or acquired. Careful site characterization will benefit site selection for specific measurements and will help us to extrapolate results and place them in a broader context. Low intensity measurements easily made by a few investigators with simple equipment may begin as soon as a general study locale is selected. Early measurements (for example soil chamber and branch enclosure gas exchange studies) will be useful for guiding further study design.

Modelling efforts should be included in design of experiments. Experimental and modelling groups (and they may overlap!) should discuss variables to be measured to include those necessary to drive models at various scales. Local, regional, and global models may be coupled. Resources should be explicitly provided for reviewing data from previous experiments and model building. Recent remote sensing should be processed before major field work begins. A goal for precision in the estimate of trace gas fluxes should be established and sensitivity studies should be conducted to evaluate measurement needs.

An experimental program

Given available resources we feel that two collaborative research projects would best supplement individual efforts. One of these activities studies the effects of rapid changes of a savanna system to pasture and intensive agriculture. A second activity aims to understand the effects of deforestation and conversion of humid tropical forest to agriculture and pasture. Combined efforts to study the changes caused when tropical forests are flooded to form reservoirs will be discussed in the future.

Each major field project must include measurements at a variety of scales. Enclosures of soil and vegetation will be useful at primary phases for evaluation of fluxes and at all phases of work for detailed connections to process studies. Tower based measurements of trace gas concentration gradients and eddy correlation fluxes will quantify gas exchange on larger scales. The possibility of incorporating eddy accumulation and/or conditional sampling approaches exists.

Field measurements

Measurements of trace gas fluxes and associated variables may be undertaken at a variety of scales. Several techniques exist for trace gas flux estimates. Enclosure methods are used for studies of soil–atmosphere, vegetation–atmosphere, and even insect–atmosphere fluxes. Enclosure methods have a long history of use because they provide numerous advantages for investigators; they are inexpensive, portable, and easily adapted to many field situations. Because of their small scale of operation, enclosure measurements of trace gases are easily linked to process level studies. For some long lived species (e.g., N_2O) they are the only currently feasible methods for flux estimation. For reactive species, enclosure methods may give only a small part of a more complex picture because gases emitted from one part of the system may react and quickly disappear in another. Studies of reactive gases require atmospheric measurements and models. Tower and aircraft based eddy correlation flux measurements are an example of methods which may be used to estimate fluxes at larger scales. Tethered balloons have been used for gradient measurements of gas concentration as well as to provide useful micrometeorological data.

To understand the development of soil changes over time, soil characteristics such as organic matter content, total nitrogen, phosphorus and cations, soil texture, moisture, and structural characteristics need to be measured along chronosequences. Distribution of organic matter and quality of organic matter may be expected to change with vegetation. Root mycorchizal associations, and root turnover under the various land uses should be examined in relation to microbial activity.

Changes in microbial processes can be measured using in situ isotope studies examining net and gross turnover of N through the microbial pool. Emissions of nitric and nitrous oxides, methane, CO_2 and CO can be measured using chamber methods. Laboratory and field experiments should be considered to elucidate important biological and physical controlling factors.

Removal or alteration of natural vegetation effects will influence canopy exchange characteristics as well as soil processes. Measurements of leaf area distribution and canopy structural characteristics will be required. Ozone uptake ought to be addressed in both humid forest and savanna systems. Ammonium release or uptake by vegetation will also be examined, especially in the savanna systems. Measurement strategies will include atmospheric gradient approaches as well as leaf enclosure methods.

To understand transports of trace gases into and out of the rain forest canopy we must not only consider traditional flux-gradient relations appropriate to heat and water transport at forest top, but also the important role of intermittent large eddies that may evacuate deep layers in the canopy. The large eddies in turn may be elements of dry convection in the mixed layer. Two to three times per day in humid forests, the entire canopy can be totally flushed by outflows from raining clouds. In consequence, it may be insufficient only to monitor gradients or eddy fluxes of trace gases. It is probably necessary to observe column concentrations within and above the forest so that the equivalent transport in these evacuation events may be estimated. One anticipates that these effects become more important for substances whose source levels are near to the forest floor.

Where vegetation is uniformly low, the situation is rather different. In principle, if a pasture, a field or a grassland savanna has sufficiently long fetch and ground vegetation is low ($<$ c. 0.5 m) standard micrometeorological techniques such as Bowen-ratio type gradient techniques suffice to make flux estimates of many quantities. Eddy correlation approaches are also possible. However, the role of edge effects on eddy correlation measurements in smaller than optimal measurement areas needs more study. The problems of measurement of trace gas fluxes using micrometeorological techniques leads to consideration of the importance of the spatial pattern of the pasture-forest patchwork. We need quantitative estimates of this scale of spatial variation, perhaps from high resolution satellite images. Studies are needed to determine how the surface pattern affects mixing in the convective boundary layer. This may depend upon the areal distribution of heterogeneous patches and the height of the convective boundary layer, a determinant of the sizes of large eddies that transport gases into the lower atmosphere. Field research on this topic and studies based on available databases on surface sensible heat flux are necessary to understanding lateral heterogeneities. In the field, small towers adequate for performing measurements over low vegetation are portable. Portable towers could be moved into different sized patches. Other surrogate measurements (cup anemometers, low frequency gas concentration) would supplement intensive micrometeorological approaches.

Field observations can be split into three categories, and instruments chosen for each task have different characteristics.

1) Instruments for ground-based studies, monitoring efforts to extend over periods of several years. These instruments should be inexpensive (many should be deployed), simple in construction (requiring little maintenance), and self contained (they should not require, for example, reference gases or pumps). Sensors that measure soil temperature, solar radiation, and horizontal wind speed and direction fall in this category.

2) Instruments for seasonal characterization studies. These include various types of chamber measurements, trace gases for which sampling is currently done are CO_2, CH_4, CO, NO, N_2O and NH_3. Gradient measurements of some species from towers also fall in this category. Currently extended-time measurements include concentration measurements of ozone and CO_2.

3) Instruments to be deployed during intensive observation periods (1–3 month periods) include the suite of fast response eddy correlation instruments. Trace gases for which such instruments exist (or will soon) for tower operation include CO_2, total hydrocarbons, CH_4 (experimental), O_3, NO, NO_x and NO_2 (experimental). Few of these instruments are now available. They may be employed at a small number of sites during the intensive phase of field studies. Fast response instruments currently deployed on research aircraft that are capable of contributing to aircraft estimation of eddy fluxes include CO_2, O_3, CH_4, CO. Aircraft operation can only be contemplated during intensive field phases.

Fluxes of reactive gases may not be interpretable without a moderately full picture of the chemical environment. The minimum set of measurements is atmospheric stability and depth of the mixed layer, O_3, CO_2, NO_x, CH_4, CO, solar radiation and total NMHC. These measurements can be carried out using routine techniques. Investigations of speciation of the NMHC should be carried out during intensive studies at each site so that the total NMHC composition can be represented as a typical surrogate in photochemical models. The minimum set should be expanded as and when possible to include measurements of J–NO_2 and J O_3 and the mixing ratios of PAN, organic nitrates, NO_y, HCHO, H_2O, CH_3CHO organic acids, and night-time build up of HONO and NO_3 by DOAS methods. Measurements of CCN and aerosol composition must be made during intensive research periods. The importance of inhomogeneity in the distributions of reactive species both in the gas phase and cloud water phases needs to be given appropriate study, especially in relation to corresponding reactions between phases.

Modelling

A program designed to clarify the effects of land use changes on trace gas exchange requires modelling components as well as long term measurement programs. Process modelling efforts at the ecosystem, canopy and atmospheric scales should run in parallel, and should be integrated wherever possible. Moreover, we should consider development of both detailed mechanistic models as well as more highly parameterized models that can be driven with variables available from remote sensing and ground-based data bases.

Modelling of soil processes should occur at two or more scales. Mechanistic models that are useful to assess our understanding of factors controlling microbial processes at the cellular level and diffusion of gaseous products between the soil and the atmosphere (Galbally and Johansson 1989). Characteristics of such models might include: 1) C and N impacts and distribution within the soil estimates of gross rates of microbial processes (N mineralization, nitrification, immobilization); diffusion characteristics of inorganic N and organic C in soil solution as a function of soil texture and moisture; diffusion of gases as a function of soil texture and moisture; and kinetics of a biological reactions within the soil

In parallel with development of detailed mechanistic models, other models should be developed that are amenable to integration with GIS data bases, output from meteorological models, and remote sensing. Characteristics of inputs to such models might include: climatic and weather variables; soil classes (soil texture and pH where available); land use classification; and ecosystem type (and, hence, inference about C and N cycling rates).

At this scale of resolution, mechanisms of soil processes may be fairly simplistic. In contrast to more detailed mechanistic models, trace gas flux from these models can be used as inputs to canopy or atmospheric models, for extrapolation to regional scales, and for comparison with tower and aircraft based estimates. This will require development of geographically referenced databases of land use types. Construction of databases should begin early in the experiments.

An aim of the observational expeditions and modelling efforts is to obtain regional estimates of trace gas fluxes and assess sensitivity of these fluxes to changes in the boundary conditions. Model calculations are done on well-defined, arbitrary grids; to include processes on scales smaller than this grid parameterizations are used. In general, the degree of difficulty of parameterization is proportional to the nonlinearity of the subgrid-scale processes being considered. In observational work, we perform the same procedure in the process of "scaling-up" observations from the local-scale (chambers, tower observations, aircraft observations) to the regional scale. Each method has its own characteristic length and time scales, analogous to the model grid scale. We do not necessarily know the degree to which the processes are linear. We seek techniques to find the function (linear or otherwise) that allows this extrapolation. Independent estimates of the extrapolated results must be found.

A first approach – one may obtain characteristic trace gas fluxes over well-defined surfaces using chambers. If care has been taken to adequately describe fluxes as a function of ecosystem types that are readily distinguished by satellite derived surface classes, one can compose the summed suite of ecosystem fluxes to apply to the region. A similar procedure could be applied to observations obtained from micrometerological towers. In this case, the "grid area" would be much larger, an area determined by prevailing wind and stability conditions, known as the "foot print" (100–300 m radius).

New approaches to analyzing data from research aircraft are needed. Categorization of fluxes by surface type along the flight path may provide; 1) a bridge to data obtained on chamber or tower scales and 2) definition of the characteristic scales of the changes in surface type and the "response" in terms of boundary layer variability of trace gas variation. We note that aircraft observations may provide one of the few estimates of these characteristic scales which will not be identified for each species of interest. For example, methane, CO_2, and sensible heat flux may have very different source distributions and perhaps the extrapolation function should be different for each. We believe that approaches for determining these functions empirically are needed.

Regional flux estimates are the output from regional (numerical) transport models. It is in the context of diagnostic transport use of these models that distributions of trace gas concentrations, as well as fluxes, may be employed. Sensitivity tests that involve altering the spatial distribution of source fluxes, constrained by the observations give another perspective on the significance of the observational uncertainties on the regional emission (deposition) estimates. We also recommend sensitivity studies of model results to the complexity of subgrid scale parameterization be performed, with the aim of finding the simplest adequate scheme.

The degree of integration among models at various scales (ecosystem, canopy, atmospheric chemistry, etc.) will depend on the gas of interest. For example, soil process models that estimate N_2O and CH_4 emissions may not require integration with canopy transport models to permit extrapolation to regional and global scales. Model estimates can be tested by comparison with flux estimates from tower and aircraft-based measurements as technological developments permit. For reactive gases models of soil emissions can provide inputs for canopy models; canopy models (which include chemistry and biological exchange) then estimate ecosystem-atmosphere exchange, which interact directly with boundary layer models of chemistry. For example, soil emissions of NO_x hypothetically may be modelled as a

Table 2. A timetable for savanna and humid forest activities.

Year	Savanna activities	Humid forest activities
1991	Site selection, develop details of experimental plan	Site selection, develop details of experimental plan
1991	Instrument intercomparison, sensitivity studies	Begin measurements, modelling, and remote sensing
1992	Interpretation meeting	Begin measurements, modelling, remote sensing
1993	Intensive campaign	Interpretation meeting, sensitivity studies
1994	Intensive campaign	
1995	Final interpretations	Intensive campaign
1996		Intensive campaign
1997		Final interpretations

function of soil nitrogen characteristics, soil texture, soil temperature, and moisture and diffusion path length. Flux estimates that are output from this model can then be input to canopy models, which incorporate chemical oxidation and plant physiological characteristics. Outputs from these models can be considered inputs to boundary layer models that describe chemical transformations and transport.

Models at all levels must be studied with sensitivity techniques to provide indications of which are the most important variables to be measured. There is need for extensive interactions between the modelling community concerned with trace gas exchange and those concerned with global chemical transport models to ensure that models developed in the IGAC study of Biosphere-Atmosphere Trace Gas Exchange in the Tropics (BATGE) interface with global models.

Remote sensing

A number of remote sensing approaches which may be useful to this study are currently available or in development. Land use classification based on thematic mapper or SPOT sensor data are crucial to selection of specific field sites, description of the heterogeneity or "patchiness" of regions, and for measurements of areal extent as a bases of extrapolation. While these sensors will easily differentiate broad land-cover classes, detection of different agricultural systems, degree of degradation, and stage of succession in forest regeneration may be difficult. Remote sensing studies must be initiated to examine these detection capabilities.

These sensors, as well as the AVHRR, can also be used to estimate light absorption by canopies, especially in the dry forest and savanna regions. Models are currently under development to predict carbon, energy and water exchange using remote sensing drivers.

Experimental sensors such as radar (including SIR–C and SAT) are being studied for use in measurements of canopy structure, foliar distribution, and vegetation classification. Research is currently being funded by NASA in a number of tropical countries. Efforts should be made to collaborate with these groups. Likewise, high spectral resolution sensors are currently being tested (e.g., AVIRIS and SCIAMACHY); these hold potential for spectral characterization of such variables as canopy chemical characteristics and trace gas concentrations.

An international consortium uniting USA, Japan, and European countries is planning a late 1990's launch of two polar-orbiting platforms (EOS) designed, to estimate a variety of physical, biological, and chemical parameters on Earth. NASA is currently funding a number of interdisciplinary studies designed to use these sensors in multi-disciplinary studies. We strongly suggest coordination and collaboration with those projects in tropical systems. The BATGE program may be able to provide ground truth data for some of the remotely sensed data discussed here.

Timetable

The forest and savanna programs would run nearly in parallel with the savanna program running a step ahead. This course is chosen because we feel that methodological problems arising from complex air transport in the heterogeneous environment arising from forest clearance will be more intractable in the forest region. Experience gained in the savanna work may be shared with forest investigators. In addition, a number of projects are currently underway in the savanna regions (e.g., DECAFE) and an international effort dealing with savanna systems is already in place (i.e., RSSO). A timetable for humid forest and savanna zone activities is given in Table 2.

Selection of experimental sites and elaboration of experimental designs should begin as soon as possible. We recommend that a meeting be held in Feb 1991 at the time of the SCOPE conference on tropical trace gas exchange. Ground based field studies, modelling studies, and evaluation of remote sensing data should begin immediately in both programs. An instrument intercomparison for trace gas species measurement is scheduled for the savanna in 1991 along with model sensitivity studies which will evaluate measurement schemes in light of available data. The product of the sensitivity studies should be a "figure of merit" level of precision for all measurements, a plan for comparing measurements at different scales, and a scheme for including data in models. A plan for managing the expedition database must also be developed at this time.

In 1992 the savanna group should hold its first work-

Table 3. Members of the Biosphere-Atmosphere Trace Gas Exchange in the Tropics (BATGE) Coordinating Committee.

Convenors:
P. Matson (USA)
C. Johansson (Sweden)

Coordinating committee:
A. F. Bouwman (The Netherlands)
R. Conrad (Germany)
B. Cros (Congo)
R. Delmas (France)
I. Galbally (Australia)
M. Keller (USA)
J. Levine (USA)
P. Sanchez (USA)
E. Sanhueza (Venezuela)
S. Wofsy (USA)
W.-X. Yang (China)

shop for data interpretation and finalize plans for 2 major field campaigns to occur from 1992 to 1994 in wet and dry seasons. Time will be reserved for evaluation and interpretation of results between the 2 field campaigns. The savanna group should conclude its activities in 1995. The forest program should plan sensitivity studies in 1993 with intensive campaigns in 1994 through 1996 and conclusion in 1997.

Conclusions

The project we consider here includes 2 large scale activities planned for 5 to 7 years duration; this project greatly expands the scope of the BATGE project proposed in the original IGAC document. The membership of BATGE provisionally has been augmented (see Table 3). Subcommittees have been assigned the task of identifying possible humid forest and savanna zone field sites for long term investigation. Several groups have also provisionally agreed to participate in the instrument intercomparison planned for 1992.

These activities require considerable resources including scarce equipment (research aircraft) and scarce personnel (an interdisciplinary team of scientists). We consider the scarcity of personnel perhaps the greatest limitation to our continuing work. Because this work is planned to take place in developing tropical countries, we feel that scientists from those countries should be involved to the maximum extent possible. Equipment and other resources should be sought internationally when unavailable from local governmental sources. Students should be trained and practicing scientists from related fields encouraged to join our activities. We feel that the future of global change research depends upon establishing a broad scientific base the world-over rather than in a but a few industrialized nations.

Acknowledgements – We are grateful to D. Ojima for his editorial assistance and especially for his patience.

References

Bartlett, K. B., Crill, P. M., Sebacher, D. I., Harriss, R. C., Wilson, J. O. and Melack, J. M. 1988. Methane flux from the central Amazon flood plain. – J. Geophys. Res. 93: 1571–1582.

Chatfield, R. B. and Delany, A. C. 1990. Convection links biomass burning to increased tropical ozone – however, models will tend to overpredict O_3. – J. Geophys. Res. 95: 18473–18488.

Cros, B., Delmas, R., Nganga, R., Clairac, B. and Fontan, J. 1988. Seasonal trends of ozone in equatorial Africa: experimental evidence of photochemical formation. – J. Geophys. Res. 93: 8355–8366.

Crutzen, P. J. and Gidel, L. T. 1983. A two dimensional model of the atmosphere 2: The tropospheric budget of the anthropogenic chlorocarbons, CO, CH_4, CH_3Cl, and the effect of various NO_x sources on tropospheric ozone. – J. Geophys. Res. 88: 6641–6661.

Galbally, I. E. and Johansson, C. 1989. A model relating laboratory rates of nitric oxide production and field measurements of nitric oxide emission from soils. – J. Geophys. Res. 94: 6473–6480.

Goreau, T. J. and de Mello, W. Z. 1988. Tropical deforestation: some effects on atmospheric chemistry. – Ambio 17: 275–281.

Johansson, C., Rodhe, H. and Sanhueza, E. 1988. Emission of NO in a tropical savanna and a cloud forest during the dry season. – J. Geophys. Res. 93: 7180–7192.

Kaplan, W. A., Wofsy, S. C., Keller, M. and da Costa, J. M. 1988. Emission of NO and deposition of O_3 in a tropical forest system. – J. Geophys. Res. 93: 1389–1395.

Keller, M., Goreau, T. J., Wofsy, S. C., Kaplan, W. A. and McElroy, M. B. 1983. Production of nitrous oxide and consumption of methane by forest soils. – Geophys. Res. Letters 10: 1156–1159.

– , Kaplan, W. A. and Wofsy, S. C. 1986. Emissions of N_2O, CH_4, and CO_2 from tropical forest soils. – J. Geophys. Res. 91: 11791–11802.

– , Jacob, D. J., Wofsy, S. C. and Harriss, R. C. 1991. Effects of tropical deforestation on global and regional atmospheric chemistry. – Climatic Change 19: 139–158.

Logan, J. A. 1985. Tropospheric ozone: seasonal behavior, trends, and anthropogenic influence. – J. Geophys. Res. 90: 10463–10482.

Luizao, F., Matson, P., Livingston, G., Luizao, R. and Vitousek, P. 1989. Nitrous oxide flux following tropical land clearing. – Global Biogeochem. Cycl. 3: 375–382.

Seiler, W. and Conrad, R. 1987. Contributions of tropical ecosystems to the global budget of trace gases especially CH_4, H_2, CO, and N_2O. – In: Dickinson, R. E. (ed.), The geophysiology of Amazonia. Wiley, New York, pp. 133–160.

Ecological Bulletins 42: 164–173. Copenhagen 1992

Trace gas exchange in rice cultivation

Heinz Rennenberg, Reiner Wassmann, Hans Papen and Wolfgang Seiler

Rennenberg, H., Wassmann, R., Papen, H. and Seiler, W. 1992. Trace gas exchange in rice cultivation. – Ecol. Bull. (Copenhagen) 42: 164–173.

Rice cultivation constitutes a source of numerous atmospheric trace gases including CH_4, N_2O, and reduced sulfur compounds. Sufficient information to estimate the source strength of rice cultivation on a global scale is only available for CH_4. Though these estimates still have a high degree of uncertainty, rice cultivation has been identified as one of the most important individual sources of atmospheric CH_4. Trace gas emission in rice cultivation is the net result of interconnected microbial processes and plant-microbial interactions. This review summarizes the present knowledge on production, consumption, and transport of atmospheric trace gases in rice paddies. The available field data on the diurnal and seasonal variations, regional differences, and the effect of agricultural practices are discussed on the basis of the biological and physico-chemical processes involved.

H. Rennenberg, R. Wassmann, H. Papen and W. Seiler, Fraunhofer Inst. für Atmosphärische Umweltforschung, Kreuzeckbahnstr. 19, D-8100 Garmisch-Partenkirchen, Germany.

Introduction

Methane (CH_4) and nitrous oxide (N_2O) are thought to be the most important radiatively active trace gases in the atmosphere next to carbon dioxide (CO_2). Atmospheric concentrations of these trace gases (at present: CH_4: 1.7 ppm; N_2O: 310 ppb) are increasing with time at a rate of 1% (CH_4) and 0.25% (N_2O) yr^{-1}. The contribution of this increase to the supposed global warming during the past 100 years is estimated to be 19% (CH_4) and 4% (N_2O). As the rate of increase of atmospheric CH_4 exceeds that of CO_2 by a factor of two, its effect on climate may become even more important in future (Bouwman 1990a).

Although changes in global sink strength cannot be excluded (Melillo et al. 1989) the increase in the atmospheric CH_4 and N_2O concentrations is mainly attributed to an increase in source strengths. For both gases the biosphere is thought to be the predominant source: agricultural and natural wetlands, landfills, ruminants and other biogenic sources contribute at more than 70% to the total annual atmospheric CH_4 burden (300–700 Tg yr^{-1}); emissions from cultivated (3 Tg N_2O-N yr^{-1}) and natural soils (6 Tg N_2O-N yr^{-1}) form the major source in the global N_2O budget (Bouwman 1990b). Among the biogenic sources of CH_4 and N_2O identified, ecosystems in which anaerobic conditions prevail are of particular importance. Under these conditions CH_4, N_2O and other trace gases are produced as intermediate and end products in numerous microbial processes, e.g. methanogenesis, denitrification, dissimilatory sulfate reduction, etc. (Zehnder and Stumm 1988).

Anaerobic conditions are frequently found in the waterlogged soils of natural wetlands, landfills, and rice paddies. With 1.45×10^6 km^2 the land area used for rice cultivation constitutes 10% of the total cultivated land and is close to the land area covered by natural wetlands. Irrigated and rainfed rice paddies, in which water saturation leads to continuously anoxic soil conditions during major parts of the vegetation period, make up c. 86% of the total rice growing area (Schütz et al. 1990a). Therefore, rice paddies are thought to be a significant source of atmospheric trace gases, especially of CH_4. This article summarizes the present understanding of the source strength of rice paddies for atmospheric trace gases on the basis of the production-, consumption- and transport-processes involved.

Significance of rice cultivation as a source of atmospheric trace gases

Rice cultivation as a significant source of atmospheric CH_4 was first discussed by Koyama (1964). From lab-

Table 1. Global annual methane emission from rice cultivation as estimated by different authors.

Reference	Estimate (Tg CH_4 yr^{-1})
Koyama 1964	190
Ehhalt and Schmidt 1978	280
Cicerone and Shetter 1981	59
Khalil and Rasmussen 1983	95
Seiler et al. 1984	35 – 59
Blake 1984	142 – 190
Crutzen 1985	120 – 200
Holzapfel-Pschorn and Seiler 1986	70 – 170
Cicerone and Oremland 1988	60 – 170
Schütz et al. 1989b	50 – 150
Aselmann and Crutzen 1989	30 – 75[a]
	60 – 140[b]
Schütz and Seiler 1989	70 – 170
Schütz et al. 1990a	50 – 150
Wang et al. 1990	60 – 120
Neue et al. 1990	25 – 60
Bouwman 1990b	53 – 114

[a] calculated from mean emission rates.
[b] calculated from temperature dependency of emission rates.

oratory experiments on CH_4 production with soil samples taken from rice paddies in Japan he estimated a global source strength of 190 Tg CH_4 yr^{-1} (1 Tg=10^{12} g). Ehhalt and Schmidt (1978) extrapolated these data to the middle of the 1970s and calculated even higher CH_4 emission rates of 280 Tg yr^{-1}. This value accounted for more than 50% of the total global CH_4 emission into the atmosphere. Since these early reports on the source strength of rice paddies for CH_4, annual global emission rates estimated from field measurements and from model calculations have been reported by several authors (Table 1). Extrapolation of the first data from in situ measurements of CH_4 emission rates obtained by closed chamber techniques in a Californian rice paddy (Cicerone and Shetter 1981) indicated much lower values (59 Tg CH_4 yr^{-1}) than those reported from previous estimates. In situ measurements performed in rice paddies of southern Spain (Seiler et al. 1984) confirmed these low rates of CH_4 emission (see Table 1). However, the data base for both calculations consisted of relatively few measurements making the estimate of global annual emission rates highly uncertain. In addition, the field site in Spain was influenced by the penetration of water from the Mediterranean Sea into the ground water of the paddy field. This seawater influx may have reduced CH_4 production.

A more reliable estimate of the annual global CH_4 emission from rice paddies was obtained by an improved closed chamber technique, that allowed continuous records of flux data of the entire inundation period of an Italian rice paddy (Holzapfel-Pschorn and Seiler 1986). From these data a mean rate of CH_4 emission of 16 mg m^{-2} h^{-1} and a global source strength of 70–170 Tg CH_4 yr^{-1} were calculated (Holzapfel-Pschorn and Seiler 1986). These measurements were continued for an additional three years, so that diurnal and seasonal fluctuations as well as differences between individual years in rice cultivation could be evaluated (Schütz et al. 1989b). As a result an average emission rate of 32 mg CH_4 m^{-2} h^{-1} was calculated. Applying this rate of CH_4 emission and considering the global distribution of rice paddies as well as the actual soil temperatures in the individual rice growing regions, a global annual emission rate of 100±50 Tg CH_4 per year was calculated (Schütz et al. 1989b). A major uncertainty in this estimate derives from methodological constraints. Chamber measurements disturb the natural environment of the ecosystem producing the gas. This is particularly important for CH_4 emission because of the influence of the rice plant on production and emission processes. However, by frequently opening the boxes a significant influence on plant growth and development was prevented. A further uncertainty in this estimate is that Italian rice paddy fields may not be representative for the most important rice growing areas of the world. Approximately 90% of the global paddy rice cultivation is located in Asia (Schütz et al. 1990a).

Data on the emission of CH_4 from rice paddies in Asia only recently became available. Intensive field studies were performed during a period of almost three years at a site located near Hangzhou in one of the most important rice growing areas of the PR China in which multiple cropping is applied (Schütz et al. 1990a, Wang et al. 1990). In 1988 an average emission rate of c. 7.3 mg CH_4 m^{-2} h^{-1} was measured during the early vegetation period (April to June); during the late vegetation period (Aug to Sept) average emission rates were much higher (25.5 mg CH_4 m^{-2} h^{-1}). Even higher fluxes of CH_4 from rice paddies (60 mg CH_4 m^{-2} h^{-1}) have been reported from sporadic measurements in the Sichuan province, another important rice growing area of PR China (Khalil et al. 1990). Extremely low rates of CH_4 emission were found in Japanese rice paddies: investigations of CH_4 emissions from different soil types in Japan yielded low emission rates for alluvial soil (2.8–9.6 mg CH_4 m^{-2} h^{-1}), Andosol (0.2–4.1 mg CH_4 m^{-2} h^{-1}), and peaty soil (16.3 mg CH_4 m^{-2} h^{-1}) (Yagi and Minami 1990). Though the data from the field studies in Hangzhou have not been completely analyzed, the available results from this study and from other investigations in Asia provide a global annual estimate of 100±50 (Schütz et al. 1990a) or 80±20 (Wang et al. 1990) Tg CH_4 yr^{-1}. These estimates of the source strength of rice cultivation for the emission of CH_4 are consistent with most of the recent calculations from other authors (Table 1). Obviously, the data base available for global calculations of the CH_4 emission from rice cultivation has considerably improved during recent years. Still a significant uncertainty remains, because the data available from Asian rice paddies are only representative for a minor part of this most important rice growing area.

In situ measurements of N_2O fluxes from rice paddies into the atmosphere were first reported by Denmead and Simpson (1979). During the first week after flooding, the emission corresponded to an average value of 88 μg N m^{-2} h^{-1}. Minami (1987) reported much lower N_2O emission rates from Japanese rice paddies with an average value of c. 3 μg N m^{-2} h^{-1}. Khalil et al. (1990) were unable to find a significant exchange of N_2O between Chinese rice paddies and the atmosphere. Considerable emissions of N_2O from surface water draining of rice paddies were reported by Minami and Fukushi (1984); the N_2O losses from drainage water amounted to 200–1190 μg N m^{-2} h^{-1}. Evolution of several volatile sulfur compounds from rice soils was reported by Minami and coworkers. The main sulfur compounds produced in Japanese rice soils were identified as H_2S, MSH, DMS, COS, CS_2 and DMDS (Takai and Asami 1962, Minami and Fukushi 1981a,b, Minami 1982). Other atmospheric trace gases emitted from rice cultivation include H_2 (Schütz et al. 1988), CO (Conrad et al. 1988, Khalil et al. 1990), C_2H_4, C_2H_6, C_3H_6, C_3H_8, $CHCl_3$ (Khalil et al. 1990), NO (Galbally et al. 1987) and NH_3 (Trevitt et al. 1988). Based on few sporadic measurements Khalil et al. (1990) observed a deposition of C_2H_2 and several chlorohydrocarbons. For none of these compounds does a sufficient data base exist that would enable reasonal estimates of fluxes on a global scale.

Mechanisms of production, consumption and exchange of trace gases in rice cultivation

Rice paddies are characterized by a high degree of spatial heterogeneity. This structure of the ecosystem is reflected by coexisting microbial communities of entirely different metabolic activites. Anaerobic and aerobic zones represent antagonistic sources and sinks. Their vicinity in rice paddies lead to an extensive exchange of compounds following source-sink concentration gradients. Volatile compounds can escape from this internal cycling via emission into the atmosphere.

Trace gas emission in rice cultivation is the result of a complex system of physico-chemical processes and plant-microbial interactions. Root exudates and root litter constitute substrates not only for microorganisms that produce trace gases in the anaerobic part of the soil, but also for bacteria that consume the trace gases produced this way. Transport of oxygen inside the plant to the roots and its release to the root surface provides the aerobic environment required for the growth of microorganisms that consume trace gases produced in the anaerobic part of the soil. The rice plant itself constitutes an important gas-exchange path between the soil and the atmosphere (Schütz et al. 1991). The relative contribution of these processes as well as their interrelation in trace gas emission from rice cultivation are far from being understood. Most of our present knowledge comes from studies on the production, consumption and transport of CH_4 in rice paddies. Studies on the interaction of the processes involved in CH_4 emission with those involved in the emission of other trace gases are almost entirely lacking.

Trace gas production

Methane production by methanogenic bacteria takes place as a terminal process of a food chain, in which organic matter is ultimately degraded under anaerobic conditions to the gaseous end products CO_2 and CH_4 (Conrad 1989). The strictly anaerobic methanogens that belong to the Archaebacteria can only utilize a limited number of substrates. They can be subdivided according to their substrate requirements in obligate chemolithotrophic, "quasi"-chemolithotrophic, and methylotrophic methanogens (Papen and Rennenberg 1990). In obligate chemolithotrophic methanogens CH_4 is exclusively produced from carbon dioxide and dihydrogen; "quasi"-chemolithotrophic methanogens are capable of using formic acid and carbon monoxide as additional substrates. Methylotrophic methanogenic bacteria produce methane from several methylated compounds, including methanol, acetate, trimethylamine, and dimethylsulfide (Kiene et al. 1986, Vogels et al. 1988). All these reactions are connected with strongly negative values of the standard free energy change (Vogels et al. 1988). As a consequence of the narrow spectrum of compounds that can be utilized for methanogenesis organic matter has first to be decomposed by numerous microbial species in consecutive fermentation processes to compounds that fulfill the substrate requirements of methanogenic bacteria. Therefore, the quality of the primary organic substrates strongly influence the rate of CH_4 production.

Next to the supply of organic matter, temperature is the most important factor controlling methane production in rice cultivation (Conrad et al. 1987). As the majority of isolated methanogens show temperature optima of 30–40°C (Vogels et al. 1988), CH_4 production in rice paddies appears to proceed under suboptimal temperatures in regions, where the local soil temperature seldomly exceeds 30°C. This view is consistent with the finding of Yamane and Sato (1961) that CH_4 formation is at its maximum at 40°C and is small at 20°C in waterlogged paddy soils. It can therefore be assumed that diel and seasonal changes in temperature have a significant effect on CH_4 production in rice paddies. At sufficient substrate supply a temperature increase from 17°C to 30°C was shown to result in a 2.5- to 3.5-fold increase in CH_4 production (Conrad et al. 1987). A recent comparison of different paddy soils showed an even higher increase in CH_4 production rates, when the temperature was elevated from 25°C to 35°C. As the production of methanogenic substrates by fermentation processes can be more temperature-sensitive than the CH_4 production itself (Conrad et al. 1987), these high rates of increase in CH_4 production with temperature may partially be the consequence of enhanced availability of substrates for methanogenesis.

In situ rates of CH_4 production in paddy soils have only recently been reported (Schütz et al. 1989a). In the Italian rice paddy studied CH_4 was not produced in the oxic soil surface layer of 1 to 2 cm depth. Considerable CH_4 production was found in the anoxic soil down to 25 cm depth, covering the entire zone of the rice roots. A preferred depth of CH_4 production could not be established; the contribution of deeper soil zones was not determined. In two consecutive years (1985, 1986) CH_4 production integrated over 25 cm depth increased during the season, though the number of methanogenic bacteria did not change. From this finding it may be concluded that CH_4 production is not limited by the density of the bacterial population, but by the availability of substrates. Maximum rates of CH_4 production of more than 300 ml CH_4 m^{-2} h^{-1} (215 mg CH_4 m^{-2} h^{-1}) were measured in Aug. Obligate chemolithotrophic methanogens, using CO_2 and H_2 as substrates, contributed to the total CH_4 production at 30 to 50%; acetate was an equally important precursor of CH_4 production (Schütz et al. 1989a).

While CH_4 production by methanogenic bacteria takes place exclusively under anaerobic conditions, the production of N_2O is observed in both, anaerobic and aerobic environments. The most important microbial processes involved in the production of N_2O are denitrification (e.g. Tiedje 1988), chemolithotrophic nitrification (Ritchie and Nicholas 1972, Bremner and Blackmer 1978), and heterotrophic nitrification (Papen et al. 1989, Rosswall et al. 1989). In denitrification NO_3^-/NO_2^- are reduced under anaerobic conditions as terminal electron acceptors to the gaseous products N_2O and/or N_2. This reduction is strictly coupled to electron transport phosphorylation. More than 80% of the NO_3^-/NO_2^-N is reduced to N_2O plus N_2 by respiratory denitrification (Tiedje 1988). Whether chemolithotrophic ammonia oxidizers of the *Nitrosomonas*-type are able to produce under aerobic conditions N_2O during the oxidation of ammonia to NO_2^- (Ritchie and Nicholas 1972, Bremner and Blackmer 1978) is a controversial issue. Poth and Focht (1985) provided evidence that *Nitrosomonas* under oxygen-limited conditions will use NO_2^- as an electron acceptor. Therefore, the observed N_2O production may be the result of nitrifier-denitrification. Robertson and Tiedje (1987) reported that more than 50% of the observed N_2O emission from forest soil cores could not be accounted for by denitrification and chemolithotrophic nitrification. Therefore, yet unidentified microbial processes must be considered as biogenic sources of atmospheric N_2O. Papen et al. (1989) provided presumptive evidence that heterotrophic nitrification may be one of these processes. As a broad spectrum of different bacterial and fungal genera are able to catalyze heterotrophic nitrification, a high potential for N_2O production by this process must be considered (Papen and Rennenberg 1990). Other microbial processes that may be a source of N_2O include assimilatory nitrate reduction, dissimilatory nitrate reduction to ammonium, and N_2O production by nitrate reductase of *Enterobacteriaceae* from nitrite (Tiedje 1988).

The relative contribution of these processes to the overall rate of N_2O production and emission in rice paddies has not been established. Garcia (1974, 1975) used the velocity of N_2O consumption as a measure for denitrification activity in Senegalese rice paddies. With this method he estimated a denitrification activity of 1–5.6 μg N g^{-1} h^{-1}. However, this value may be an underestimation of the denitrification activity since the N_2O consumption method will only reflect denitrifiers, which contain a N_2O reductase. In future studies the model developed by Betlach and Tiedje (1981) has to be taken into account, which predicts that any factor reducing the overall rate of denitrification (e.g., nitrate concentration, availability of the C-sources, oxygen concentration) would result in an increase in the N_2O fraction of the denitrification products. Therefore, decreased rates of denitrification may result in enhanced rates of N_2O production.

Volatile sulfur compounds can be produced in the soil by chemical and biological processes (Bremner and Steele 1978). The most important microbial process producing H_2S is the reduction of sulfate under strictly anaerobic conditions. This process is catalyzed by the obligate anaerobic sulfate reducing bacteria (e.g. Widdel 1988). Decomposition of reduced sulfur containing organic matter results in the production of H_2S and methanethiol (MSH). As pointed out by Widdel (1988) the degradation of biomass with sulfate as the terminal electron acceptor yields 57-times more H_2S than could be produced by degradation of reduced sulfur compounds to H_2S in the same amount of biomass. H_2S may also be produced from COS in cyanobacteria by the carbonic anhydrase activity (Rennenberg 1991). Though an appreciable population of cyanobacteria is abundant in many rice paddies, it is unknown whether significant amounts of COS are available for H_2S production by this reaction. In addition to H_2S and MSH, DMS (dimethylsulfide), COS and CS_2 are produced by microbial processes under aerobic and anaerobic conditions; the microorganisms and the processes responsible have not yet been elucidated. In marine ecosystems DMS is produced from dimethylsulfoniopropionate by phytoplankton (Taylor and Kiene 1989); whether this reaction is also involved in the production of DMS in terrestrial ecosystems is unknown (Rennenberg 1991). Volatile reduced sulfur compounds may also be produced during methanogenesis. DMS, e.g., can be converted to H_2S and CH_4 by methylotrophic methanogenic bacteria (Kiene et al. 1986). Takai and Asami (1962), Minami and Fukushi (1981a,b) and Minami (1982) reported the emission of H_2S, DMS, MSH, COS, and CS_2 from paddy soils supplemented with cysteine, cystine, or methionine. However, the processes involved in the production of these compounds were not studied. Whether the rice plant itself is a source of atmospheric sulfur as observed for many other plant species (Rennenberg 1991) has not been investigated.

Trace gas consumption

Oxidation of CH_4 has mostly been concluded from differences between laboratory estimates of CH_4 production and field measurements of CH_4 emission. This approach has led to the assumption that as much as 80% of the CH_4 produced in rice cultivation is oxidized and, thus, is not emitted into the atmosphere (Holzapfel-Pschorn et al. 1985). Another estimate suggests a value closer to 60% (Sass et al. 1990). Methane oxidation based on in situ measurements of CH_4 production and field measurements of CH_4 emission was found to be highly dependent on the season (Schütz et al. 1989b). During the early season, when rice plants were entirely submerged and CH_4 emission was mainly by ebullition, less than half of the CH_4 produced was emitted. Later in the season, when the rice plants had developed their vascular transport system and most of the CH_4 emission proceeded via the plant, up to 90% of the CH_4 produced was oxidized. Obviously enhanced CH_4 production in rice paddies coincides with enhanced CH_4 oxidation. This phenomenon may be explained 1) by a prolonged residence time of CH_4 when CH_4 transport proceeds through the plant rather than by ebullition, and/or 2) by a stimulation of the activity of CH_4 oxidizers at enhanced rates of CH_4 production. Obviously other processes, e.g. O_2 transport to the rhizosphere, are also involved in the observed high rates of CH_4 oxidation. CH_4 emission undergoes not only seasonal, but also diurnal changes. It may therefore be assumed that the ratio of CH_4 production to CH_4 oxidation does not remain at a constant value during a 24 h period. However, measurements that enable an estimation of the diurnal changes of CH_4 oxidation in rice paddies have not been reported.

Methanotrophic bacteria are thought to be aerobic organisms, though there is indirect evidence that anaerobic oxidizers do exist (Oremland and Capone 1988). Attempts to isolate bacteria that oxidize CH_4 in the absence of O_2, have so far not been successful. From radioisotope studies and from investigations on the sulfate and CH_4 distribution in marine sediments it has been assumed that sulfate reducing bacteria play a role in anaerobic CH_4 oxidation. However, inhibitors of sulfate-reducing bacteria had no significant effect on CH_4 oxidation (cf. Galchenko et al. 1989).

CH_4 oxidation can take place in the aerobic microenvironments of the rhizosphere and in the oxic part of the upper soil horizons. It is initiated by the copper containing enzyme methane mono-oxygenase, converting CH_4 to methanol (Oremland and Capone 1988). Selective inhibitors of this reaction are copper-chelating agents; the effect of these inhibitors on methane mono-oxygenase can be reversed by addition of copper-II-ions. The use of selective inhibitors may provide a means for the direct measurement of CH_4 oxidation in future studies. Methanotrophic bacteria can grow with CH_4 as their sole source of energy. In laboratory studies oxidation of CH_4 by a variety of organisms, including nitrifying bacteria such as the dominant terrestrial species *Nitrosomonas europaea,* is competitively inhibited by nitrogen compounds, especially by ammonium (Whittenburry et al. 1970, Hyman and Wood 1983, Jones and Morita 1983). It may therefore be assumed that nitrogen containing fertilizers may reduce CH_4 oxidation thereby enhancing CH_4 emission. As nitrogen containing fertilizers reduced rather than enhanced CH_4 emission from rice paddies (Schütz et al. 1989b), it appears as if nitrifiers do not significantly participate in CH_4 oxidation in paddy soils.

In contrast to CH_4 that is predominantly consumed by aerobic processes, N_2O can only be consumed by denitrification under anaerobic/microaerobic conditions. The enzyme responsible for N_2O consumption, nitrous oxide reductase, is only present in part of the denitrifier population. Thus, N_2O production and N_2O consumption may occur at different microsites in anaerobic soils. Still, N_2O produced under anaerobic conditions has a greater chance to be consumed than N_2O that might have been produced under aerobic conditions.

The consumption of volatile sulfur compounds is mediated by chemical and biological processes. In soils a major fraction of the hydrogen sulfide formed undergoes rapid chemical conversion to pyrite framboids and cristals (Meyer 1977). Under aerobic conditions sulfide can chemically be oxidized to sulfate. Still H_2S can also be consumed by biological processes, i.e. bacterial photosynthesis, sulfide assimilation in algae and plants, and biological sulfide oxidation by thiobacilli. High populations of unidentified sulfur bacteria were found in Japanese paddy soils (cf. Watanabe and Furusaka 1980). This finding is surprising, as the anaerobic conditions required for photosynthetic sulfur bacteria are usually not met in the surface soil of flooded rice fields. The role of these microorganisms as a sink of H_2S produced in the anaerobic zones of the soil by, e.g., sulfate reducing bacteria has not been investigated. Cyanobacteria, algae, and plants are able to utilize H_2S as a source of reduced sulfur for protein synthesis (De Kok 1990). H_2S assimilation is thought to be catalyzed by cysteine synthase, though the apparent K_M of the enzyme for sulfide measured in vitro was found to be several orders of magnitude higher than the intracellular sulfide concentration (cf. De Kok et al. 1989). The uptake of COS by cyanobacteria (cf. Rennenberg 1991), as well as the metabolism of DMS by methylotrophic methanogenic bacteria (Kiene et al. 1986) constitute additional processes of comsumption of reduced sulfur compounds, though another volatile sulfur-compound, namely H_2S, is produced in these reactions.

Trace gas exchange

CH_4 produced in the anoxic sediment of the rice paddy may be released into the atmosphere by three different pathways: 1) It may diffuse from the supersaturated sediment along the sediment/water and the water/air interface according to the concentration gradient. 2) When the partial pressure of CH_4 exceeds the hydrostatic pressure in the supersatured sediments, gas bubbles will be formed and released to the water surface by ebullition. 3) CH_4 may enter the root of the rice plant and may move by diffusion in the aerenchyma of the plant to the atmosphere. A recent review by Schütz et al. (1991) provides a detailed characterization of these paths of CH_4 exchange between the paddy soil and the atmosphere.

In Italian rice paddies the relative contribution of these gas exchange processes varied with the season. During the early season, CH_4 was predominantly emitted by ebullition. Depending on the developmental stage of the plant, the contribution of ebullition to the emission of CH_4 decreased and transport through the rice plant increased during tillering (Schütz et al. 1989a). These field observations are consistent with recent morphological studies and laboratory experiments (Butterbach, unpubl. data): an appreciable aerenchyma is not present in young rice plants, but is developed during tillering. When rice plants were inserted into an apparatus separating the root and the shoot, gas transport from the root to the shoot increased by more than an order of magnitude during plant development. During the reproductive phase of the rice plant CH_4 emission proceeded by more than 90% through the plants' aerenchyma (Cicerone and Shetter 1981, Schütz et al. 1989a). Diffusion of dissolved CH_4 across the water/atmosphere interface was in the range of less than 1 to 5% of the total flux throughout the season (Holzapfel-Pschorn et al. 1986, Schütz et al. 1989a). A minor contribution of diffusion to the overall emission of CH_4 from rice paddies has also been found in other studies (e.g. Denmead and Freney 1992).

In these studies transport through the rice plant was estimated from the CH_4 emission in the presence of plants as compared to the CH_4 emission when the plants were cut below the water surface. As cutting the plants may considerably disturb the entire ecosystem, the plants' role in the exchange of CH_4 between the soil and the atmosphere should be re-evaluated in future studies by the use of less rigorous methods, e.g. the enclosure of those parts of the rice plant from which CH_4 is released.

As trace gas emission from rice paddies is mediated by physical and biophysical processes it can be assumed that the paths of emission established for CH_4 are not different from those for other trace gases. In recent studies by Mosier et al. (1990) the presence of young rice plants only slightly enhanced the flux of N_2+N_2O to the atmosphere. This finding is consistent with the observation that a considerable aerenchyma is not yet developed in young rice plants. Investigations of the exchange processes involved in the emission of other trace gases in rice cultivation have not been reported.

Interpretation of field data on the basis of the processes involved

Diurnal variations

Significant diurnal changes in the rate of CH_4 emission from rice paddies with maximum daily emission rates during the late afternoon were observed during field measurements in Italy (Schütz et al. 1989b). These diurnal variations were attributed to changes in soil temperature. A more intensive analysis of these data and additional laboratory studies on the effect of temperature on CH_4 emission have recently been reported (Schütz et al. 1990b). The diel changes of CH_4 fluxes in Italian rice paddies were found to be significantly correlated with diel temperature changes in a distinct soil depths. In May and June as well as in Aug the soil depth with the best correlation was 1–5 cm, in June and July a better correlation was found at 10–15 cm soil depth. From these data it may be concluded that temperature dependent changes in the processes involved in CH_4 production, consumption and/or transport are responsible for the diurnal fluctuations in the CH_4 emission observed in the field. The process(es) determining the CH_4 flux may then be localized in distinct soil depths showing correlation between temperature and CH_4 emission. However, temperature dependent changes alone are insufficient to explain the diurnal variation in the CH_4 emission measured, as the diel amplitudes in temperature and in CH_4 emission exhibited entirely different patterns during the season. The diel amplitudes of CH_4 emission were high during the early and late vegetation period and smaller in between. Shading of the soil by the developing rice plants caused the diurnal temperature amplitudes in the soil to decrease during the season.

Recent results from rice paddies in PR China have confirmed this assumption (Schütz et al. 1990a, Wang et al. 1990). The first rice crop growing between the end of April and the end of June showed two diel maxima in CH_4 emission, one at noon and another during the night. The second rice crop growing between the beginning of Aug and the beginning of Oct showed only one maximum in CH_4 emission that was measured during the night. These patterns of CH_4 emission do obviously not correlate with diel changes in soil temperature. It may be assumed that temperature-independent diel changes in O_2 and carbohydrate supply to the roots may modulate the O_2 concentration in the rhizosphere and the exudation by the roots, thereby affecting CH_4 production and oxidation. However, experimental evidence for such an assumption has so far not been provided.

As far as diurnal changes in CH_4 emission are caused by changes in temperature, the temperature-dependent processes responsible for these changes have not been identified. The following mechanisms may be considered in this respect. Temperature may directly effect CH_4 production and the growth of methanogenic bacteria. It may change the activity and the population of microorganisms producing methanogenic substrates. Methanogenic and methanotrophic bacteria may be differently affected by changes in temperature. Changes in soil temperature will alter root respiration and may influence the O_2 diffusion into the rhizosphere and, as a consequence, CH_4 oxidation.

Seasonal variations

The rate of CH_4 emission from rice cultivation changes significantly during the season. In Italian rice paddies, the avarage daily emission rate showed a first maximum in May shortly after flooding during tillering of the rice plants. A second maximum was observed in July during the reproductive stage of the rice plants. A third maximum was found during two out of three consecutive vegetation periods in late Aug (Schütz et al. 1989b). Similar seasonal variations of CH_4 emission were observed in Chinese (Schütz et al. 1990a, Wang et al. 1990) and Japanese rice paddies.

The seasonal patterns of CH_4 emission observed by several authors correlated poorly with seasonal changes in temperature (Cicerone et al. 1983, Seiler et al. 1984, Schütz et al. 1990b), but similarities in the general trends were found (Schütz et al. 1990b). Therefore, it appears that other factors than temperature modulate the seasonal changes of CH_4 emission from rice cultivation. From the published literature the availability of substrates for methanogenesis appears to be the most important factor determining seasonal changes in CH_4 emission. The first maximum of emission, observed shortly after flooding, has been attributed to the degradation of organic matter present in the soil. This view is consistent with the findings that 1) additional input of rice straw into soils enhanced this maximum of CH_4 emission (e.g. Schütz et al. 1989b) and 2) an early peak in CH_4 emission from soils with low organic matter content is not observed (Sass et al. 1990). The second maximum of CH_4 emission is assumed to be caused by intensive root exudation and enhanced availabitiy of root litter that may be used as substrates for methanogenesis either directly or subsequently to degradation processes. However, experimental evidence for a developmentally determined exudation of organic compounds by the roots of rice plants is lacking. The third maximum of CH_4 emission observed in some investigations has been attributed to the decomposition of plant litter and dying plant material. Though this assumption is feasible, experimental support has so far not been provided. As seasonal patterns of CH_4 emission determined on the same spot varied considerably during consecutive years (Schütz et al. 1989b), other factors than substrate availability for methanogenesis may also be involved in seasonal changes of CH_4 emission.

Fertilizer treatment

Application of fertilizers can significantly alter the emission of CH_4 from rice paddies. The published literature on the effects of different fertilizer treatments are partially inconsistent or even contradictory. Organic amendments to the soil are generally thought to increase the rate of CH_4 emission as a result of an enhanced availability of methanogenic substrates. Addition of rice straw, e.g., resulted in a drastic increase in methane emission in Italian (Schütz et al. 1989b) and Japanese (Minami 1992) rice paddies. Also when compost was introduced into the soil, enhanced rates of CH_4 emission were observed (Minami 1992), whereas first results from Chinese rice paddies did not provide evidence for an effect of organic manure (animal excrements, rape seed cake) on CH_4 emission (Schütz et al. 1990a).

Application of mineral fertilizers has lead to much more complex results. Addition of $CaCN_2$ enhanced CH_4 emission from Italian rice paddies (Holzapfel-Pschorn and Seiler 1986), but counteracted the stimulating effect of rice straw (Schütz et al. 1989b). Cicerone and Shetter (1981) observed a 5-fold increase in CH_4 emission upon fertilization with $(NH_4)_2SO_4$, but the same type of fertilizer resulted in a 50% decrease in CH_4 emission in studies by Schütz et al. (1989b). In the same investigation surface applied instead of incorporated $(NH_4)_2SO_4$ resulted in an increased rather than a decreased rate of CH_4 emission (Schütz et al. 1989b). A combination of $(NH_4)_2SO_4$ and rice straw enhanced CH_4 emission compared to the application of straw alone (Schütz et al. 1989b). In Chinese rice paddies application of K_2SO_4 did not significantly affect CH_4 emission rates (Schütz et al. 1990a). Urea, when incorporated into the soil, decreased CH_4 emission; when applied in combination with straw CH_4 emission was enhanced above the values measured with straw alone (Schütz et al. 1989b). Whereas an amendment of 80 kg N-fertilizer ha^{-1} only slightly altered CH_4 emission compared to an unfertilized plot, addition of 100 kg N-fertilizer ha^{-1} (in the form of ammonium-phosphate) decreased CH_4 emission by an order of magnitude (Minami 1992).

For most of these phenomena a clear cut explanation can not be given, because a couple of questions have not been addressed in the experimental approaches applied. E.g., under which conditions the effect of $(NH_4)_2SO_4$

fertilizer is an effect of the sulfate added, under which conditions it is an effect of N-fertilization? In which way the type of fertilizer applied, the method of application, and the actual amount of fertilizer used interact with the soil type and the history of the soil? Obviously more research is needed to improve the present understanding of the effect of fertilizer treatment on CH_4 emission from rice paddies and the underlying processes.

Emission of N_2O from rice paddies is also affected by fertilizer treatment, especially if N-containing fertilizers are applied. Smith et al. (1982) observed that only 0.01–0.05% of the urea-N added was emitted as N_2O-N from paddy fields. Minami (1987) reported much higher losses of fertilizer-N (0.33 and 0.55%) by N_2O emission, when NH_4-containing fertilizers were used. In a recent study of Bhadrachalam et al. (1992) 5% of the fertilizer N apllied in the form of urea was lost from a rice field within two years. As 4 to 6% of this loss was attributed to N_2O, this amounts as a total loss of approximately 0.25% of fertilizer N by N_2O emission. The latter emissions are relatively high as compared to other crops (Bouwman 1990b); therefore, N_2O losses from NH_4-containing fertilizers in paddy fields may be a considerable factor in the global atmospheric N_2O budget.

Conclusions

Recent field measurements of the CH_4 emission from rice paddies have provided a data-set that enables a reasonal estimate of the source strength of rice paddies in the global atmospheric CH_4 budget. A major uncertainty in this estimate is, however, the small amount of data available from the most important rice growing areas in Asia. Data on the emission of other trace gases from rice cultivation are scarce and insufficient even for a rough global estimate on an annual scale.

Though it is essential to strongly intensify continuous field measurements of the emission of CH_4 and other trace gases from rice paddies, available manpower and costs will limit the number of sites for such an approach. Therefore, field measurements alone will not provide the information required for accurate calculations of the trace gas exchange in rice cultivation. For a reliable extrapolation based on a limited number of field measurements, a better understanding of the individual processes involved in the production, the consumption, and the transport of trace gases and their interactions are required. Therefore, future research has to focus on field measurements and mechanistic studies as well. Both types of investigations should be performed in close collaboration with model developments to ensure that those data are compiled that are required for present and future model calculations.

Acknowledgement – The authors thank K. Butterbach, Fraunhofer Inst. für Atmosphärische Umweltforschung, Garmisch-Partenkirchen, for providing unpubl. data.

References

Aselmann, I. and Crutzen, P. J. 1989. Freshwater wetlands: global distribution of natural wetlands and rice paddies, their net primary productivity, seasonality and possible methane emissions. – J. Atm. Chem. 8: 307–358.

Betlach, M. R. and Tiedje, J. M. 1981. Kinetic explanation for accumulation of nitrite, nitric oxide and nitrous oxide during bacterial denitrification. – Appl. Environ. Microbiol. 42: 1074–1084.

Bhadrachalam, A., Chakravorti, S. P., Banerjee, N. K., Mohanty, S. K. and Mosier, A. R. 1992. Denitrification in intermittently flooded rice fields and N-gas transport through rice plants. – Ecol. Bull. (Copenhagen) 42: 183–187.

Blake, D. R. 1984. Increasing concentration of atmospheric methane. Ph.D. thesis, – Univ. of California, Irvine, CA.

Bouwman, A. F. 1990a. Soils and the greenhouse effect. – Wiley, Chichester, pp. 25–32.

– 1990b. Exchange of greenhouse gases between terrestrial ecosystems and the atmosphere. – In: Bouwman, A.F. (ed.), Soils and the greenhouse effect. Wiley, Chichester, pp. 61–127.

Bremner, J. M. and Blackmer, A. M. 1978. Nitrous oxide: emission from soils during nitrification of fertilizer nitrogen. – Science 199: 295–296.

– and Steele, C. G. 1978. Role of microorganisms in the atmospheric sulfur cycle. – Adv. Microbiol. Ecol. 2: 155-201.

Cicerone, R. J. and Shetter, J. D. 1981. Sources of atmospheric methane: measurements in rice paddies and a discussion. – J. Geophys. Res. 86: 7203–7209.

– , Shetter, J. D. and Delwiche, C. C. 1983. Seasonal variation of methane flux from a Californian rice paddy. – J. Geophys. Res. 88: 11022–11024.

– and Oremland, R. S. 1988. Biogeochemical aspects of atmospheric methane. – Global Biogeochem. Cycl. 2: 299–327.

Conrad, R. 1989. Control of methane production in terrestrial ecosystems. – In: Andreae, M. O. and Schimel, D. S. (eds), Exchange of trace gases between terrestrial ecosystems and the atmosphere. Wiley, Chichester, pp. 39–58.

– , Schütz, H. and Babbel, M. 1987. Temperature limitation of hydrogen turnover and methanogenesis in anoxic paddy soil. – FEMS Microbiol. Ecol. 45: 281–289.

– , Schütz, H. and Seiler, W. 1988. Emission of carbon monoxide from submerged rice fields into the atmosphere. – Atm. Environ. 22: 821–823.

Crutzen, P. J. 1985. The role of the tropics in atmospheric chemistry. – In: Dickinson, R. (ed.), Geophysiology of Amazonia. Wiley, Chichester, pp. 107–132.

De Kok, L. J. 1990. Sulfur metabolism in plants exposed to atmospheric sulfur. – In: Rennenberg, H., Brunold, Ch., De Kok, L. J. and Stulen, I. (eds), Sulfur nutrition and sulfur assimilation in higher plants. Fundamental, environmental, and agricultural aspects. SPB Acad. Publ., The Hague, pp. 111–130.

– , Stahl, K. and Rennenberg, H. 1989. Fluxes of atmospheric hydrogen sulphide to plant shoots. – New Phytol. 112: 533- 542.

Denmead, O. T. and Freney, J. R. 1992. Transfer coefficients for water-air exchange of ammonia, carbon dioxide and methane. – Ecol Bull. (Copenhagen) 42: 31–41.

– and Simpson, J. R. 1979. Nitrous oxide emission during denitrification in a flooded field. – Soil Sci. Soc. Am. J. 43: 716–718.

Ehhalt, D. H. and Schmidt, U. 1978. Sources and sinks of atmospheric methane. – Pageoph. 116: 452–464.

Galbally, I. E., Freney, J. R., Muirhead, W. A., Simpson, J. R., Trevitt, A. C. F. and Chatte, P. M. 1987. Emission of nitrogen oxides (NO_x) from a flooded soil fertilized with urea: relation to other nitrogen loss processes. – J. Atm. Chem. 5: 343–365.

Galchenko, V. F., Lein, A. and Ivanov, M. 1989. Biological sinks of methane. – In: Andreae, M. O. and Schimel, D. S. (eds), Exchange of trace gases between terrestrial ecosystems and the atmosphere. Wiley, Chichester, pp. 59–71.

Garcia, J. L. 1974. Réduction de l'oxyde nitreux dans les sols de rizieres du Sénegal: mesure de l'activité denitrifiante. – Soil Biol. Biochem. 6: 79–84.

– 1975. Evaluation de la dénitrification dans les rizieres par la methode de réduction de N_2O. – Soil Biol. Biochem. 7: 251–256.

Holzapfel-Pschorn, H., Conrad, R. and Seiler, W. 1985. Production, oxidation and emission of methane in rice paddies. – FEMS Microbiol. Ecol. 31: 343–351.

– , Conrad, R. and Seiler, W. 1986. Effects of vegetation on the emission of methane from submerged paddy soil. – Plant and Soil 92: 223–233.

– and Seiler, W. 1986. Methane emission during a cultivation period from an Italian rice paddy. – J. Geophys. Res. 91: 11803–11814.

Hyman, M. and Wood, P. M. 1983. Methane oxidation by *Nitrosomonas europaea*. – Biochem. J. 212: 31–37.

Jones, R. D. and Morita, R. Y. 1983. Methane oxidation by *Nitrosococcus oceanus* and *Nitrosomonas europaea*. – Appl. Environ. Microbiol. 45: 401–410.

Khalil, M. A. K. and Rasmussen, R. A. 1983. Sources, sinks and seasonal cycles of atmospheric methane. – J. Geophys. Res. 88: 5131–5144.

– , Rasmussen, R. A., Wang, M. X. and Ren, L. 1990. Sources of methane in China: rice fields, biogas pits, cattle, urban areas, and wetlands. – In: Newman, L., Wang, W. and Kiang, C. S. (eds), Global and regional atmospheric chemistry. Proceedings of the international conference on global and regional environmental atmospheric chemistry, Beijing, May 3–10, 1989.

Kiene, R. P., Oremland, R. S., Catena, A., Miller, L. G. and Capone, D. G. 1986. Metabolism of reduced methylated sulfur compounds in anaerobic sediments and by a pure culture of an estuarine methanogen. – Appl. Environ. Microbiol. 52: 1037–1045.

Koyama, T. 1964. Biogeochemical studies on lake sediments and paddy soils and the production of hydrogen and methane. – In: Miyaki, T. and Koyama, T. (eds), Recent researches in the field of hydrosphere, atmosphere, and geochemistry. Maruzen, Tokyo, pp. 143–177.

Melillo, J. M., Steudler, P. A., Aber, J. D. and Bowden, R. D. 1989. Atmospheric deposition and nutrient cycling. In: Andreae, M. O. and Schimel, D. S. (eds), Exchange of trace gases between terrestrial ecosystems and the atmosphere. Wiley, Chichester, pp. 263–280.

Meyer, B. 1977. Sulfur, energy, and environment. – Elsevier, Amsterdam, pp. 152–155.

Minami, K. 1982. Volatilization of sulfur from paddy soils. – JARQ 15: 167–171.

– 1987. Emission of nitrous oxide (N_2O) from agro-ecosystems. – JARQ 21: 22–27.

– and Fukushi, S. 1981a. Detection of carbonyl sulfide among gases produced by the decomposition of cysteine in paddy soils. – Soil Sci. Plant Nutr. 27: 105–109.

– and Fukushi, S. 1981b. Volatilization of carbonyl sulfide from paddy soils treated with sulfur-containing substances. – Soil Sci. Plant Nutr. 27: 339–345.

– and Fukushi, S. 1984. Methods for measuring N_2O flux from water surface and N_2O dissolved in water from agricultural land. – Soil Sci. Plant Nutr. 30: 495–502.

Mosier, A. R., Mohanty, S. K., Bhadrachalam, A. and Chakravorti, S. P. 1990. Influence of rice plants on the evolution of N_2 and N_2O from the soil to the atmosphere. – Mitteil. Dt. Bodenk. Ges. 60: 115–120.

Neue, H. U., Becker-Heidmann, P. and Scharpenseel, H. W. 1990. Organic matter dynamics, soil properties, and cultural practices in rice lands and their relationship to methane production. – In: Bouwman, A. F. (ed.), Soils and the greenhouse effect. Wiley, Chichester, pp. 457–466.

Oremland, R. S. and Capone, D. G. 1988. Use of "specific" inhibitors in biogeochemistry and microbial ecology. – Adv. Microb. Ecol. 10: 285–383.

Papen, H., von Berg, R., Hinkel, I., Thoene, B. and Rennenberg, H. 1989. Heterotrophic nitrification by *Alcaligenes faecalis:* NO_2^-, NO_3^-, N_2O, and NO production in exponentially growing cultures. – Appl. Environ. Microbiol. 55: 2068–2072.

– and Rennenberg, H. 1990. Microbial processes involved in emissions of radiatively important trace gases. – In: Transactions of the 14th int. congress of soil science. Int. Soc. of Soil Sci., Kyoto, vol. II, pp. 232–237.

Poth, M. A. and Focht, D. D. 1985. ^{15}N kinetic analysis of N_2O production by *Nitrosomonas europaea:* an examination of nitrifier denitrification. – Appl. Environ. Microbiol. 49: 1134–1141.

Rennenberg, H. 1991. The significance of higher plants in the emission of sulfur compounds from terrestrial ecosystems. – In: Sharkey, T. D., Holland, E. A. and Mooney, H. A. (eds), Trace gas emissions from plants. Academic Press, New York, pp. 217–260.

Ritchie, G. A. F. and Nicholas, D. J. D. 1972. Identification of the sources of nitrous oxide produced by oxidative and reductive processes in *Nitrosomonas europaea*. – Biochem. J. 126: 1181–1191.

Robertson, G. P. and Tiedje, J. M. 1987. Nitrous oxide sources in aerobic soils: nitrification, denitrification, and other biological processes. – Soil Biol. Biochem. 19: 187–193.

Rosswall, T., Bak, F., Baldocchi, D., Cicerone, R. J., Conrad, R., Ehhalt, D. H., Firestone, M. K., Galbally, I. E., Galchenko, V. F., Groffman, P. M., Papen, H., Reeburgh, W. S. and Sanhueza, E. 1989. What regulates production and consumption of trace gases in ecosystems: biology or physicochemistry? – In: Andreae, M. O. and Schimel, D. S. (eds), Exchange of trace gases between terrestrial ecosystems and the atmosphere. Wiley, Chichester, pp. 73–95.

Sass, R. L., Fischer, F. M., Harcombe, P. A. and Turner, F. T. 1990. Methane production and emission in a Texas agricultural wetland. – Global Biogeochem. Cycles 4: 47- 68.

Schütz, H., Conrad, R., Goodwin, S. and Seiler, W. 1988. Emission of hydrogen from deep and shallow freshwater environments. – Biogeochem. 5: 295–311.

– and Seiler, W. 1989. Methane flux measurements: methods and results. – In: Andreae, M. O. and Schimel, D. S. (eds), Exchange of trace gases between terrestrial ecosystems and the atmosphere. Wiley, Chichester, pp. 209–228.

– , Seiler, W. and Conrad, R. 1989a. Processes involved in formation and emission of methane in rice paddies. – Biogeochem. 7: 33–53.

– , Holzapfel-Pschorn, A., Conrad, R., Rennenberg, H. and Seiler, W. 1989b. A three years continuous record on the influence of daytime, season, and fertilizer treatment on methane emission rates from an Italian rice paddy field. – J. Geophys. Res. 94: 16405–16416.

– , Seiler, W. and Rennenberg, H. 1990a. Soil and land use related sources and sinks of methane (CH_4) in the context of the global methane budget. – In: Bouwman, A. F. (ed.), Soils and the greenhouse effect. Wiley, Chichester, pp. 269–285.

– , Seiler, W. and Conrad, R. 1990b. Influence of soil temperature on methane emission from rice paddies. – Biogeochem., (in press).

– , Schröder, P. and Rennenberg, H. 1991. Role of plants in regulating the methane flux to the atmosphere. – In: Sharkey, T. D., Holland, E. A. and Mooney, H. A. (eds), Trace gas emissions from plants. Academic Press, New York, (in press).

Seiler, W., Holzapfel-Pschorn, A., Conrad, R. and Scharffe,

D. 1984. Methane emission from rice paddies. – J. Atm. Chem. 1: 241–268.

Smith, C. J., Delaune, R. D. and Patrick Jr., W. H. 1982. Nitrous oxide emission following urea-N fertilization of wetland rice. – Soil Sci. Plant Nutr. 28: 161–171.

Takai, Y. and Asami, T. 1962. Formation of methyl mercaptan in paddy soil. – Soil Sci. Plant Nutr. 8: 40–44.

Taylor, B. F. and Kiene, R. P. 1989. Microbial metabolism of dimethyl sulfide. – In: Saltzman, E. S. and Cooper, W. J. (eds), Biogenic sulfur in the environment. Am. Chem. Soc., ACS Symp. Series 393, pp. 202–221.

Tiedje, J. M. 1988. Ecology of denitrification and dissimilatory nitrate reduction to ammonium. – In: Zehnder, A. J. B. (ed.), Biology of anaerobic microorganisms. Wiley, New York, pp. 179–244.

Trevitt, A. C. F., Freney, J. R., Denmead, O. T., Zhu Zhau Liang, C. G.-X. and Simpson, J. R. 1988. Water-air transfer resistance for ammonia from flooded rice. – J. Atm. Chem. 6: 133–147.

Vogels, G. D., Keltjens, J. T. and van der Drift, C. 1988. Biochemistry of methane production. – In: Zehnder, A. J. B. (ed.), Biology of anaerobic microorganisms. Wiley, New York, pp. 707–770.

Wang, M. X., Dai, A., Shen, R. X., Wu, H. B., Schütz, H., Rennenberg, H. and Seiler, W. 1990. CH_4 emission from a Chinese rice paddy field. – J. Sci. China, (in press).

Watanabe, I. and Furusaka, C. 1980. Microbial ecology of flooded rice soils. – Adv. Microbial Ecol. 4: 125–168.

Whittenbury, R., Phillips, K. C. and Wilkinson, J. F. 1970. Enrichment, isolation and some properties of methane-utilizing bacteria. – J. Gen. Microbiol. 61: 205–218.

Widdel, F. 1988. Microbiology and ecology of sulfate- and sulfur-reducing bacteria. – In: Zehnder, A. J. B. (ed.), Biology of anaerobic microorganisms. Wiley, New York, pp. 469–586.

Yagi, K. and Minami, K. 1990. Effects of organic matter applications on methane emission from japanese paddy fields. In: Bouwman, A. F. (ed.), Soils and the greenhouse effect. Wiley, Chichester, pp.467–473.

Yamane, I. and Sato, K. 1961. Effect of temperature on the formation of gases and ammonium-nitrogen in the water-logged soils. – Sci. Rep. Res. Inst. Tohoku Univ. D (Agr.) 12: 31–46.

Zehnder, A. J. B. and Stumm, W. 1988. Geochemistry and biogeochemistry of anaerobic habitats. – In: Zehnder, A. J. B. (ed.), Biology of anaerobic microorganisms. Wiley, New York, pp. 1–38.

Ecological Bulletins 42: 174–182. Copenhagen 1992

Agronomic practices affecting methane fluxes from rice cultivation

Heinz-Ulrich Neue

Neue, H.-U. 1992. Agronomic practices affecting methane fluxes from rice cultivation. – Ecol. Bull. (Copenhagen) 42: 174–182.

Only few experimental data are available on trace gas emission from ricefields. Methane is the most important greenhouse gas released from ricefields to the atmosphere. Methane emission from ricefields have been estimated to account for 20% of global methane emission. Major rice systems, soil processes upon flooding, and agronomic practices are described and discussed in view of methane emission. There is a wide scope of mitigation options without sacrificing rice yields, especially through water management, fertilization and rice cultivars. The complex interactions between methane production, oxidation, and emission in ricefields have to be further elucidated before efficient and acceptable mitigating technologies can be developed.

H.-U. Neue, Int. Rice Res. Inst., P.O. Box 933, Manila, Philippines.

Introduction

Rice is grown under a wider variety of climatic, soil, and hydrological conditions than any other crop. It is grown from the equator to latitudes as far as 50°N and 40°S, and from sea level to altitudes of >2,500 m. The temperature may be as low as 4°C during the seedling stage and as high as 40°C at flowering. Rice is grown irrigated in arid areas, and in rainfed areas with only 500 mm rain per season. Rice is cultivated as a dryland crop and in soils that are submerged to >1 m. It is the only major crop grown on flooded soils.

Anaerobic decay of organic matter in wetland ricefields is a major source for increasing methane fluxes to the atmosphere. Nitrogen transformation in rice soils could result in nitrous oxide emission. Both gases significantly enhance global warming and affects various chemical reactions in the atmosphere. Ricefields may account for 50% of methane emission from agricultural activities or 20% of global methane emissions (US-EPA 1990). The harvested wetland rice area and irrigation will likely increase to meet the demand for rice in the next decades. Mitigation options are required to stabilize atmospheric methane concentrations.

Since only few experimental data are available on trace gas emission from ricefields the objective of this paper is to give an overview of major rice environments, properties of rice soils, and agronomic practices, and to review and deduce effects of agronomic practices on trace gas fluxes from rice cultivation.

Rice environments

Because the water regime during the growing season is the most discriminating physical factor, ricelands can be ordered into two main systems: wetland rice and dryland rice. Neue (1989) outlined a comprehensive classification of rice environments by applying water source and floodwater depth as diagnostic criteria. Further differentiation is made by modifiers related to climate, landforms, fieldforms, floodwater, soil, and cropping systems (Table 1). The total area of dryland rice harvested is c. 20 mill ha and that of wetland rice is c. 120 mill ha (Table 2). In less developed countries where rice is the staple food, irrigated riceland supply, on an average, c. 70% of the total rough rice production (IRRI 1989).

Rice soils

Soil reduction

Flooding an air-dry soil sets in motion a series of physical, chemical, and biological processes that profoundly influence the stable and dynamic properties of soils. The limited O_2 diffusion into the soil, the supply of biodegradable C, and the activity of the edaphon are the key to most of the characteristic biochemical and chemical processes in flooded soils. The chemistry of flooded soils has recently been reviewed by Neue (1988). These processes include soil reduction and associated electrochemical changes; N immobilization and

Table 1. Classification of rice environments (adapted from Neue 1989).

The diagnostic characteristics							
Water source	Irrigation		No irrigation (pluvial, phreatic, surface flow, tidal)				
Floodwater depth (cm)	0–5	5–25	0–25	25–50	50–100	>100	<0
Rice environment	Shallow irrigated rice	Medium irrigated rice	Shallow rainfed rice	Medium rainfed rice	Deep-water rice	Very deepwater rice	Non flooded rice
Land ecosystem	Wetland						Dryland

The modifiers	
Climate	
Zone	Temperate, subtropical, tropical (low altitude, high altitude)
Temperature regime at different growth stages	Low (<20°C), high (>35°C), pattern of daily means, minimum and maximum
Rainfall pattern	Even, monomodal, bimodal, long rainfall season (>4 months), short rainfall season (<4 months)
Day length (growing season)	Short (<12 h), long (>12 h)
Landforms	Inland valley, alluvial fan, piedmont plains, floodplains (river, lacustrine, coastal), tidal land, terraces, undulating, rolling, hilly, steep slopes
Fieldforms	Not levelled, levelled, unbunded, bunded (with levees)
Floodwater	
Floodwater regime	Assured supply and control of floodwater, controlled floodwater, irregular prolonged flooding, irregular brief floodings, tidal flooding, flash-floods, consecutive days of flooding (<30, 30–60, 60–90, 90–120, >120) per season
Floodwater quality	pH, EC, particulates (organic, inorganic), solutes (organic, inorganic), microbiology
Soil	
Soil moisture regime	Epiaquic[a], aquic, peraquic, udic, perudic, ustic, xeric, aridic
Water holding capacity in the root zone (water available for plant)	Low (<140 mm), medium (140–180 mm), high (180–220 mm), very high (>220 mm)
Percolation rate/d^{-1}	Low (<2 mm), medium (2–10 mm), high (>10 mm)
Soil epipedon	Anthropic, histic, mollic, ochric, umbric, vertic[a], salic[a], natric[a], andic[a]
Epipedon reaction	Sulfuric[a], allic, dysic, acid, euic, nonacid, calcareous, alkaline[a]
Subsurface horizon	Albic, argillic, agric, calcic, cambic, kandic, oxic, natric, salic, spodic, sulfuric, sulfidic[a]
Other characteristics	Depth, texture/skeleton, mineralogy, amorphous colloids, plinthite, structure, underlying pan layer, organic C, CEC
Nutrient def/tox	N, P, K, Zn, S, Si, Cu, Fe, Mn, Al, B, Na
Cropping system	1. Monoculture: single rice, double rice, triple rice 2. Multiple culture: intercropping (sorjan[b], others) crop rotation (seasonal, periodical)
Length of growing period of rice crop (days)	Very short (<100), short (<101–120), medium (121–140), long (>140)
Land preparation	Zero tillage, dry tillage, wet tillage, puddled
Cultural practices	1. Transplanting: random, controlled 2. Direct seeding: dry seeding, wet seeding, water seeding (broadcasting, drilling, dibbling)
Fallow system	Flood fallow, dry fallow

[a] Terms do not strictly follow Soil Taxonomy.
[b] Sorjan (Indonesian) is a cropping system in which flooded rice and upland crops are grown simultaneously in alternate ditches and raised beds.

fixation; production of an array of organic compounds, especially organic acids; and release of NH_4^+, CO_2, CH_4, and H_2S. Fig. 1 illustrates the changes of some soil characteristics upon flooding.

The magnitude of reduction is determined by the amount of easily degradable organic substrates; their rate of decomposition; and the amounts and kinds of reducible nitrates, iron and manganese oxides, sulfates, and organic compounds. The pattern of the decreasing Eh after flooding is controlled by decomposition rates of organic matter and the content of nitrates and Mn/Fe oxides. The most important redox buffer systems in rice soils are Fe(III) oxyhydroxides/Fe(II) and organic compounds stabilizing the Eh of the soil solution somewhat between +100 and −100 mV (Ponnamperuma 1984). Corresponding Eh values of the bulk soil may be up to 150 mV lower.

Although the reduction of flooded soils proceeds stepwise in a thermodynamic sequence, the given oxidation-reduction systems (Ponnamperuma 1972, Patrick and Reddy 1978) are only partially applicable to field conditions. The mineral phases present in soils are not pure and often unknown, and a large portion of reduced Fe^{2+} and Mn^{2+} ions are held at ion exchange sites (Tsuchiya et al. 1986). Changes in pH and activities of reactants and products can also alter the order of

Table 2. Harvested rice area by region and rice ecosystem 1988[a].

Region	Rice ecosystem (1000 ha)				
	Irrigated	Rainfed	Deepwater	Upland	Total
East Asia	34,253	1,908	0	824	36,985
South East Asia	12,763	15,058	4,263	3,946	36,030
South Asia	18,062	21,450	8,062	7,046	54,620
Latin America	2,634	483	0	4,794	7,911
Africa	1,007	1,315	152	2,986	5,460
Others	3,081	0	0	0	3,081
World	71,800	40,214	12,477	19,596	144,087

[a] World Rice Facts 1989. Int. Rice Res. Inst. (1990).

redox reactions. As a consequence, reduction potentials of a given redox reaction span a fairly wide range.

Chemical reactions that are favored thermodynamically are not necessarily favored kinetically. In soils, the catalysis of redox reactions is mediated by microbial organisms.

Katyal (1977) demonstrated the acceleration and intensification of Eh and pH change with the addition of plant residues. The effect of vetch, which has a low C–N ratio was greater than that of rice straw (Yu 1985). Changes were more pronounced when organic substrates were added to soil low in organic matter (Nagarajah et al. 1989).

Depending on texture, mineralogy, soil and aquatic fauna, and land preparation, flooding increases or decreases soil bulk density. Increasing the soil bulk density of flooded soils retards organic matter decomposition,

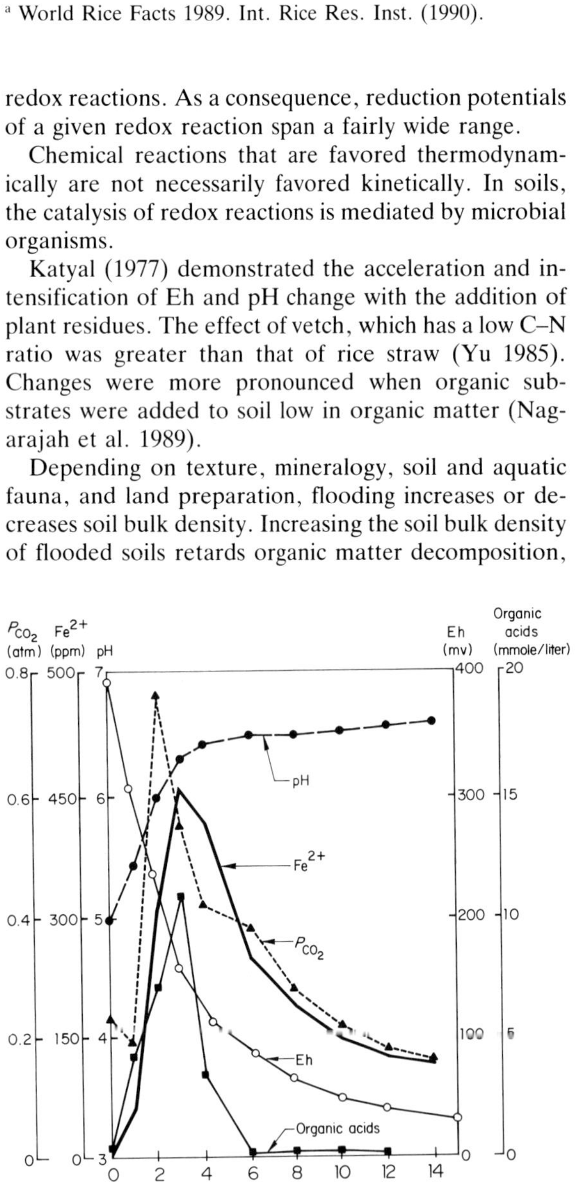

Fig. 1. Kinetics of pH, Eh, P_{CO_2}, water-soluble Fe^{2+} and organic acids in the soil solution of a flooded Ultisol at 30°C, (adapted from Ponnamperuma 1985).

increases the concentration and residue time of organic acids, and reduces the speed of Eh and pH changes.

pH change and partial pressure of CO_2

The increase in pH of acid soils is initially brought about by soil reduction of Fe- oxyhydroxides conforming to

$$pE = 17.8 + pFe^{2+} - 3\ pH$$

in the soil solution (Ponnamperuma 1984).

The pH decrease of sodic and calcareous soils and the final regulation of the pH rise in acid soils are the results of CO_2 accumulation. The pH values at steady state of flooded alkaline, calcareous and acid soils are highly sensitive to the partial pressure of CO_2. Carbon dioxide that accumulates in large amounts profoundly influences the chemical equilibria of almost all divalent cations (Ca^{2+}, Mg^{2+}, Fe^{2+}, Mn^{2+}, Zn^{2+}) in flooded soils.

Up to 2.6 t CO_2 ha^{-1} is produced in the puddled layer during the first few weeks of flooding (IRRI 1964). After addition of organic substrates, the partial pressure of CO_2 in a flooded soil may reach a peak of almost 100 KPa (Ponnamperuma 1985, Neue and Bloom 1989). Typical values in flooded soils range from 5 to 20 kPa (Kundu 1987, Patra 1987). Carbon dioxide concentrations greater than 15 KPa retard root development, leading to wilting and reduced nutrient uptake (Dent 1986).

Organic matter decomposition

The optimum temperature range for anaerobic decomposition of rice straw was found to be 30–35°C (Acharya 1935). Submergence of soils retards initial decomposition of rice straw in the field only slightly, compared with the situation in upland soils (Neue and Scharpenseel 1987). In tropical flooded rice soils, decomposition of the remaining, more resistant metabolites and residues is similar to that in upland soils, with half-lives of c. 2 years if the following conditions are met: soil is intensively puddled each cropping season; soil temperature of the puddled layer is 30–35°C; neutral pH; low soil bulk density and wide soil-water ratio; shallow floodwater; high and balanced nutrient supply; no long-lasting accumulation of organic acids; permanent supply of energy-rich photosynthetic aquatic and benthic biomass; high diversity of micro- and macroorganisms that provide successive fermentation down to CO_2, CH_4, H_2, and NH_3; supply of O_2 into the reduced layer by rice root excretion and oligochaete population; and diurnal oversaturation of the floodwater with O_2 due to photosynthetic aquatic biomass enhancing the aeration function of oligochaetes.

Decomposition is retarded in puddled soils with low and imbalanced nutrient supply, high bulk density, and low biological diversity and activity, as demonstrated in an Aeric Paleaquult of Northeast Thailand (Snitwongse et al. 1988).

If the biological activity is restricted to bacteria, as in laboratory experiments, the decomposition of rice straw in flooded soils is highly retarded. In laboratory incubation studies, only 7–18% of the incorporated straw was mineralized after 100 days following an order according to soil pH ranging from 4.9 to 6.6 (Capistrano 1988).

Humus of soils that are alternately flooded and drained differ from that of dryland soil (Kuwatsuka et al. 1978, Tsutsuki and Kuwatsuka 1978, Tsutsuki and Kumada 1980). In flooded soils, the degree of humification is less; the content of H_2 and N, higher; the degree of unsaturation, lower; and the contents of carboxyl and phenolic groups, lower; meanwhile, the alcoholic and methoxyl groups are higher. Humification indices given as the ratio of nonhumified to humified materials (Sequi et al. 1986) are high (low humification) in topsoils (0–10 cm) of flooded rice soils and decrease with depth. Electrophoresis of the humified materials displays bands only in the low pH range, revealing mainly low molecular weight, highly oxidized substances in the range of fulvic acids (Nobili, pers. comm.).

In general, waterlogging of soils (continuously reduced) is equated with accumulation of poorly decomposed plant materials and peat formation. But ricefields, even if waterlogged, are different because of soil, water, and crop management. Pedogenesis in tropical ricefields may be more appropriately related to the category of Bonneau's (1982) wet soils with high temperature in all seasons, which show rapid mineralization and weak humification. Based on long-term yield trials in the Philippines, a relation among soil C content, N fertilizer rate, and rice grain yield was established (Neue 1985, Smith et al. 1987). The optimum C content in puddled and flooded soils was found to be 2–2.5%.

Fermentation products

Anaerobic fermentation produces an array of organic substances, many of them transitory and not found in well-aerated soils; Ponnamperuma (1984) listed various gases, hydrocarbons, alcohols, carbonyls, volatile fatty acids, nonvolatile fatty acids, phenolic acids, and volatile S compounds.

If organic substrates are added at the beginning of flooding, the production of volatile acids occurs only in the initial stage of flooding. Short-term H_2 evolution immediately follows the disappearance of O_2 in the first days after flooding. Thereafter, CO_2 production increases; and finally, with decreasing CO_2, CH_4 formation increases (Takai et al. 1956, Neue and Scharpenseel 1984).

With increasing temperature up to 35°C, decomposition starts earlier and is more vigorous in every case. At high temperatures, the formation of CO_2 and CH_4 occurs earlier and is stronger, but that of H_2 and organic acids is smaller in amount, although it also occurs earlier (Yamane and Sato 1961). The period of occurrence and the amount of gaseous products and volatile acids depend largely on temperature and reducing conditions.

At soil temperatures found in flooded tropical soils, CO_2 and CH_4 formation occur sooner and in larger amounts than in cooler climates (Tsutsuki and Ponnamperuma 1987). The annual mean soil temperature in flooded tropical ricefields is similar to the mean maximum air temperature (Neue 1988). The amount of CH_4 found in the soil solution and in gas bubbles of flooded soils may be up to eight times higher than that of CO_2 after the initial stage of flooding. The change in favor of CH_4 is likely to be caused by assimilation of $C0_2$ and precipitation of carbonates rather than reduction of CO_2 to CH_4, but the controlling processes still need elucidation. According to Takai (1970), the bulk of CH_4 is formed through decarboxylation of acetic acid, which would result in a 1:1 ratio of CO_2 and CH_4 formation.

The ratio of CO_2 and CH_4 formation is regulated by the ratio of the oxidizing capacity (amount of reducible O_2, NO_3^-, Mn^{4+}, and Fe^{3+}) to the reducing capacity (Takai 1961). The actual oxidizing capacity is influenced highly by O_2 diffusion from the atmosphere, floodwater, and plant roots; the soil bulk density (soil-water ratio); and fertilization. Production of CH_4 is stimulated under an O_2–free N atmosphere, and higher accumulations of volatile acids are found in soils with higher bulk density (lower soil-water ratios).

Methane production is lower in reduced environments if sulfate concentrations are high as in saline soils. Inhibitory effects and interactions with sulfate-reducing bacteria are given as possible reasons (Mitsch and Gosselink 1986). Competition for H_2 and toxicity of sulfide are possible mechanisms.

Consecutive addition of organic substrates through plant growth and photosynthetic biomass production in the floodwater maintains the fermentation chain and CH_4 production.

Methane emission

Factors influencing CH_4 emission from wetland ricefields have recently been reviewed by Schütz et al. (1989a), Bouwman (1990), and Neue et al. (1990). In considering these factors, it is important to distinguish between processes influencing CH_4 production and oxidation in the soil and processes influencing its escape from the soil once produced. CH_4 production is highly influenced by the soil temperature, the redox potential (Eh), the soil pH, the carbon supply to methanogenic bacteria, toxic substances, and competition with sulfate-reducing bacteria. Holzapfel-Pschorn and Seiler (1986) reported marked increases in CH_4 emission when soil temperature increased from 20 to 25°C. Redox potentials (corrected to pH 7) of −0.15 to −0.19 volts are necessary for methanogenesis, and it is favored by a neutral or slightly above neutral pH. Incorporation of sulfate and nitrate depresses CH_4 production (Jacobsen et al. 1981).

The influence of the application of mineral fertilizer on the in situ CH_4 emission rates was first studied by Cicerone and Shetter (1981) who observed a fivefold increase of CH_4 emission after fertilization with ammonium sulfate. Extended field experiments in Italy indicated that the influence of fertilizers on the CH_4 emission rates is rather complicated and needs further intensive studies (Cicerone and Shetter 1981, Schütz et al. 1989b). In general, the CH_4 emission rates are found to be strongly dependent on the type, the rate, and the application mode of fertilizers. Fertilization with deeply incorporated urea or ammonium sulfate decreased CH_4 emission rates by c. 50%. But preliminary data obtained from field experiments in China did not indicate a significant effect of mineral fertilizer (e.g., KCl, K_2SO_4) nor of organic amendments (animal manure) on CH_4 emission rates (Schütz et al. 1990). Application of rice straw greatly enhanced CH_4 emission, whereas effects of compost application have been found negligible.

Sebacher et al. (1986) found that CH_4 emission increased in tundra wetlands with floodwater depths up to 10 cm; greater depths decreased CH_4 emission. Floodwater depth greatly influences photosynthetic aquatic biomass production and aquatic fauna activity in wetland ricefields, both of which may be important in CH_4 production and escape. Oxidation of CH_4 in the thin oxidized floodwater-soil interface may greatly reduce CH_4 emission (Bont et al. 1978). Holzapfel-Pschorn et al. (1986) reported that 67% of the CH_4 produced during a rice crop growing season was oxidized and only 23% was emitted to the atmosphere. Factors influencing CH_4 escape from the soil are related to microbial CH_4 oxidation, and processes controlling ebullition, diffusion, and transport through the plant (Schütz et al. 1989a). In rice soils, release of CH_4 into the atmosphere is >90% mediated through the rice plant (Seiler 1984). At ripening, rice plants release more CH_4 (Bont et al. 1978). By cutting off the emerged parts of the rice plants, actual CH_4 emission was reduced to a value that corresponds to 10% of the CH_4 emission rate in rice paddies with intact plants (Holzapfel-Pschorn et al. 1986). Unvegetated ricefields emitted c. 50% of the CH_4 released from planted fields, and the emission was almost exclusively due to ebullition (Schütz et al. 1989b).

Significant variations in CH_4 emission have been observed under different water regimes. High percolation rates (20 mm d^{-1}) reduced methane emission to almost zero (Yagi et al. 1990).

The CH_4 emission rates show significant diurnal and seasonal variations which, in addition, seemed to be dependent on climate and field management. In Italy, maximum daily CH_4 emission rates were observed during late afternoon when topsoil temperatures reached maximum values (Schütz et al. 1989b). The CH_4 emission rates in ricefields of China exhibited two different diurnal patterns for the early and the late rice vegetation period (Schütz et al. 1990). During the early vegetation period, two maximum CH_4 emission rates occurred, one at noon and one during the night. In the late vegetation period, there was only a nocturnal maximum in the diurnal course of CH_4 emission.

In addition to the high diurnal variations, specific seasonal maxima of CH_4 emission have been monitored. Three seasonal maxima were found in Italy: the first one, shortly after flooding; the second one, during the reproductive phase of the rice plants; and the third one, in two of the three-year observation time, in the late stage of the vegetation period (Schütz et al. 1990). In China, the early rice crop also showed three maxima, whereas the late rice crop showed only two (Schütz et al. 1990).

Three seasonal maxima were also observed in Japan while the early maximum was absent in Texas, USA. In California rice fields Cicerone et al. (1983) found two peaks of methane emission. Emission rates were extremely high at the second peak during the last month of the growing season.

The first seasonal maximum seems to reflect the initial methane production upon flooding. The following maxima seem to be induced by increased root exudation, root degradation and/or cycles of decay of aquatic biomass.

Agronomic practices

Various rice culture systems and agronomic practices have developed to suit the physical, biological, and socioeconomic conditions of different regions and environments. Effects of agronomic practices on growth and yield of rice are fairly well documented but interrelationships to CH_4 emission are poorly understood. Major agronomic practices are described in view of mitigation/adaptation options that may be available without sacrificing rice yields.

Land preparation

Tillage operations vary according to water availability, soil texture, topography, rice culture, and resources available. Kawaguchi and Kyuma (1977) found that 40% of the tropical rice soils they studied had at least 45% clay. Soils with such high clay content have poor structure and are hard when dry. Since hand- and animal-powered tillage is still common in most Asian countries and the principle forms of mechanization are only the 10–15 hp tiller (hand tractor), wet tillage is the preferred land preparation. This comprises soaking land until the soil is saturated, plowing, puddling and harrowing. One third of the total water required for a rice crop is needed for the wet field preparation.

According to De Datta (1981), the advantages of wet tillage are: improved weed control; ease of transplanting; establishment of reduced soil condition which improves soil fertility and fertilizer management; reduced draft requirement; reduced water percolation; reliability of monsoon rains by the time land preparation is completed; higher fertilizer efficiency, especially for N-fertilizer.

Flooding and puddling a soil provides an ideal growth medium for rice by supplying abundant water, buffering soil pH near neutral, enhancing N fixation and carbon supply, and increasing diffusion rates, mass flow, and availability of most nutrients. Standing water stabilizes the soil moisture regime, moderates soil temperature, and prevents soil erosion. Flooding a ricefield results in better growth and produces higher grain yields.

A moisture stress of 50 kPa (slightly above field capacity) may reduce grain yield to 20–25% of the yield of continually flooded treatments (De Datta 1981). Water stress at any growth stage may reduce yield but the rice plant is most sensitive to water deficit during the reproductive stage, causing a high percentage of sterility (Yoshida 1981). Water deficits during the vegetative stage may reduce plant height, tiller number, and leaf area which may also highly reduce yields if plants do not recover before flowering. Duration of a moisture stress is more important than the growth stage at which the stress occurred.

If initial crop growth is essential to prevent damage by early monsoonal rainfall and subsequent floods, such as in deepwater rice ecosystems, tillage and seeding are done in dry soils. Less than 200 mm of rainfall during the planting month leads to dryland preparation and seeding. In other rainfed rice ecosystems, labor constraints associated with seedbed preparation, land preparation, and transplanting may force farmers to adopt dryland tillage and seeding to ensure timely crop establishment. Where large power units can be employed because of available capital, dryland tillage is commonly practiced. In the USA and Australia, rice is generally sown in dry soil which is flooded after crop establishment. Upland ricelands are never flooded and tillage is, of course, the same as for other upland crops.

Dryland tillage and dry seeding causes initial aerobic decomposition of incorporated organic substrates slows down the decrease of the redox potential, and shortens the anaerobic phase. As a result methane production will be delayed and likely to be lower. Minimum or zero tillage should have similar effects. However, these land preparation and seeding techniques may require new rice cultural types, different modes of fertilization, higher powered tillage implements, and alternative seeding and weeding techniques.

Seeding and transplanting

Transplanted rice is the major system of rice culture in most of tropical Asia. Direct seeding of pregerminated rice in wet prepared soils becomes popular in areas with good water control and if manpower is lacking or becoming expensive. Pregerminated seeds are generally broadcast onto puddled fields without standing water (saturated soil moisture) and the field is flooded after crop establishment. The crop duration (vegetative phase) is shorter for direct-seeded rice, avoiding delay due to seedbed preparation, transplanting, and reduced initial growth because of the transplanting shock. Weed control and possible moisture stresses because of insufficient water control are the main obstacles in direct wet-seeded rice. The yield potential of direct wet-seeded rice is similar to that of transplanted rice (De Datta 1981).

The advantages of transplanting rice seedlings are: lower seed requirement (increased tillering); save seed establishment (control of pests and fertilization); variable schedule of field establishment is possible without risk; less requirements for floodwater control; and tolerance to biotic and abiotic stresses increases with seedling age.

Methane fluxes should vary between direct wet-seeded and transplanted rice because: crop duration is shorter in direct seeded rice; soil surface is aerated for 7–14 days after land preparation in direct-seeded rice; growth pattern and canopy development differ; and transplanting causes additional soil disturbances.

Fertilization

The most deficient nutrient for high wetland rice yields is N, followed by P, K, and Zn. In general, K and P are applied basally during the final land preparation. Zn may be added by seed treatments, dipping seedling roots in ZnO solution or broadcasting of Zn salts at the time Zn deficiency symptoms occur.

Potassium chloride is the principal fertilizer source of K and superphosphate is the primary source of P fertilizer. On acid rice soils, phosphate rock may be applied. The most common source of N fertilizer in wetland rice is urea, followed by ammonium-containing fertilizers such as ammonium sulfate. Most farmers use N fertilizer in two or three split applications, with part during final land preparation or shortly after planting and the remainder as topdressing at later growth stages, especially at the early panicle initiation stage.

Studies on fertilizer use and crop management to minimize N losses (volatilization of NH_3, nitrification, followed by denitrification) and to increase the efficiency of fertilizer have recently been reviewed (De Datta 1981, De Datta and Patrick 1986, De Datta 1987). For basal application, ammonium-containing or ammonium-producing (urea) N fertilizers are recommended (De Datta 1981) to minimize denitrification losses. Volatilization losses can be reduced by incorporation or deep placement of N fertilizer without standing water at final harrowing. Broadcasting basal N fertilizer into floodwater results in extensive N losses (as ammonia) to the atmosphere due to high pH values as a result of the algal assimilation or alkaline irrigation water (Fillery and Vlek 1986).

Nitrogen transformation in rice soils results in N_2O emissions. Denitrification, a multistep reduction of nitrate to N_2, creates N_2O as an intermediate product.

The development of the thin, oxidized surface layer followed by reduced soil favors the simultaneous occurrence of nitrification (ammonium oxidation) and denitrification in flooded rice soils. Both processes can also occur at the soil-root interface.

Denitrification is recognized as a major cause of N loss from flooded rice soils but only few studies on direct denitrification measurements from flooded ricefields have been published (Buresh and Austin 1988, Mosier et al. 1989). The production of N_2 and N_2O was shown to be a significant loss mechanism of fertilizer N but N_2O emissions into the atmosphere accounted only 1–2%. Frequent flooding and drying commonly occurring in rainfed rice should result in much higher N_2O emission. The factors and processes influencing N_2O emission suggest that CH_4 and N_2O production are inversely related. CH_4 production seems to be highest in flooded soils with expanding clays and high organic matter content, the very same soils that, if not flooded, are major producers of N_2O. In changing agronomic practices to reduce the fluxes of greenhouse gases, it is important to assess the trade-off between CH_4 and N_2O emission.

Reports on the influence of the mineral fertilizer application (source, mode, and rate) on CH_4 production and emission are inconsistent. The complex interrelationships of fertilization on the biochemistry (CH_4 production) of flooded soils and on plant growth (plant-mediated emission) still have to be elucidated.

It is evident that organic amendments of flooded soils increase CH_4 production and emission (Schütz et al. 1989b, 1990) by lowering the Eh and providing carbon sources. But methane production and fluxes should differ according to quantity and quality of organic substrates (C–N ratio, cellulose content, degree of humification). Application of rice straw (wide C–N ratio) increased CH_4 emission significantly while addition of compost (humified, low C–N ratio) showed negligible effects.

Pest control

Control of pests (insects, diseases, weeds, snails, birds, rats) of rice ranges from varietal resistance to cultural, biological, and chemical control. Application of pesticides to the floodwater, soil surface, or into the soil may have significant effects on methane fluxes, especially if it adversely affects the aquatic and soil flora and fauna. Many cultural practices, such as management of cropping pattern, crop residue, tillage, water, fertility, weeding, or plant spacing/population control pests in rice and affect methane fluxes at the same time. For example, the mechanical disturbance of the soil during weeding of transplanted rice (2–3 times per season by hand or small implements) increases the release of gases trapped in the soil. Weeds become an additional source of methane since they are commonly returned to the reduced soil. Furthermore, aquatic weeds may provide a more efficient pathway of CH_4 to the atmosphere than rice plants, as indicated by Holzapfel-Pschorn et al. (1986).

Water control

Water control (irrigation and drainage) is one of the most important factors in rice production. In many ricefields crops suffer from either too much or too little water because of irregular rainfall and landscape pattern.

During the monsoonal rainy season in tropical Asia, ricelands are naturally flooded. Excess water is a serious constraint in river, lacustrine, and coastal floodplains. Bunding of ricefields, and its collateral levelling of sloping land, considerably changes the water regime of that land. The overall effect is that uncontrolled runoff of water is minimized and more water, whether from natural sources or from irrigation, is retained on or in the soil. Bunding and levelling of ricefields is a perfect measure of erosion protection and allows efficient water harvesting and water conservation.

The following parameters of floodwater regimes are important for rice growth and methane fluxes: duration and depth of flooding; regularity of flooding as determined by climate, relief, and regime of rivers; and degree to which flooding is controlled by irrigation, flood protection, and drainage.

A ricefield may have two distinct flooding regimes in a given year, especially in pronounced wet-dry monsoonal climates. A ricefield may be naturally shallow to moderately deep flooded during the rainy season and fall dry or becomes shallow flooded by irrigation water during the dry season. In deepwater rice areas, floodwater depth may increase up to 4 meters for most of the growing season after dry-seeded rice is established at the onset of the wet season.

In subtropical East Asia (China, Korea, Japan), percolation rates between 8 and 20 mm d^{-1} are considered essential if paddy soils are to be highly productive (Si-Tu Soong and Zhang Weih 1985). CH_4 emission is negatively correlated to percolation rates. Yagi et al. (1990) found that CH_4 emission was reduced to almost zero at a percolation rate of 20 mm d^{-1}.

Water control is a prerequisite for high yield technology in rice cultivation and will likely be a key issue for reducing methane fluxes. Intermittent drainage and aeration are promising mitigation candidates but resulting N_2O production has to be minimized.

Rice cultivars

Since the major pathway of methane from the rice soil to the atmosphere is the aerenchyma of rice plants, differences among the 80,000 rice accessions conserved in IRRI's germplasm bank should exist. The well-developed air spaces in leaf blades, leaf sheath, culms and roots provide an efficient gas exchange between the atmosphere and the anaerobic soil. Oxygen diffusion from rice roots via the aerenchyma constitutes an important part of the root-oxidizing power aside from enzymatic oxidation due to hydrogen peroxide production. Large cultivar differences in the amount of oxygen released from rice roots have already been found (Yoshida 1981) and breeding for rices that emit less CH_4 is highly feasible.

Conclusion

Breeding rices for lower methane emission is a highly desirable approach because farmers could easily adapt this new technology. But the degree to which plant breeders may succeed in this attempt may be limited. Economically feasible soil, water, and crop management practices have to be developed at the same time, in order to maintain rice productivity and reduce methane emission. Although there is a wide scope of mitigation options, current understanding of the complex interaction between methane production, oxidation and emission to the atmosphere is insufficient. Only an integrated, interdisciplinary approach will contribute to efficient and acceptable mitigating technologies and to a better understanding of processes controlling methane fluxes.

References

Acharya, C. N. 1935. Studies on the anaerobic decomposition of plant materials. II. Some factors influencing the anaerobic decomposition. – Biochem. J. 29: 953–960.

Bonneau, M. 1982. Soil temperature. – In: Bonneau, M. and Souchier, B. (eds), Constituents and properties of soils. Academic Press, London, pp. 366–371.

Bont, J. A. M. de, Lee, K. K. and Bouldin, D. F. 1978. Bacterial oxidation of methane in a rice paddy. – Ecol. Bull. 26: 91–96.

Bouwman, A. F. 1990. The role of soils and land use in the greenhouse effect. – In: Bouwman, A. F. (ed.), Soils and the greenhouse effect. Wiley, New York. (in press)

Buresh, R. J. and Austin, E. R. 1988. Direct measurement of dinitrogen and nitrous oxide flux in flooded rice fields. – Soil Sci. Soc. Am. J. 52: 681–688.

Capistrano, R. F. 1988. Decomposition of ^{14}C-labelled rice straw in 3 submerged soils under controlled laboratory conditions. – M. Sci. thesis, Univ. of Philippines at Los Baños, Laguna, Philippines.

Cicerone, R. J. and Shetter, J. D. 1981. Sources of atmospheric methane: measurements in rice paddies and a discussion. – J. Geophys. Res. 96: 7203–7209.

– , Setter, J. D. and Delwich, C. C. 1983. Seasonal variation of methane flux from a California rice paddy. – J. Geophys. Res. 88: 11022–11024.

De Datta, S. K. 1981. Principles and practices of rice production. – Wiley, New York.

– 1987. Nitrogen transformation processes in relation to improved cultural practices for lowland rice. – Plant Soil 100: 47–69.

– and Patrick, W. H. Jr. (eds). 1986. Nitrogen economy of flooded rice soils. – Martinus Nijhoff Publ., Dordrecht.

Dent, D. 1986. Acid sulfate soils: a baseline for research and development. – ILRI Public. 39, Wageningen.

Fillery, I. R. P. and Vlek, P. L. G. 1986. Repraisal of the significance of ammonia volatilization as a N-loss mechanism in flooded rice fields. – In: Datta, S. K. and Patrick, W. H. Jr. (eds). Nitrogen economy of flooded rice soils. Martinus Nijhoff Publ., Dordrecht.

Holzapfel-Pschorn, A., Conrad, R. and Seiler, W. 1986. Effects of vegetation on the emission of methane from submerged paddy soil. – Plant Soil 92: 223–233.

– and Seiler, W. 1986. Methane emission during a cultivation period from an Italian rice paddy. – J. Geophy. Res. 91: 11803–11814.

IRRI (Int. Rice Res. Inst.). 1964. Annual report for 1963. – Manila, Philippines.

– 1989. Towards 2000 and beyond. – Manila, Philippines.

– 1990. World Rice Facts, 1989. – Manila, Philippines.

Jacobsen, P., Patrick, W. H. Jr. and Williams, B. G. 1981. Sulfide and methane formation in soils and sediments. – Soil Sci. 132: 279–287.

Katyal, J. C. 1977. Influence of organic matter on chemical and electrochemical properties of some flooded soils. – Soil Biol. 9: 259–266.

Kawaguchi, K. and Kyuma, K. 1977. Paddy soils in tropical Asia: their material, nature and fertility. – The Univ. Press of Hawaii, Honolulu, HI, USA.

Kundu, D. K. 1987. Chemical kinetics of aerobic soils and rice growth. – Ph.D. thesis, Indian Agric. Res. Inst., New Delhi, India.

Kuwatsuka, S., Tsutsuki, K. and Kumada, K. 1978. Chemical studies on humic acids. I. Elementary composition of humic acid. – Soil Sci. Plant Nutr. 23: 337–347.

Mitsch, W. J. and Gosselink, J. G. 1986. Wetlands. – Van Nostrand Reinhold Comp., New York.

Mosier, A. R., Chapman, S. L., Freney, J. R. 1989. Labelled dinitrogen in atmosphere, floodwater and porewater of a flooded rice field fertilized with 15-N urea. – Fert. Res. 19: 127–136.

Nagarajah, S., Neue, H.-U. and Alberto, M. C. R. 1989. Effect of Sesbania, Azolla and rice straw incorporation on the kinetics of NH_4, K, Fe, Mn, Zn and P in some flooded rice soils. – Plant Soil 116: 37–48.

Neue, H.-U. 1985. Organic matter dynamics in wetland soils. – In: Wetland soils: characterization, classification and utilization. Int. Rice Res. Inst., Manila, Philippines, pp. 109–122.

– 1988. Holistic view of chemistry of flooded soil. – In: Proceedings of the first int. symp. on paddy soil fertility, 6–13 Dec. 1988. Chiang-mai, Thailand, IBSRAM, Bangkok, pp. 21–56.

– 1989. Rice growing soils: constraints, utilization and research needs. – FFTC Book Series No. 39: 1–14, Food and Fertilizer Technol. Center, Taipei.

– , Becker-Heidmann, P. and Scharpenseel, H. W. 1990. Organic matter dynamics, soil properties, and cultural practices in rice lands and their relationship to methane production. – In: Bouwman, A. F. (ed.), Soils and the greenhouse effect. Wiley, New York, (in press).

– and Bloom, P. R. 1989. Nutrient kinetics and availability in

flooded soils. – In: Progress in irrigated rice research. Int. Rice Res. Inst., Manila, Philippines, pp. 173–190.
– and Scharpenseel, H. W. 1984. Gaseous products of the decomposition of organic matter in submerged soils. – In: Organic matter and rice. Int. Rice Res. Inst., Manila, Philippines, pp. 311–328.
– and Scharpenseel, H. W. 1987. Decomposition pattern of ^{14}C-labelled rice straw in aerobic and submerged rice soils of the Philippines. – Sci. Total Environ. 62: 431–434.
Patra, P. K. 1987. Influence of water regime on the chemical kinetics of soils and rice growth. – Ph.D. thesis. Indian Agric. Res. Inst., New Delhi, India.
Patrick, W. H. and Reddy, C. N. 1978. Chemical changes in rice soils. – In: Soils and rice. Int. Rice Res. Inst., Manila, Philippines, pp. 361–380.
Ponnamperuma, F. N. 1972. The chemistry of submerged soils. – Adv. Agron. 24: 29–96.
– 1984. Effects of flooding on soils. – In: Kozlowski, T. T. (ed.), Flooding and plant growth. Academic Press, New York, pp. 9–45.
– 1985. Chemical kinetics of wetland rice soils relative to soil fertility. – In: Wetland soils: characterization, classification, and utilization. Int. Rice Res. Inst., Manila, Philippines, pp. 71–89.
Schütz, H., Seiler, W. and Conrad, R. 1989a. Processes involved in formation and emission of methane in rice paddies. – Biogeochem. 7: 33–53.
– , Holzapfel-Pschorn, A., Conrad, R., Rennenberg, H. and Seiler, W. 1989b. A three-year continuous record on the influence of daytime, season and fertilizer treatment on methane emission rates from an Italian rice paddy field. – J. Geophys. Res., (in press).
– , Seiler, W. and Rennenberg, H. 1990. Soil and land use related sources and sinks of methane (CH_4) in the context of the global methane budget. – In: Bouwman, A. F. (ed.), Soils and the greenhouse effect. Wiley, New York, (in press).
Sebacher, D. I., Harris, R. C., Barlett, K. B., Sebacher, S. M. and Grice, S. S. 1986. Atmospheric methane sources: Alaskan tundra bogs, an alpine fen and a subartic bored marsh. – Tellus 38: 1–10.
Seiler, W. 1984. Contribution of biological processes to the global budget of CH_4 in the atmosphere. – In: Kleig, M. J. and Reddy, C. A. (eds), Current perspectives in microbial ecology. Am. Soc. of Meteorol., pp. 468–477.
Sequi, P., De Nobili, M., Leita, L., and Cerciguani, G. 1986. A new index of humification. – Agrochem. 30: 175–179.
Si-Tu Soong and Zhang Weih. 1985. Subsurface drainage of lowland rice fields in China. – In: Soil physics and rice. Int. Rice Res. Inst., Manila, Philippines, pp. 351–366.
Smith, J., Neue, H.-U. and Umali, G. 1987. Soil nitrogen and fertilizer recommendations for irrigated rice in the Philippines. – Agric. Syst. 24: 165–181.
Snitwongse, P., Pongpan, S., Neue, H.-U. 1988. Decomposition of ^{14}C-labelled rice straw in a submerged and aerated rice soil in northeastern Thailand. – In: Proceedings of the First Int. Symp. on Paddy Soil Fertility, 6–13 Dec. 1988. Chiang-mai, Thailand, IBSRAM, Bangkok, pp. 461–480.
Takai, Y. 1961. Reduction and microbial metabolism in paddy soils (3). – Nogyo Gijitsu (Agro. Tech.) 19: 122–126, (in Japanese with English summary).
– 1970. The mechanism of methane fermentation in flooded soils. – Soil Sci. Plant Nutr. 16: 238.
– , Koyama, T. and Kamura, T. 1956. Microbial metabolism in reduction process of paddy soil. Part I. – Soil Plant Food 2: 63–66.
Tsuchiya, K., Wada, H. and Takai, Y. 1986. Leaching of substances from paddy soils. 4. Water solubilization of inorganic components in submerged soils. – Jpn. J. Soil Sci. Plant Nutr. 57: 593–597.
– and Kumada, K. 1980. Chemistry of humic acids. – Fert. Sci. 3: 93–171, (in Japanese with English summary).
– and Kuwatsuka, S. 1978. Chemical studies on soil humic acids. II. Composition of oxygen-containing functional groups of humic acids. – Soil Sci. Plant Nutr. 24: 547–560.
– and Ponnamperuma, F. N. 1987. Behavior of anaerobic decomposition products in submerged soils. Effects of organic material amendment, soil properties, and temperature. – Soil Sci. Plant Nutr. 33: 13–33.
US-EPA. 1990. Methane emissions and opportunities for control. Workshop results of intergovernmental panel on climate change, response strategies working group, coordinated by Japan Environment Agency and U.S. Environmental Protection Agency, Washington, Sept. 1990.
Yagi, K., Minami, K. and Ogawa, Y. 1990. Effects of water percolation on methane emission from paddy fields. – Res. Rep. Div. Environ. Planning, Nat. Inst. Agro-Environ. Sci. 6: 105–112.
Yamane, I. and Sato, K. 1961. Effect of temperature on the formation of gases and ammonium nitrogen in the waterlogged soils. – Rep. Inst. Agric. Res. Tokoku Univ. 12: 1–10.
Yoshida, S. 1981. Fundamentals of rice crop science. – Int. Rice Res. Inst., Manila, Philippines.
Yu, T. 1985. Physical chemistry of paddy soils. – Springer, Berlin.

Ecological Bulletins 42: 183–187. Copenhagen 1992

Denitrification in intermittently flooded rice fields and N-gas transport through rice plants

A. Bhadrachalam, S. P. Chakravorti, N. K. Banerjee, S. K. Mohanty and A. R. Mosier

Bhadrachalam, A., Chakravorti, S. P., Banerjee, N. K., Mohanty, S. K. and Mosier, A. R. 1992. Denitrification in intermittently flooded rice fields and N-gas transport through rice plants. – Ecol. Bull. (Copenhagen) 42: 183–187.

Field studies were conducted in dry season, intermittently flooded rice *Oryza sativa* L. to quantify nitrogen (N) loss due to denitrification. Nitrogen gas flux (N_2 + N_2O) and N_2O flux were monitored using a closed chamber collection method and mass spectrometric and gas chromatographic analytical techniques throughout two growing seasons in ^{15}N-labeled urea fertilized rice fields. A nitrification inhibitor, coated calcium carbide (CCC), was used during the second season to determine the impact of limiting nitrification on denitrification. Nitrogen gas flux was detectable one day after urea addition at rice transplanting. About 99% of the N gas evolution occurred within 3 weeks of fertilization. Peak emissions occurred 8–10 days after transplanting. Six and 94% of the N evolved as N_2O and N_2 during the first year and 4 and 96% during the second year, respectively. Inclusion of rice plants inside the collection chambers increased measured N gas flux by c. 30%. Adding CCC at transplanting eliminated N_2 and N_2O production from the added urea. During both years N_2 + N_2O emissions from these rice fields amounted to only c. 5% of the fertilizer N applied.

A. Bhadrachalam, S. P. Chakravorti and S. K. Mohanty, Cent. Rice Res. Inst., Cuttack, India. – N. K. Banerjee, Indian Agric. Res. Inst., New Delhi India. – A. R. Mosier, USDA/ARS, P.O. Box E, Ft Collins, CO 80522, USA.

Introduction

Low urea fertilizer N use efficiency in rice is commonly due to loss of 40–70% of the added N from the system (De Datta 1981, Craswell et al. 1984). Ammonia volatilization and N_2 emission following nitrification-denitrification are the important loss mechanisms when leaching and runoff are eliminated (Fillery and De Datta 1986, Freney et al. 1990). Ammonia volatilization has been directly measured at a number of locations in rice fields (Fillery et al. 1986, Freney et al. 1990), but N loss due to denitrification has been quantified indirectly. Only few and very recent studies have measured denitrification directly in rice fields (Buresh and Austin 1988, Mosier et al. 1989, Buresh and De Datta 1990, Lindau et al. 1990).

The purpose of the studies reported here was to directly quantify N_2 and N_2O emissions from urea fertilized, intermittently flooded rice. Specific objectives within this goal were to evaluate the effect on N gas flux of length of time that soil cover gas collection chambers remained in place, the effect of time of day that gas flux measurements were made, the effect of a nitrification inhibitor, coated calcium carbide (CCC) (Banerjee and Mosier 1989) on denitrification, and the effect of including rice plants inside the gas collection chambers on N gas emissions from the saturated soils.

Methods

Field plots

Year 1. The field plots were located at the Central Rice Research Institute, Cuttack, India, experimental field. The soil is a Aeric Fluvaquept that is only slowly permeable to vertical water movement. The soil contained 0.63% C and 0.08% N, with a cation exchange capacity of 13.8 m Eq 100 g^{-1}, pH 5.9. In the center of a 5×20 m field plot, a 1 m^2 plot was isolated by driving a rectangular steel confinement 30 cm into the soil. This small plot was covered before urea (100 kg N ha^{-1}) was surface applied to the entire field plot by hand and four-week old rice seedlings were hand planted (10 cm apart

Table 1. Total N gas flux (N_2 + N_2O + N oxides) measured at different times during the first 22 days[1] after fertilizing and transplanting.

Time after transplanting	Total N gas flux afternoon (1400–1600 h)				evening (1800–2200 h)				overnight (1800–0900 h)			
	1 h		2 h		between		within[2]		between		within	
	mean	SD	mean	SD	mean	SD	mean	SD	mean	SD	mean	SD
Days					g N ha^{-1} d^{-1}							
1	16	21	15	8	10	0	3	0	14	7	14	6
2	38	16	21	9	19	6	27	1	19	2	21	1
3	29	2	32	5	39	4	40	1	35	2	39	13
4	59	18	67	21	100	27	78	4	51	9	68	20
5	210	120	220	78	180	20	210	18	110	26	130	52
6	270	84	270	98	260	57	290	88	210	40	280	93
7	540	180	560	130	340	20	550	150	310	35	510	95
8	500	91	470	100	200	170	210	240	150	25	250	190
9	–[3]	–	490	69	–	–	–	–	750	71	980	110
10	–	–	560	150	–	–	–	–	500	85	580	11
11	–	–	380	52	–	–	–	–	410	28	390	1
12	–	–	300	160	–	–	–	–	270	5	410	36
13	–	–	–	–	–	–	–	–	190	1	270	53
14	–	–	170	42	–	–	–	–	160	34	220	33
15	–	–	190	50	–	–	–	–	240	11	270	14
16	–	–	140	23	–	–	–	–	130	17	–	–
17	–	–	71	16	–	–	–	–	102	17	–	–
18	–	–	60	30	–	–	–	–	74	10	–	–
19	–	–	47	4	–	–	–	–	34	3	–	–
20	–	–	–	–	–	–	–	–	12	1	–	–
21	–	–	20	1	–	–	–	–	18	16	–	–
22	–	–	42	0	–	–	–	–	12	12	–	–

[1] Data are not given for the remainder of the study, even though gas samples were collected for the afternoon 2 h period and overnight throughout the cropping.
[2] Gas collection chambers placed within the rice row enclosed 3 plants.
[3] No samples were collected at this time.

with 20 cm row spacing). Inside the small plot an equivalent amount of N (99 atom% ^{15}N-urea) was uniformly surface applied and rice seedlings were planted into the soil slurry. The standing water was c. 3 cm deep at planting.

Year 2. A second year's experiment was conducted in field plots as described above with the following exceptions: Two 1 m^2 plots were established within a 100 m^2 experimental plot. Both were fertilized with ^{15}N-urea as for year 1 and the second 1 m^2 plot was additionally amended with CCC (Banerjee and Mosier 1989) at the equivalent rate of 20 kg CaC_2 ha^{-1}, by mixing the CCC with the urea and hand spreading the urea-CCC mixture as done for the urea alone.

Gas flux measurements

Measurement of total N gas flux (N_2 + N_2O) and N_2O was begun about four h after urea was applied. The gasses were sampled using closed rectangular chambers 12×14×40 cm, made from plate glass. Two holes were made in the top of each chamber to accommodate rubber septa through which a needle could be inserted to withdraw a gas sample by syringe. Three replicate chambers were inserted into the saturated soil so that a 9 cm gas headspace existed. The serum stoppers were left out while the chambers were installed to alleviate a pressure buildup. An initial gas sample was withdrawn from each chamber by syringe immediately after installation. The gas was immediately injected into an evacuated 10 cm^3 glass vial, and the tube's rubber septum immediately coated with silicone sealant. The following sampling schedule was performed for the days indicated in Table 1 and Fig. 1: essentially daily during the first 4 wk after fertilizing, and less frequently the remainder of the cropping season. At c. 1400 h, triplicate chambers were again installed between the rice rows and gas samples were removed after 0, 1 and 2 h. Then the chambers were removed from the plot. At c. 1800 h, triplicate chambers were installed between the rice rows and within the rice row, enclosing three rice plants. These chambers were left in place overnight. Samples were withdrawn after 0, 4 and 15 h.

Based upon the information obtained from year 1, year 2 gas samples were collected each afternoon and sampled after 0 and 2 h. Sampling chambers were replaced each evening and sampled after 0 and 15 h. During year 2, plants were always included within the gas sampling chamber.

Gas samples in the glass vials were shipped to Ft. Collins, CO, USA, where the gas was analyzed for N_2O

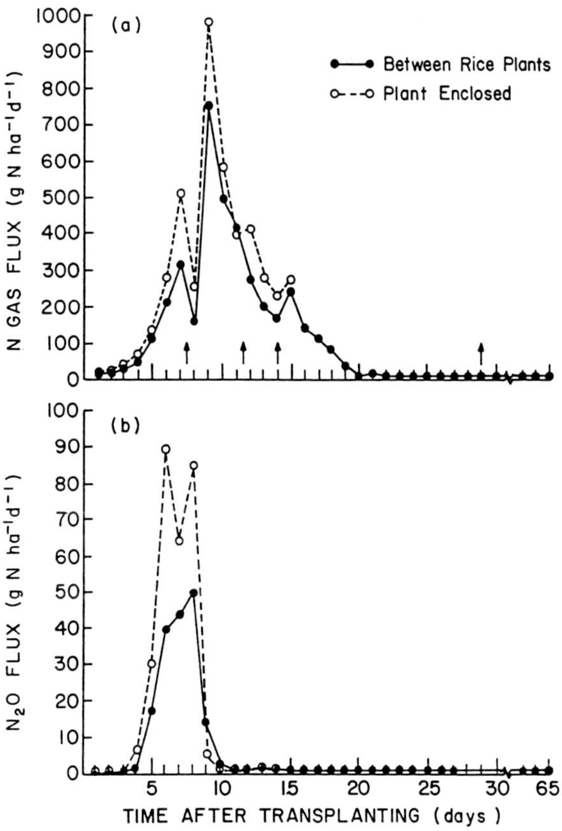

Fig. 1. (a) Total N gas flux (N_2 + N_2O + N oxides) from an intermittently flooded rice plot, using the overnight collection period, throughout year 1 growing season. Arrows indicate when the plot was reflooded. (b) Nitrous oxide gas flux from the rice plot.

by gas chromatography (Mosier and Mack 1980) and for total N gas (included N_2, N_2O and NO) by mass spectrometry (Mosier et al. 1986). The flux of total N gas and N_2O was calculated from the change in gas concentration within the sampling chamber with time (Mosier et al. 1989).

Results and discussion

Total N gas

Increases in N-gas flux were detected one day after urea fertilization and remained above detectable levels (2–5 g N ha^{-1} d^{-1}, for the following 20 d in year 1 and 24 d during year 2 (Figs. 1a and 2a). During this time the gas flux varied between 2 and 980 g N ha^{-1} d^{-1} during year 1 and 2 to 550 during year 2. The N gas emissions were <10 g N ha^{-1} d^{-1} the remainder of the cropping season during both yr. The total N gas production (N_2 + N_2O + NO) was c. 5% of the fertilizer N added during both cropping seasons. Dinitrogen emissions from the urea + CCC plot (Fig. 2a) were <100 g N ha^{-1} over the growing season compared to 5100 g N ha^{-1} in the plot treated with urea alone. These data confirm that CCC effectively eliminates N loss by denitrification (Banerjee et al. 1990) in urea-fertilized rice soils.

Gas flux measurements were made during each afternoon for 1 and 2 h between the plant rows and for 4 h and overnight within and between plant rows (Table 1) during year 1. From these measurements we found, first, that the flux measurements calculated from 1 and 2 h gas samplings during the afternoon were not different. This indicates that only a time zero and 1 or 2 h sample was needed. Secondly, the gas fluxes determined from separate chamber settings in the afternoon, 1 or 2 h measurements, and the nighttime sampling 4 h after installing the chambers (between the plant rows) were the same. A third finding from the nighttime samples, was that the N gas flux measured after 4 h and overnight were the same from between or from within the rice row. The fact that the midday, 4 h and overnight N flux rates were not different suggests that possible gas pressure increases inside the chamber during the long collec-

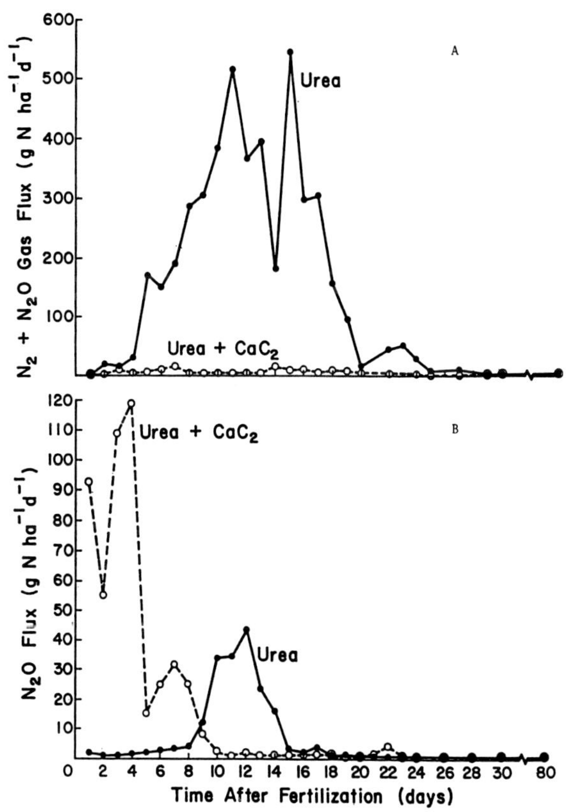

Fig. 2. (A) Total N gas flux (N_2 + N_2O + N oxides) from intermittently flooded rice plots that were urea or urea + coated calcium carbide treated, using the overnight gas collection period, throughout the year 2 growing season. (B) Nitrous oxide gas flux from urea and urea + coated calcium carbide treated plots.

tion periods did not appreciably alter the N-gas evolution. Mosier et al. (1989), in an Australian flooded rice field, also found that N_2 emissions did not vary with time of measurement during each day. This was surprising in view of the results of Galbally and Chalk (1987) which showed a marked diel cycle of production and removal of floodwater nitrite and nitrate.

A fourth finding was that the N gas fluxes were c. 30% greater when plants were inside the gas collection enclosure during the overnight measurements (Table 1 and Fig. 1a). Because the floodwater depth was either shallow or nonexistent during the course of the study it is surprising that the plants had a measurable effect on N gas emissions from the intermittently flooded rice field. The observation that even very young rice plants facilitate the efflux of N gasses from saturated soils was, however, confirmed in recent greenhouse studies (Mosier et al. 1990).

Nitrous oxide

The N gas evolution measured from the ^{15}N-fertilized plots without CCC totaled c. 5 kg N ha^{-1} in both cropping seasons. Of this total c. 6% was N_2O (Fig. 1b) in the first year and c. 4% (Fig. 2b) during the second year. The amount of N_2O was relatively large compared to N_2O efflux from flooded rice (Freney et al. 1981, Mosier et al. 1989) which is probably indicative of the different oxidative status of the continuously and intermittently flooded soils. During year 1; 6, 7, and 8 d after transplanting, N_2O emissions comprised c. 25% of the total N gases evolved. On day 9, the N_2O emissions declined to only c. 1% of the total and were lower thereafter. In year 2, N_2O emissions from the urea treated plot peaked on the 12th day after transplanting and totaled c. 200 g N ha^{-1} compared to 300 g N ha^{-1} during year 1.

Adding CCC to the rice field with the urea altered the pattern of N_2O emission. The reason for this is that sufficient acetylene was produced during the initial few days of the experiment to block the conversion of N_2O to N_2 by denitrification that normally occurs in saturated soils and as occurred in the plot treated only with urea. Sufficient native nitrate was apparently still present in the soil to provide substrate for the relatively large (Fig. 2b) N_2O production for 8 d following transplanting. Mass spectrometric analyses showed that the N_2O produced during this time did not contain elevated ^{15}N content. This indicates that it did not arise from the added highly ^{15}N enriched urea, therefore must have been produced from N already in the soil.

Conclusions

Denitrification does not appear to be a primary N loss mechanism in the intermittently flooded rice systems investigated in these studies. Although N losses, c. 5 kg N ha^{-1}, were several times greater than those reported by Buresh and Austin 1988, Buresh and De Datta 100 in Philippine rice fields, the total impact of the measured N losses on rice growth should be minimal. Adding C_2H_2 by CCC to limit nitrification virtually eliminated N_2 production through denitrification. These studies did establish that N gas emissions from saturated soils is facilitated by the rice plants. These data and those presented by Mosier et al. (1990) suggest that even young rice plants serve to transport N gases produced in the soil into the atmosphere. This gas transport system was first demonstrated for methane in laboratory studies by Raimbault et al. (1977) and in the field by Cicerone and Shetter (1981). It is evident that the rice plants must be included inside gas collection chambers in order to quantity N and other gas emissions from rice fields.

Acknowledgments – This research was conducted as a part of the Indo/US Science and Technology Initiative on N-fertilizer use efficiency supported by the Indian Council of Agricultural Research, US-AID, and USDA-ARS.

References

Banerjee, N. K. and Mosier, A. R. 1989. Coated calcium carbide as a nitrification inhibitor in upland and flooded soils. – J. Ind. Soil Sci. 37: 306–313.

– , Mosier, A. R., Uppal, K. S. and Goswami, N. N. 1990. Encapsulated calcium carbide to reduce denitrification losses from urea-fertilized flooded rice. – Mitt. der Deuts. Bodenk. Ges., (in press).

Buresh, R. J. and Austin, E. R. 1988. Direct measurement of dinitrogen and nitrous oxide flux in flooded rice fields. – Soil Sci. Soc. Am. J. 52: 681–688.

– and De Datta, S. K. 1990. Denitrification losses from puddled rice soils in the tropics. – Biol. Fert. Soil. 9: 1–13.

Cicerone, R. J. and Shetter, J. D. 1981. Sources of atmospheric methane: measurements in rice paddies and a discussion. – J. Geophys. Res. 86: 7203–7209.

Craswell, E. T., De Datta, S. K., Werarathna, C. S. and Vlek, P. L. 1984. Fate and efficiency of nitrogen fertilizers applied to wetland rice. 1. The Philippines. – Fert. Res. 6: 49–63.

De Datta, S. K. 1981. Principles and practices of rice production. Wiley, New York.

Fillery, E. R. P. and De Datta, S. K. 1986. Ammonia volatilization from nitrogen sources applied to rice fields: 1. Methodology, ammonia fluxes and nitrogen-15 losses. – Soil Sci. Soc. Am. J. 50: 80–86.

Fillery, I. R. P., Simpson, J. R. and De Datta, S. K. 1986. Contribution of ammonia volatilization to total nitrogen loss after application of urea to wetland rice fields. – Fert. Res. 8: 193–202.

Freney, J. R., Denmead, O. T., Watnabe, I. and Craswell, E. T. 1981. Ammonia and nitrous oxide losses following applications of ammonium sulfate to flooded rice. – Aust. J. Agric. Res. 32: 27–34.

– , Trevitt, A. C. F., De Datta, S. K. Obcemea, W. N. and Real, J. G. 1990. The interdependence of ammonia volatilization and denitrification as nitrogen loss processes in flooded rice fields in the Philippines. – Biol. Fert. Soils. 9: 31–36.

Galbally, I. E. and Chalk, P. M. 1987. Emission of nitrogen

oxides (NO_x) from a flooded soil fertilized with urea: relation to other nitrogen loss processes. – J. Atmos. Chem. 5: 343–365.

Lindau, C. W., DeLaune, R. D., Patrick, W. H. Jr. and Bollich, P. K. 1990. Fertilizer effects on dinitrogen, nitrous oxide, and methane emissions from lowland rice. – Soil Sci. Soc. Am. J. 54: 1789–1794.

Mosier, A. R. and Mack, L. 1980. Gas chromatographic system for precise, rapid analysis of N_2O. – Soil Sci. Soc. Am. J. 44: 1121–1123.

– , Guenzi, W. D. and Schweizer, E. E. 1986. Soil losses of dinitrogen and nitrous oxide from irrigated crops in northeastern Colorado. – Soil. Sci. Soc. Am. J. 50: 344–348.

– , Chapman S. L and Freney J. R. 1989. Determination of dinitrogen emission and retention in floodwater and porewater of a lowland rice field fertilized with ^{15}N urea. – Fert. Res. 19: 127–136.

– , Mohanty, S. K., Bhadrachalam, A. and Chakravorti, S. P. 1990. Evolution of dinitrogen and nitrous oxide from the soil to the atmosphere through rice plants. – Biol. Fert. Soils. 9: 61–67.

Raimbault, M., Rinaudo, G., Garcia, J. L. and Boreau, M. 1977. A device to study metabolic gases in the rice rhizosphere. – Soil Biol. Biochem. 9: 193–196.

Ecological Bulletins 42: 188–194. Copenhagen 1992

Factors controlling ammonia and nitrous oxide emissions from flooded rice fields

John R. Freney and Owen T. Denmead

Freney, J.R. and Denmead, O.T. 1992. Factors controlling ammonia and nitrous oxide emissions from flooded rice fields. – Ecol. Bull. (Copenhagen) 42: 188–194.

We have determined the importance of several cultural and environmental factors on ammonia volatilization and nitrous oxide emission from rice fields in experiments conducted in Australia, China and the Philippines. Ammonia loss has been determined in the field by micrometeorological methods. Investigations in which ammonium bicarbonate, ammonium sulfate or urea were incorporated into the soil or broadcast into the floodwater indicate that the important variables controlling ammonia loss are the ammoniacal nitrogen concentration, the pH and temperature of the floodwater, and windspeed; other factors exert their influences through effects on those primary variables. Nitrogen losses from volatilization have varied from 0 to 56% of the nitrogen supplied. Our studies have also shown that some ammonia is emitted to the atmosphere from the leaves of rice plants, but these emissions appear to represent only a small part of the nitrogen budget of the crop, probably less than 5% of the nitrogen applied.
Nitrous oxide emissions from temperate and tropical rice fields were studied with chamber techniques. Investigations in two cultural systems, a rotation involving pasture legumes and bare fallow, which produced an appreciable nitrate build-up before the field was flooded, and a system in which a rice field received applications of urea or ammonium sulfate some weeks after flooding, suggest that the emission of nitrous oxide is controlled by the time of flooding relative to nitrate production.

J. R. Freney, Div. of Plant Industry, CSIRO, GPO Box 1600, Canberra, ACT 2601, Australia. – O. T. Denmead, Centre for Environ. Mechanics, CSIRO, GPO Box 821, Canberra, ACT 2601, Australia.

Introduction

It is imperative that yields of food crops be increased throughout Asia to feed its rapidly increasing population. Rice constitutes nearly 20% of the total food grain production worldwide and is the dominant food grain (70%) for Asia (Stangel 1979). Consequently, rice must feature prominently in any strategy designed to supply food to Asia's increasing population.

Nitrogen (N) is the chief nutrient limiting production in modern rice varieties (Brady 1979); grain yield can be increased markedly by applications of fertilizer N (Patnaik and Ras 1979). There is a widespread move to urea as the major form of N produced and used, because of its relatively low manufacturing cost and low transportation cost per unit of N. Urea now accounts for 69% of the total fertilizer N consumed in Asia (Stangel and Harris 1987).

However, there is concern about the efficiency of urea as a N source for flooded rice. Usual farmers' practices in Asia result in recoveries of only 30–40% of the N applied and recoveries rarely exceed 60% (De Datta 1987). Recent research suggests that the main reason for poor recoveries is gaseous loss of the applied N to the atmosphere by either ammonia (NH_3) volatilization or denitrification (Fillery and Vlek 1986, De Datta 1987, Simpson and Freney 1988).

The relative importance of these two loss processes varies from region to region, but the reasons for the variation are not known. Before appropriate strategies can be designed to reduce losses, the importance of each process needs to be determined in different localities, and the controlling factors understood. To achieve these ends, studies need to be carried out at a large number of locations with different soil types, environmental conditions, fertilizer types and management

Table 1. Some characteristics of the experimental sites.

Characteristic	Deniliquin	Griffith	Dan Yang	Fengqui	Los Baños	Calauan	Mabitac
Location	Australia	Australia	China	China	Philippines	Philippines	Philippines
Coordinates	35°S 145°E	34°S 146°E	32°N 119°E	35°N 114°E	35°N 121°E	14°N 121°E	16°N 121°E
Soil classification	Chromic pelloxerert	Typic pelloxerert	Typic humaquept	Thermic eutrochrept	Typic tropaquent	Aeric tropaquept	Vertic tropaquept
Soil texture	Clay	Clay	Clay	Loam	Clay	Clay	Silty clay
pH (1:5, soil:water)	5.8	8.4	5.4	8.8	6.7	6.0	4.6
Total N(g kg^{-1})	1.1	1.1	1.7	0.5	1.7	4.0	2.1
Cation exchange capacity [c mol (Na^+)kg^{-1}]	20	36	25	8	40	48	26

practices. Because environmental conditions have a large effect on NH_3 loss, the results of such studies are most useful when the losses are assessed with techniques that do not change the microenvironment (Denmead 1983).

This paper summarizes studies carried out by the authors, in association with colleagues in Australia, China and the Philippines, into the extent of N losses from rice fields through volatilization, denitrification, loss from plant leaves, and nitrous oxide (N_2O) emissions and the influences of several cultural and environmental factors on them.

Methods

Gaseous N losses from fertilizers applied to flooded rice *Oryza sativa* L. have been determined in Australia, China and the Philippines at the locations given in Table 1. The experimental areas in Asia were puddled to disperse the clay and reduce water infiltration. In Australia, work has been conducted with crops which were dry-seeded and flooded approximately four weeks after emergence, and crops which were water-seeded shortly after flooding.

Emissions of NH_3 have been determined by nondisturbing micrometeorological methods based on gradient-diffusion or mass balance approaches (Denmead 1983). In both methods, atmospheric NH_3 concentrations have been determined by chemical trapping (Denmead et al. 1976, 1977, Leuning et al. 1985), followed by elution of the NH_3 with distilled water and measurement of its concentration with an NH_3 gas sensing electrode and millivolt meter (Orion 1983). Ammoniacal N in water samples was also determined with an NH_3 electrode, urea by a diacetyl monoxime method (Mulvaney and Bremner 1979) and pH with a glass electrode-millivolt meter assembly.

Nitrous oxide emissions were measured with emission chambers employing both closed and open air-flow systems. The atmospheric N_2O concentrations within the chambers during enclosure were measured with nondispersive infrared gas analyzers fitted with N_2O detectors. Gases which could interfere with the measurement (water vapour, carbon dioxide, carbon monoxide, nitric oxide and nitrogen dioxide) were removed from the air-stream by trapping. The emission of N_2O was calculated from the rate of change of the N_2O concentration of air flowing through the chamber (Denmead 1979, Denmead et al. 1979).

Results and discussion

Ammonia volatilization

In our studies, the loss of N by NH_3 volatilization following applications of urea, ammonium bicarbonate or ammonium sulfate, has varied from 0 to 56% of the N applied, depending on fertilizer practice and environmental conditions (Leuning et al. 1984, Simpson et al. 1984, 1985, Cai et al. 1986, 1988, Simpson and Freney 1988, Simpson et al. 1988, Trevitt et al. 1988a, b, Zhu et al. 1988, Cai et al. 1989, De Datta et al. 1989, Mosier et al. 1989, Zhu et al. 1989). In the following sections we examine the ways in which agronomic, chemical and physical factors have influenced those losses.

Ammonia volatilization from a water solution proceeds because of a difference in NH_3 concentration between the air in equilibrium with the water surface, ϱ_o, and the ambient air, ϱ_a (Koelliker and Miner 1973, Denmead et al. 1982, Leuning et al. 1984, Denmead and Freney 1992). Increasing windspeed increases the rate of volatilization by promoting more rapid transport away from the water surface and more rapid mixing in the liquid phase, so that ϱ_o is closer to the equilibrium concentration of the bulk water, ϱ_w. These influences can be incorporated into a bulk transfer equation of the form

$$F = V_b (\varrho_w - \varrho_a) \quad (1)$$

where F is the vertical flux density of NH_3 and V_b is a bulk transfer coefficient, a function of the horizontal windspeed, u, and the height at which ϱ_a is measured (Denmead and Freney 1992). The NH_3 gas concentration in equilibrium with the liquid phase (ϱ_w) is deter-

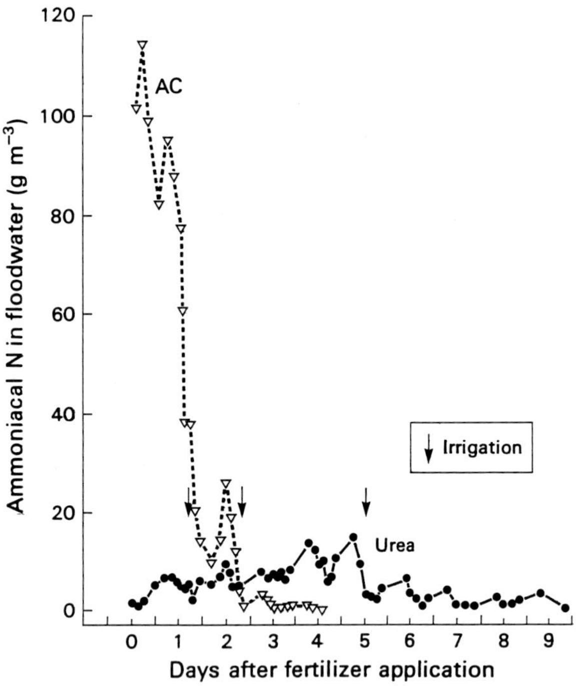

Fig. 1. Ammoniacal N concentrations in the floodwater after applications of urea and ammonium bicarbonate (AC) and incorporation into the soil at transplanting. From Zhu et al. (1989).

mined from the floodwater ammoniacal N concentration ($[NH_4^+ + NH_3]_{soln}$), its temperature (T, K) and its pH by the relationships

$$[NH_3]_{soln} = \frac{[NH_4^+ + NH_3]_{soln}}{1 + 10^{(0.09018 + 2729.92/T - pH)}} \tag{2}$$

and

$$\varrho_w = [NH_3]_{soln}\, 10^{(1.6937 - 1477.7/T)} \tag{3}$$

(Emerson et al. 1975, Hales and Drewes 1979, Leuning et al. 1984).

Thus, measurements of floodwater temperature, pH and ammoniacal N concentration, together with windspeed above the floodwater and ϱ_a can be used to calculate the rate of NH_3 emission. We have found this a very useful framework for examining volatilization and have used it successfully for measuring and modelling NH_3 loss from extensive areas and small plots (Leuning et al. 1984, Freney et al. 1985, Denmead et al. 1987, Trevitt et al. 1988a, Denmead and Freney 1992).

Our studies have shown also that other factors exert their influences on volatilization through effects on the primary variables listed above: ammoniacal N concentration, pH, temperature, and windspeed. Illustrations of these interactions are given below.

Fertilizer composition

Ammonia volatilization is influenced markedly by fertilizer composition, but conclusions regarding the efficiencies of the three main nitrogenous fertilizers used in Asia, ammonium bicarbonate, ammonium sulfate and urea, are difficult to make because comparisons of NH_3 loss from them have not been made at the same time in the same environment. Ammonium bicarbonate and urea have been compared under the same conditions in two trials in China (Cai et al. 1986, Zhu et al. 1989). In these trials, the fertilizers were broadcast into the floodwater and incorporated by harrowing just before transplanting. When ammonium bicarbonate was applied, the ammoniacal N concentration and pH of the floodwater increased. Because both these also increase ϱ_w, Eq. (2) and (3), NH_3 volatilization commenced immediately after application, and proceeded at high rates, up to 2.3 kg N ha^{-1} h^{-1}, during the first few periods of observation (Cai et al. 1986, Zhu et al. 1989). The high rates of emission did not persist because the ammoniacal N concentration of the floodwater decreased rapidly (Fig. 1), and emission virtually ceased 3 days after application. When urea was applied, the ammoniacal N concentration of the floodwater did not increase until some of the urea was hydrolyzed at the soil surface as a result of reaction with the enzyme urease. Consequently, little NH_3 was lost from the urea-treated areas during the first few days after application, and the maximum rate of NH_3 emission was not reached until 4 or 5 days after. By this time, the ammoniacal N concentration of the floodwater was high. Losses were sustained for at least 7 days.

Ammonia losses from the calcareous site in China (Fengqiu, Table 1, Zhu et al. 1989) were much larger than those from the acidic site (Dan Yang, Table 1, Cai et al. 1986). At Fengqiu, the losses from the ammonium bicarbonate and urea treatments amounted to 39% and 30% of the applied N, respectively, compared with only 18% and 9% at Dan Yang. This difference can be attributed to the comparatively low pH values of the water at Dan Yang (6.4–7.8) and the high pH of the water at Fengqiu (7.8–10.6), and consequent differences in ϱ_w. The pH differences may be due partly to differences in the pH of the soils and irrigation water, but they are more likely to be due to differences in the growth of photosynthetic organisms at the two sites (Roger et al. 1987). At Dan Yang, overcast conditions prevailed during most of the experimental period and little photosynthetic biomass was evident. However, at Fengqiu the days were bright and sunny and a prolific growth of algae ensued. The resulting increase in pH increased the proportion of ammoniacal N present as NH_3 gas and increased the potential for NH_3 loss, Eq. (2) and (3).

Emissions of NH_3 following applications of ammonium sulfate have been much lower, from no detectable emissions (Denmead et al. 1979) to 5% (Freney et al. 1981), but these results cannot be compared directly with those discussed above. The studies with ammonium sulfate were made at different sites, in different seasons and the fertilizer was applied in different ways. In the studies of Denmead et al. (1979), the fertilizer was drilled into the soil before flooding and in those of Freney et al. (1981), the ammonium sulfate was broadcast into the floodwater and not incorporated. These various results illustrate that the effects of fertilizer composition on NH_3 loss are confounded by differences in floodwater properties, environmental conditions and management.

Fertilizer rate

When urea has been applied at different rates, ammoniacal N has accumulated in the floodwater in amounts roughly proportional to the amounts of urea applied. The rate of application has had little effect on floodwater pH values, and obviously could not affect the other important variables controlling NH_3 loss: floodwater temperature and windspeed. Consequently, following Eq. (2) and (3), NH_3 loss has been linearly proportional to the rate of urea applied (r=0.99, Simpson et al. 1988). In those experiments, the mean loss over a 10-day period following application was 17% of the applied N.

Method of fertilizer application

The effect of method of application has been studied at two sites in China (Cai et al. 1986, Zhu et al. 1989) and four sites in the Philippines (Freney et al. 1990). Ammonia was lost at high rates following the broadcasting of urea into the floodwater 10 days after transplanting rice seedlings. At one of the sites in the Philippines a mean of 52% of the applied N was volatilized as NH_3 during the first 8 days after urea application. Ammonia loss was reduced slightly (to a mean of 43%) by broadcasting the fertilizer into 0.05 m deep floodwater and incorporating it into the soil with a rotary harrow before transplanting. Removal of the floodwater before application of the urea onto the saturated soil and incorporation by harrowing considerably reduced NH_3 loss (to 9%).

The effects of those different management treatments on NH_3 loss can be related to effects on the ammoniacal N concentration of the floodwater (Simpson et al. 1988). Incorporation of urea into the soil in the absence of floodwater presumably increased the contact of the fertilizer with the soil resulting in faster hydrolysis of urea, greater reaction of the NH_3 and ammonium ions so produced with organic matter and cation exchange sites, and a lower concentration of ammoniacal N in the floodwater. Incorporation of urea in the presence of floodwater was effective only when the cation exchange capacity of the soil was very high (Freney et al. 1990).

Time of fertilizer application

Ammonia loss from urea applied to dry-seeded rice has varied markedly depending on the time of application and stage of plant growth. The extent of the loss can be attributed to effects on the ammoniacal N concentration, pH and temperature of the floodwater, and windspeed near the floodwater surface.

Ammonia volatilization was found to be negligible when urea was applied to the surface of a dry cracking soil 2 days before permanently flooding the field. Before flooding, the urea prills remained undissolved on the soil surface and hydrolysis of the urea could not proceed. When the soil was flooded, the urea prills were washed into the cracks in the soil, which subsequently closed. This prevented the movement of soluble N into the floodwater and there was no source of NH_3 in the floodwater for volatilization to occur (Humphreys et al. 1988).

When urea is broadcast into the floodwater a few days after permanent flooding, losses can be high. In our investigations, they varied from 11 to 21% of the N applied (Simpson et al. 1984, 1985, Humphreys et al. 1988). Then, the rice plants just protruded through the surface of the floodwater, resulting in little shading to restrict photosynthetic activity of microorganisms and little attenuation of the wind. The losses were consequently associated with high floodwater pHs and high windspeeds, hence high transfer coefficients, near the water surface (Humphreys et al. 1988), both of which would tend to increase volatilization rate, following Eq. (1), (2) and (3).

Ammonia losses have been low (3 to 8% of the applied N) when urea has been applied into the floodwater at panicle initiation (Humphreys et al. 1988). At this stage of growth, the plant canopy shades the floodwater, inhibiting algal photosynthesis and consequent pH elevation, and reducing water temperatures. In addition, the plant canopy restricts air movement at the water surface. The net result is low NH_3 gas concentrations in the floodwater, low transfer coefficients, and low rates of volatilization, Eq. (1).

Floodwater depth

Ammonia loss from shallow (0.05 m) and deep (0.14 m) ponds has been studied following the broadcasting of urea into the floodwater (Freney et al. 1988). In these investigations, urea was hydrolysed at a faster rate in the shallow pond than in the deep pond. The ammoniacal N concentration in the water resulting from the hydrolysis was inversely related to the depth of water for the first 5 days after application. Each day, the water in the shallow pond warmed up faster than that in the deep pond and attained a higher maximum temperature. Thus, in the initial stages, ϱ_w was higher in the shallow pond than the deep pond, and NH_3 loss was more rapid. Towards the end of the experiment, when ammoniacal N concentrations in the two ponds were similar, NH_3 emission rates were comparable.

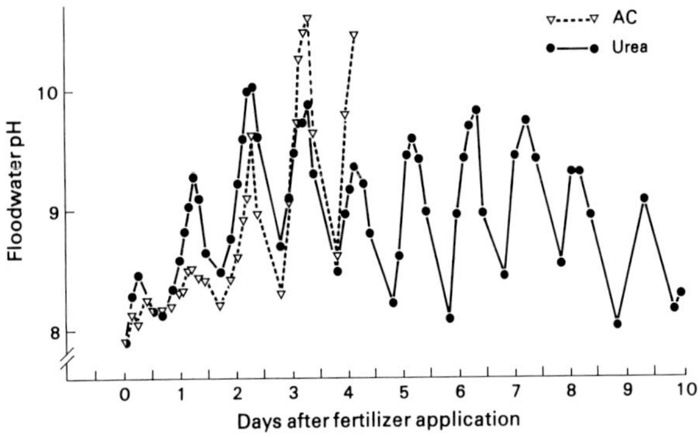

Fig. 2. Floodwater pH fluctuations after applications of ammonium bicarbonate (AC) and urea. From Zhu et al. (1989).

Algal growth

The pH of the floodwater is markedly affected by the growth and metabolism of photosynthetic microorganisms (Roger et al. 1987). During daylight hours, they fix CO_2 and excrete hydroxylions. As a result the pH of the floodwater rises dramatically during the day. At night, the microorganisms respire, CO_2 is emitted into the water and the pH of the floodwater decreases (Fig. 2). Floodwater pH values in excess of 10 have been observed during the day (Simpson et al. 1984) and the pH may decrease by 2 or 3 units during the night. These changes can have marked effects on NH_3 volatilization. It can be calculated from Eq. (1), (2) and (3) that, other things being equal, a change of pH from 8 to 10 will increase ϱ_w and the volatilization rate more than tenfold.

Overcast conditions in the atmosphere, such as existed in the field experiment at Dan Yang, China, described by Cai et al. (1986), restrict the growth of photosynthetic organisms in the floodwater and result in low floodwater pH values and low NH_3 loss rates. Reducing the amount of light penetrating the floodwater, either by shading (Simpson et al. 1988, Trevitt et al. 1988b) or by increasing the turbidity of the floodwater (Freney, Denmead, Trevitt, Muirhead and Simpson, unpubl. data), also restricts the growth of photosynthetic organisms, reduces pH values and results in lower NH_3 losses.

These results suggest that NH_3 loss might be controlled if the growth of photosynthetic organisms were restricted by adding biocides or algicides to the floodwater. Muirhead et al. (1986), Bowmer and Muirhead (1987) and Simpson et al. (1988) adopted this approach and showed that changes in dissolved CO_2 were suppressed, diurnal variation in pH values was reduced and NH_3 loss was lowered by the use of chemicals such as terbutryn and dichlorophen.

Under certain conditions the growth of photosynthetic algae can also reduce NH_3 loss. For instance, Zhu et al. (1989) found that addition of phosphate and clear sunny days resulted in a prolific growth of algae on the floodwater surface which was so dense that the transfer of NH_3 gas to the atmosphere was restricted for significant periods until the algal layer was broken up by strong winds.

Windspeed

Increasing windspeed should increase the rate of NH_3 volatilization by promoting more rapid transport of NH_3 away from the water surface. This effect has been demonstrated in flooded rice fields (Freney et al. 1981, Leuning et al. 1984), in irrigated maize (Denmead et al. 1982) and in evaporation pans (Cai 1988). Leuning et al. (1984) observed that the transfer velocity for NH_3, V_b (Eq. 1), increased linearly with windspeed up to 5 m s^{-1} and Cai (1988) found that the relationship was reasonably linear for wind speeds up to 6 m s^{-1} (r=0.86). Denmead et al. (1982), working in higher windspeeds (up to 8 m s^{-1}) found an exponential relationship between V_b and windspeed, and Cai (1988) found that above 6 m s^{-1}, V_b increased at a greater rate as windspeed increased. She found that over the whole range of wind speeds an equation of the form

$$\ln V_b = 0.249\,u - 6.12 \tag{4}$$

gave the best description of the relationship between V_b and windspeed measured at 1.2 m (r=0.95). In Eq. (4) both V_b and windspeed have units of m s^{-1}. As suggested by Denmead et al. (1982), the enhanced volatilization in high winds was probably due to better mechanical mixing of the NH_3 solution, which reduces the liquid phase resistance, and to more rapid, gas-phase transport in the air. The dependence of V_b on windspeed is discussed more in Denmead and Freney (1992).

Temperature stratification in the water

Theoretical considerations suggest that NH_3 volatilization should be dependent not only on solution chemistry, but also on the mechanics of gas transfer across the air-water interface. In principle, it is possible to describe the resistances to transfer across the interface in terms of windspeed alone (Deacon 1977, Denmead et al. 1987, Denmead and Freney 1992). However, it can happen that temperature stratification can produce significant additional resistance to NH_3 transfer in the aqueous phase (Leuning et al. 1984, Trevitt et al. 1988a). This phenomenon is marked in turbid water (Leuning et al. 1984), but it can occur also in clear water when skies are cloudy (Trevitt et al. 1988a). In both situations, it appears that solar radiation is absorbed strongly near the water surface with the consequent development of a stable temperature profile in the floodwater. As a result, mixing in the water is suppressed and NH_3 escaping from the surface is replenished only slowly.

Ammonia losses from rice plants

The concept of an NH_3 compensation point was introduced by Farquhar et al. (1980) to explain the finite partial pressure of NH_3 gas which exists in the intercellular air-spaces of plant leaves. If the ambient partial pressure is less than the compensation point, plants lose NH_3 to the air by diffusion through the stomata, and if higher, they absorb NH_3. Wetselaar and Farquhar (1980) suggested that elevated compensation points during senescence, due to the breakdown of proteins, may be responsible for substantial losses of N from plants.

Recently, we examined NH_3 losses from rice plants during senescence in the field, but found that the compensation point remained relatively low during that period and losses were quite small, amounting to only 1.5 kg N ha^{-1} over the six weeks following anthesis. If NH_3 was emitted from plants at that rate for the five months in the life of a rice crop when a significant plant cover exists, the total emission would amount to 5 kg N ha^{-1}, which is <5% of the N fertilizer applied (120 kg N ha^{-1}) and of small consequence in the N budget of the crop.

Nitrous oxide emission

In all the flooded rice systems studied, some N_2O has been emitted, although it has represented only a small part of the total N budget (0.02–1.4% of the N present in the floodwater). The extent of the loss has varied with the cultural conditions and N management.

In one system (at Deniliquin, N.S.W., Australia), the rice was grown in a rotation in which legumes were used to build up soil N between rice crops, and a bare fallow was used just before flooding to release the N for the rice crop. Prior to flooding, the surface 0.08 m soil contained 40 kg N ha^{-1} as nitrate. Rice was seeded into the flooded soil a few days after flooding. Emission of N_2O peaked at 0.01 kg N ha^{-1} h^{-1} one day after flooding and declined to undetectable levels by day 13 (Denmead et al. 1979). The emission of N_2O coincided with the disappearance of nitrate from the floodwater, and the total emission accounted for 1.4% of the nitrate initially present.

In another system (on an alkaline clay at Griffith, N.S.W., Australia), negligible amounts of N_2O were emitted when urea was broadcast into the water three days after flooding (Simpson et al. 1984). Other studies have examined N_2O emissions following applications of fertilizer N at panicle initiation. Freney et al. (1981) measured an emission of N_2O of 0.1% of the applied N when ammonium sulfate was applied at that stage and Mosier et al. (1989) measured only 0.02% when urea was applied. The former study was carried out in the tropics at Los Baños, Philippines, with higher water temperatures than those experienced in the other studies.

To summarize, the largest losses have occurred when appreciable nitrate was present before flooding. When flooding occurs before nitrate production, it appears that the conditions created favour the further reduction of the N_2O produced to dinitrogen. This is probably attributable to different relative rates of diffusion and reduction of N_2O in the different systems.

References

Bowmer, K. H. and Muirhead, W. A. 1987. Inhibition of algal photosynthesis to control pH and reduce ammonia volatilization from rice floodwater. – Fert. Res. 13: 13–29.

Brady, N. C. 1979. Opening address. – In: Nitrogen and rice. Int. Rice Res. Inst., Los Baños, pp. 1–2.

Cai, G. X. 1988. Gaseous loss of nitrogen from fertilizers applied to flooded rice. Ph.D. Thesis, Univ. of Queensland, Brisbane.

– , Zhu, Z. L., Trevitt, A. C. F., Freney, J. R. and Simpson, J. R. 1986. Nitrogen loss from ammonium bicarbonate and urea fertilizers applied to flooded rice. – Fert. Res. 10: 203–215.

– , Freney, J. R., Humphreys, E., Denmead, O. T., Samson, M. and Simpson, J. R. 1988. Use of surface films to reduce ammonia volatilization from flooded rice fields. – Aust. J. Agric. Res. 39: 177–186.

– , Freney, J. R., Muirhead, W. A., Simpson, J. R., Chen, D. L. and Trevitt, A. C. F. 1989. The evaluation of urease inhibitors to improve the efficiency of urea as a N-source for flooded rice. – Soil Biol. Biochem. 21: 137–145.

Deacon, E. L. 1977. Gas transfer to and across an air-water interface. – Tellus 29: 363–374.

De Datta, S. K. 1987. Advances in soil fertility research and nitrogen fertilizer management for lowland rice. – In: Freney, J. R., Wetselaar, R., Trevitt, A. C. F. and Simpson, J. R. (eds), Efficiency of nitrogen fertilizers for rice. Int. Rice Res. Inst., Los Baños. pp. 27–41.

– , Trevitt, A. C. F., Freney, J. R., Obcemea, W. N., Real, J. G. and Simpson, J. R. 1989. Measuring nitrogen losses from lowland rice using bulk aerodynamic and nitrogen-15 balance methods. – Soil Sci. Soc. Am. J. 53: 1275–1281.

Denmead, O. T. 1979. Chamber systems for measuring nitrous oxide emission from soils in the field. – Soil Sci. Soc.Am. J. 43: 89–95.

– 1983. Micrometeorological methods for measuring gaseous losses of nitrogen in the field. – In: Freney, J. R. and Simpson, J. R. (eds), Gaseous loss of nitrogen from plant-soil systems. Martinus Nijhoff, The Hague. pp. 133–158.

– and Freney, J. R. 1992. Transfer coefficients for water-air exchange of ammonia, carbon dioxide and methane. – Ecol. Bull. (Copenhagen) 42: 31–41.

– , Freney, J. R. and Simpson, J. R. 1976. A closed ammonia cycle within a plant canopy. – Soil Biol. Biochem. 8: 161–164.

– , Freney, J. R. and Simpson, J. R. 1979. Nitrous oxide emission during denitrification in a flooded field. – Soil Sci. Soc. Am. J. 43: 716–718.

– , Freney, J. R. and Simpson, J. R. 1982. Dynamics of ammonia volatilization during furrow irrigation of maize. – Soil Sci. Soc. Am. J. 46: 149–155.

– , Simpson, J. R. and Freney, J. R. 1977. A direct field measurement of ammonia emission after injection of anhydrous ammonia. – Soil Sci. Soc. Am. J. 41: 1001–1004.

– , Trevitt, A. C. F., Muirhead, W. A., Freney, J. R. and Simpson, J. R. 1987. Modelling ammonia loss from flooded soil systems. – In: Bacon, P. E., Evans, J., Storrier, R. R. and Taylor, A. C. (eds), Nitrogen cycling in temperate agricultural systems. Australian Soc. of Soil Sci. Inc., Riverina Branch, Wagga Wagga, N.S.W., pp. 353–359.

Emerson, K., Russo, R. C., Lund, R. E. and Thurston, R. V. 1975. Aqueous ammonia equilibrium calculations. Effect of pH and temperature. – J. Fish. Res. Bd. Can. 32: 2379–2383.

Farquhar, G. D., Firth, P. M., Wetselaar, R. and Weir, B. 1980. On the gaseous exchange of ammonia between leaves and the environment: determination of the ammonia compensation point. – Plant Physiol. 66: 710–714.

Fillery, I. R. P. and Vlek, P. L. G. 1986. Reappraisal of the significance of ammonia volatilization as an N loss mechanism in flooded rice fields. – Fert. Res. 9: 79–98.

Freney, J. R., Denmead, O. T., Watanabe, I. and Craswell, E. T. 1981. Ammonia and nitrous oxide losses following applications of ammonium sulfate to flooded rice. – Aust. J. Agric. Res. 32: 37–45.

– , Leuning, R., Simpson, J. R., Denmead, O. T. and Muirhead, W. A. 1985. Estimating ammonia volatilization from flooded rice fields by simplified techniques. – Soil Sci. Soc. Am. J. 49: 1049–1054.

– , Trevitt, A. C. F., Muirhead, W. A., Denmead, O. T., Simpson, J. R. and Obcemea, W. N. 1988. Effect of water depth on ammonia loss from lowland rice. – Fert. Res. 16: 97–107.

– , Trevitt, A. C. F., De Datta, S. K., Obcemea, W. N. and Real, J. G. 1990. The interdependence of ammonia volatilization and denitrification as nitrogen loss processes from flooded rice soils in the Philippines. – Biol. Fertil. Soils 9: 31–36.

Hales, J. M. and Drewes, D. R. 1979. Solubility of ammonia in water at low concentrations. – Atmos. Environ. 13: 1133–1147.

Humphreys, E., Freney, J. R., Muirhead, W. A., Denmead, O. T., Simpson, J. R., Leuning, R., Trevitt, A. C. F., Obcemea, W. N., Wetselaar, R. and Cai, G. X. 1988. Loss of ammonia after application of urea at different times to dry-seeded, irrigated rice. – Fert. Res. 16: 47–57.

Koelliker, J. K. and Miner, J. R. 1973. Desorption of ammonia from anaerobic lagoons. – Trans. Am. Soc. Agric. Engineers. 16: 148–151.

Leuning, R., Denmead, O. T., Simpson, J. R. and Freney, J. R. 1984. Processes of ammonia loss from shallow floodwater. – Atmos. Environ. 18: 1583–1592.

– , Freney, J. R., Denmead, O. T. and Simpson, J. R. 1985. A sampler for measuring atmospheric ammonia flux. – Atmos. Environ. 19: 1117–1124.

Mosier, A. R., Chapman, S. L. and Freney, J. R. 1989. Determination of dinitrogen emission and retention in floodwater and porewater of a lowland rice field fertilized with 15N-urea. – Fert. Res. 19: 127–136.

Muirhead, W. A., De Datta, S. K. and Gusto, R. M. 1986. Can algicides improve nitrogen use efficiency in rice by reducing ammonia volatilization. – Int. Rice Res. Inst. Spec. Semin. 9 July 1986, Los Baños.

Mulvaney, R. L. and Bremner, J. M. 1979. A modified diacetyl monoxime method for colorimetric determination of urea in soil extracts. – Commun. Soil Sci. Plant. Anal. 10: 1163–1170.

Orion. 1983. Instruction manual for ammonia electrode model 95–12-Orion Res. Inc. – Cambridge, MA.

Patnaik, S. and Ras, M. V. 1979. Sources of nitrogen for rice production. – In: Nitrogen and rice. Int. Rice Res. Inst., Los Baños. pp. 25–43.

Roger, P. A., Grant, I. F., Reddy, P. M. and Watanabe, I. 1987. The photosynthetic aquatic biomass in wetland rice fields and its effect on nitrogen dynamics. – In: Freney, J. R., Wetselaar, R., Trevitt, A. C. F. and Simpson, J. R. (eds), Efficiency of nitrogen fertilizers for rice. Int. Rice Res. Inst., Los Baños, pp. 43–68.

Simpson, J. R. and Freney, J. R. 1988. Interacting processes in gaseous nitrogen loss from urea applied to flooded rice fields. – In: Pushparajah, E., Husin, A. and Bachik, A. T. (eds), Urea technology and utilization. Malaysian Soc. of Soil Sci., Kuala Lumpur, pp. 281–290.

– , Freney, J. R., Muirhead, W. A. and Leuning, R. 1985. Effects of phenylphosphorodiamidate and dicyandiamide on nitrogen loss from flooded rice. – Soil Sci. Soc. Am. J. 49: 1426–1431.

– , Freney, J. R., Wetselaar, R., Muirhead, W. A., Leuning, R. and Denmead, O. T. 1984. Transformations and losses of urea nitrogen after application to flooded rice. – Aust. J. Agric. Res. 35: 189–200.

– , Muirhead, W. A., Bowmer, K. H., Cai, G. X. and Freney, J. R. 1988. Control of gaseous nitrogen losses from urea applied to flooded rice soils. – Fert. Res. 18: 31–47.

Stangel, P. J. 1979. Nitrogen requirement and adequacy of supply for rice production. – In: Nitrogen and rice. Int. Rice Res. Inst., Los Baños. pp. 45–69.

– and Harris, G. T. 1987. Trends in production, trade and use of fertilizers: a global perspective. – In: Freney, J. R., Wetselaar, R., Trevitt, A. C. F. and Simpson, J. R. (eds), Efficiency of nitrogen fertilizers for rice. Int. Rice Res. Inst., Los Baños. pp. 1–26.

Trevitt, A. C. F., Freney, J. R., Denmead, O. T., Zhu, Z. L., Cai, G. X. and Simpson, J. R. 1988a. Water-air transfer resistance for ammonia from flooded rice. – J. Atmos. Chem. 6: 133–147.

– , Freney, J. R., Simpson, J. R. and Muirhead, W. A. 1988b. Problems in the use of microplots for assessing the fate of urea applied to flooded rice. – Aust. J. Agric. Res. 39: 351–362.

Wetselaar, R. and Farquhar, G. D. 1980. Losses of nitrogen from the tops of plants. – Adv. Agron. 33: 263–302.

Zhu, Z. L., Cai, G. X., Simpson, J. R., Zhang, S. L., Chen, D. L., Jackson, A. V. and Freney, J. R. 1989. Processes of nitrogen loss from fertilizers applied to flooded rice fields on a calcareous soil in north-central China. – Fert. Res. 18: 101–115.

– , Zhang, S. L., Chen, D. L., Cai, G. X., Xu, Y. H., Simpson, J. R., Freney, J. R. and Jackson, A. V. 1988. Fate and efficiency of nitrogen fertilizers applied with different methods of incorporation to a calcareous paddy soil in Huang-Huai-Hai plain. – Soils (China) 20: 121–125.

Ecological Bulletins 42: 195–198. Copenhagen 1992

Measurement of dimethyl sulfide emission from lysimeter paddy fields

Ken-ichi Kanda and Katsuyuki Minami

Kanda, K. and Minami, K. 1992. Measurement of dimethyl sulfide emission from lysimeter paddy fields. – Ecol. Bull. (Copenhagen) 42: 195–198.

The closed chamber method was employed for the measurements of dimethyl sulfide (DMS) fluxes from paddy field plots (2×2.5 m, depth 1 m) into the atmosphere, from April to Oct 1989. The chamber was constructed from polycarbonate with internal height of 100 cm, covering-sectional area of 900 cm^2. Air samples were collected by using vacuumed 300 ml glass bottles. The DMS concentration of the sample was determined by using GC/FPD. Significant diurnal variations of DMS fluxes were observed in both gray lowland soil and light-colored Andosol paddy fields. DMS fluxes were higher in daytime and lower from evening through morning, and correlated to the variation of both solar radiation strength and air temperature. We also observed considerable seasonal variation of DMS fluxes from both fields during the cultivation period. The maximum DMS emission rate was measured just before heading stage of the rice plant. Annual emission rates of the DMS from paddy fields ranged from 4.1 to 7.3 mg Sm^{-2} yr^{-1}.

K. Kanda and K. Minami, Dept of Environ. Manage., Nat. Inst. of Agro Environ. Sci., Kannondai, Tsukuba, 305, Japan.

Introduction

During recent years, SO_2 emission from anthropogenic sources into the atmosphere has steadily increased at a rate of c. 2.4% yr^{-1} (Varchelyi 1985). The increase in sulfur fluxes can lead to local air pollution, acid precipitation and perhaps climate change. To determine the extent of the human impact, we need to know the amount of sulfur emission from natural sources. Although recent estimations have shown that the total natural sulfur emissions are equivalent to the current anthropogenic emissions of 80 Tg S yr^{-1} (Varchelyi 1985), there still remain large uncertainties about the chemical speciation and the magnitude of natural emissions.

Almost all sulfur gases emitted from the ocean are dimethyl sulfide (DMS). The production of DMS in the ocean has been well documented, and its emission to the atmosphere is c. 40 Tg S yr^{-1} (Andreae and Raemdonck 1983). Soil is one of the natural sulfur sources. Emission rates of carbonyl sulfide (COS), DMS and carbon disulfide (CS_2) from upland soils in Germany have been examined by Staubes et al. (1989). By contrast, very little is known about the occurrence of the sulfur gases in paddy field environments.

The purpose of this study is to gain better information about the amount of DMS emitted from paddy fields, and to elucidate the differences in DMS fluxes between different water managements using lysimeter paddy fields.

Material and methods

Experimental fields

The experiments contained four rice paddy plots (2×2.5 m, depth 1 m) at the Nat. Inst. of Agro-Environ. Sci. (NIAES), Tsukuba, Japan. Two plots contained gray lowland soil (Eutric Glelysols) and the other two contained light-colored Andosol. Before transplanting the rice plants, the plots were flooded, and mineral fertilizers were incorporated into the soil.

Water percolation rates of the plots were adjusted to 0 and 20 mm d^{-1}, respectively during the observation period by periodically opening water stopping valves

Table 1. Soil properties and the cultivation practices.

	Gray lowland soil	Light-colored Andosol
Texture	Clay loam	Loam
pH (H_2O)	6.5	5.9
Total C(%)	2.5	2.2
Total N(%)	0.15	0.18

Basal application: N: 100 kg ha^{-1} as $(NH_4)_2SO_4$, P_2O_5; 100 kg ha^{-1} as $Ca(H_2PO_4)_2$ H_2O, K_2O; 100 kg ha^{-1} as KCl. Supplement application: N, 50 kg ha^{-1} as $(NH_4)_2SO_4$. Planting density: 22.2 plants m^{-2}.

installed at the bottom of the plots. Rice plants were transplanted on 28 April and harvested on 26 Sept 1989.

Soil-, air- and water-temperature and solar radiation strength were monitored during the experimental period. Soil samples were collected from individual fields before flooding. Soil properties and the cultivation practices are listed in Table 1.

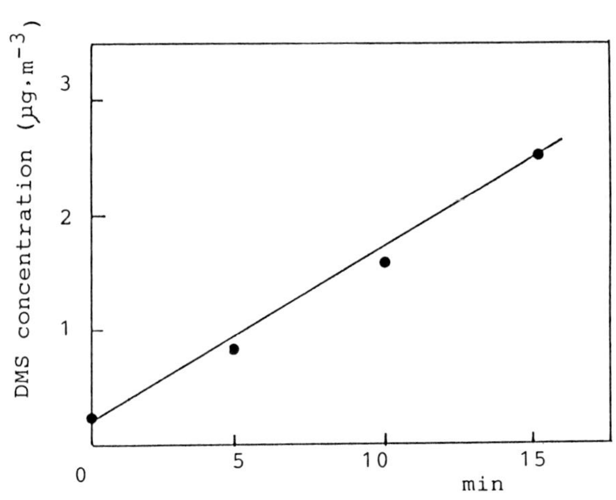

Fig. 1. Temporal increase of DMS concentration in the air within the chamber (height 100 cm, covering area of 900 cm^2). Data were obtained at lysimeter paddy fields of the NIAES.

Measurements of DMS fluxes

The sampling system proposed by Yagi and Minami (1990) was applied to the measurements of DMS fluxes. The chamber was constructed of polycarbonate with an internal height of 100 cm, covered an area of 900 cm^2, and had a small fan for mixing the air.

DMS fluxes were determined by measuring the temporal increase of the DMS mixing ratio of the air in the chamber over plots with and without rice plants. The chamber with rice plants was placed over two riceplants in the plot. The air within the chamber was evacuated into a 300 ml glass bottle. Air samples were taken over a 10 min period. All measurements were performed between 10 am and 2 pm except when measuring the diurnal variations of fluxes. Seasonal DMS measurements were carried out every one or two weeks. Measurements of the diurnal variation of DMS fluxes were carried out every 2–10 h 20 and 21 Sept 1989.

The DMS concentrations in the air sample were enriched cryogenically in the Teflon trapping tube (0.2 cm Ø, length 30 cm) packed with Chromosorb W. The DMS in the air sample was trapped in the Teflon tube and cooled by liquid nitrogen. The air sample was exhausted by use of a vacuum pump. The trapped and enriched DMS in the Teflon tube was warmed to a gaseous form using 70°C water and was analyzed by gas chromatography (Shimadzu GC-12A) equipped with a flame photometric detector and glass column (0.3 cm Ø, length 300 cm) packed with 10% 1,2,3-tris (2-cyanoethoxy) propane on 60–80 mesh Chromosorb W (AW DMCS treatment). Calibration was performed by using a standard gas generator and permeation tube (Gastec Co. Ltd).

The accuracy of this procedure was confirmed by the linear temporal increase of the DMS concentration in

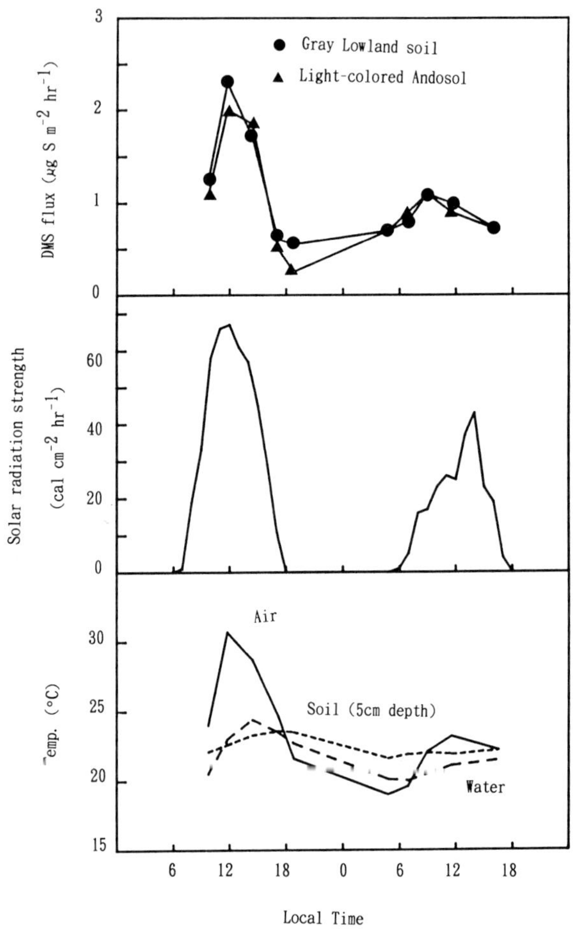

Fig. 2. Diurnal variation of DMS fluxes, solar radiation strength and temperature of air, water and soil at lysimeter paddy fields on 20-21 Sept 1989. The soil temperature was measured at 5 cm depth.

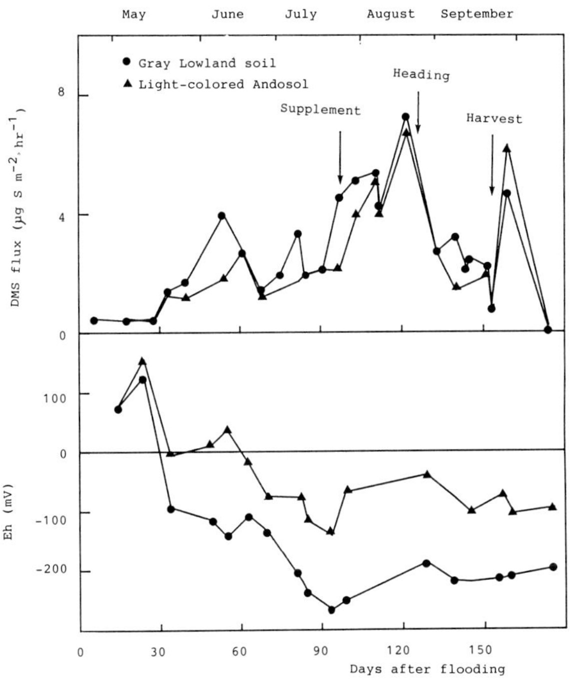

Fig. 3. Seasonal variation of DMS fluxes and soil Eh in lysimeter paddy fields. The soil Eh was measured at 5 cm depth.

the chamber. A typical example obtained at the paddy plot is presented in Fig. 1, showing the linear relationships between time and DMS concentration in the air within the chamber.

Results and discussion

The diurnal variation of DMS fluxes from the paddy field without water percolation (0 mm d^{-1}) on gray lowland soil and lightcolored Andosol is shown in Fig. 2, along with solar radiation strength and temperature of air, water, and soil. The fluxes of both soil type were higher in the daytime and lower from evening through morning. The variation in the fluxes corresponded with the variation of the solar radiation and air temperature. The DMS fluxes varied from 0.2 to 2.4 μg S m^{-2} h^{-1} during 38 h.

Staubes et al. (1989) observed diurnal variation of DMS fluxes from upland soil in Germany. They found a relationship between the diurnal change of the DMS fluxes and soil surface temperatures. Cooper et al. (1987a, 1987b) also pointed out a positive correlation between the diurnal change of DMS from a coastal wetland and the solar radiation and the diurnal change in DMS flux and soil temperature. Our measurements in paddy field plots at Tsukuba showed a positive correlation between the fluxes and solar radiation strength, the fluxes and air temperature, and slight correlation

Table 2. Diurnal variation of DMS fluxes rate on 0 mm percolation plots with or without rice plants (gray lowland soil).

Sept 1989	Time of day	With rice plants	Without rice plants
		(μ S m^{-2} h^{-1})	
12th	5	1.2	0.1
	9	2.9	1.0
	11	2.3	0
	14	1.4	0
	17	1.4	0.2
	23	1.2	−0.3
13th	6	1.3	0.2
	9	1.7	0.1
	12	2.7	0
	18	1.6	0.2

between the fluxes and soil temperature, and the fluxes and water temperature, but no difference between the fluxes and soil type. This implies that rice plants have large influence on the DMS flux. To clarify the relation between them, further detailed measurement will be needed.

The seasonal variation in DMS fluxes from the paddy fields without water percolation (0 mm d^{-1}) on gray lowland soil and light-colored Andosol is shown in Fig. 3, along with the soil Eh. Strong seasonal variation in the DMS fluxes was observed during the cultivation period. The DMS fluxes increased with the growing of the rice plants, and the maximum rate was observed just before the heading period. After the heading period, the fluxes gradually decreased. The flux variation is probably due to the change of solar radiation strength, temperature, the activity of rice plants, and perhaps other factors during the cultivation period. To elucidate the reason of the flux variation, further studies will be needed. The fluxes increased just after the harvesting. This was probably due to diffusion of DMS formed in the soil through the cut aerenchyma system of the rice

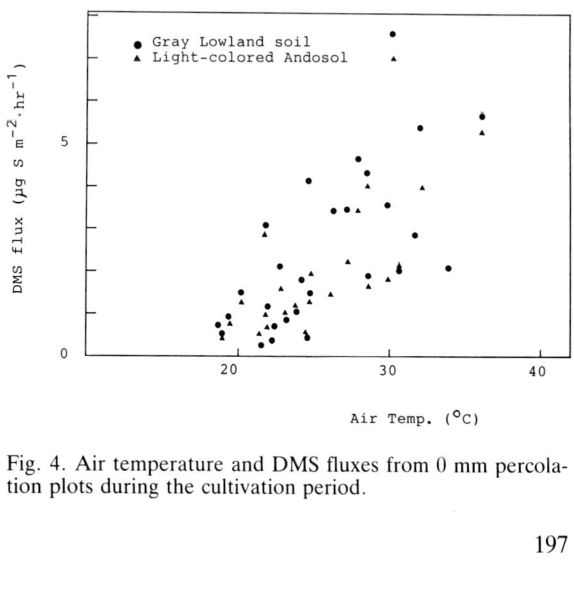

Fig. 4. Air temperature and DMS fluxes from 0 mm percolation plots during the cultivation period.

Table 3. Annual emission rates of DMS from the individual plots on the lysimeter paddy fields.

Soil	Water percolation*	Annual emission rate (mg S m^{-2})
Gray lowland	0	7.3
soil	20	4.3
Light-colored	0	6.1
Andosol	20	4.1

* Water percolation rates of the lysimeters were adjusted to 0 and 20 mm d^{-1} during the observation period by periodically opening water stopping valves installed at the bottom of the lysimeters.

plants. There was no significant difference in the DMS flux between gray lowland soil and light-colored Andosol, in spite of the difference observed in soil Eh between them.

Table 2 shows the difference of DMS fluxes from 0 mm percolation plot on gray lowland paddy soil with or without rice plants. These results suggest that DMS emission arises mostly from rice plant activity. Fig. 4 shows the relationship between air temperature and DMS fluxes from 0 mm percolation plots on gray lowland soil and light-colored Andosol during the cultivation period. The rise of air temperature tends to enhance the DMS fluxes from the rice paddy field. Fall et al. (1988) also found that the emission rate of DMS from corn, wheat and alfalfa increased with the increasing of air temperature and solar radiation under laboratory studies.

From the integration of the observed DMS emission rates, we calculated the total ammount of DMS emission from the individual plots during the experimental period. Since the amount of DMS emitted from these paddies after the experimental season was very little, the integrated value can be considered as the annual emission. The results are shown in Table 3. Emission rates of DMS from our plots ranged from 4.1 to 7.3 mg S m^{-2} yr^{-1}. Although there was no significant difference in the DMS emission rates between gray lowland soil and light-colored Andosol, the emission rates of 0 mm percolated plot was higher than the emission rates from the plot receiving 20 mm percolation.

Staubes et al. (1989) reported that the mean annual emission rate of DMS was c. 3 mg S m^{-2} for upland soils in Germany. Annual maximum emission rate of DMS from coastal wetland in Florida and Massachussetts, USA, was 2870 mg S m^{-2} (Steudler and Peterson 1984, Cooper et al. 1987). Our data on DMS emission rates from a rice paddy field were similar to that of upland soils in Germany, but lower than the emission rates from coastal wetland.

The annual DMS emission rates from the paddy fields were a small fraction of the amount of sulfur supplied as irrigation water, precipitation and fertilizer. The amount of sulfur in precipitation was c. 1 g S m^{-2} yr^{-1} in Tsukuba, Japan. In this study, the amount of sulfur from irrigation water was 5 and 38 g S m^{-2} yr^{-1} for no drainage and drainage plot, respectively. All plots were supplied 19 g S m^{-2} yr^{-1} from ammonium sulfate and superphosphate fertilizer. The amounts of sulfur supplied as fertilizer to the experiment plots are slightly higher than that of common paddy fields in Japan.

The release of the long-lived sulfur gases such as COS and CS_2 from paddy fields (Minami 1982) may have an important role in the atmospheric sulfur cycle on a global scale. Additional measurements of sulfur gas emission from rice paddy fields (especially tropical rice paddies) are needed to refine the global biogenic emission estimates.

Acknowledgements – We thank Mr. Yagi, Nat. Inst. Agro-Environ. Sci., for useful comments and the management of the plots.

References

Andreae, M. O. and Raemdonck, H. 1983. Dimethylsulfide in the surface ocean and the marine atmosphere. A global view. – Science 221: 744–747.

Cooper, D. J., de Mello, W. Z., Cooper, W. J., Zika, R. G., Saltzman, E. S., Prospero, J. K. and Savoie, D. L. 1987a. Short term variability in biogenic sulphur emissions from a Florida Spartina marsh. – Atmos. Environ. 21: 7–12.

Cooper, W. J., Cooper, D. J., Saltzman, E. S., de Mello, W. Z. Savoie, D. L., Zika, R. G. and Prospero, J. M. 1987b. Emission of biogenic sulphur compounds from several wetland soils in Florida. – Atmos. Environ. 21: 1491–1495.

Fall, R., Alberitton, D. L., Fehsenfeld, F. C., Kuster, W. C. and Goldan, P. O. 1988. Laboratory studies of some environmental variables controlling sulfur emissions from plants. – J. Atmos. Chem. 6: 341–362.

Minami, K. 1982. Volatilization of sulfur from paddy soils. – JARQ 15: 167–171.

Staubes, R., Georgii, H. W. and Ockenlmann, G. 1989. Flux of COS, DMS and CS_2 from various soils in Germany. – Tellus 41: 305–313.

Steudler, P. A. and Peterson, B. J. 1984. Contribution of gaseous sulphur from salt marshes to the global sulphur cycle. – Nature (Lond.) 311: 455–457.

Varchelyi,G. 1985. Continental and global sulphur budgets, 1. Anthropogenic SO2 emissions. – Atmos. Environ. 19: 1029–1040.

Yagi, K. and Minami, K. 1990. Effects of mineral fertilizer and organic matter applications on the emission of methane from some Japanese paddy fields. – Soil Sci. Plant Nutr. 36: 599–610.

Ecological Bulletins 42: 199–206. Copenhagen 1992

Exchange of methane and other trace gases in rice cultivation

R. Sass, O. T. Denmead, R. Conrad, J. Freney, M. Klug, K. Minami, A. Mosier, H.-U. Neue, H. Rennenberg, W. H. Su and M.-X. Wang.

Sass, R., Denmead, O. T., Conrad, R., Freney, J., Klug, M., Minami, K., Mosier, A., Neue, H.-U., Rennenberg, H., Su, W. H. and Wang, M.-X. 1992. Exchange of methane and other trace gases in rice cultivation. – Ecol. Bull. (Copenhagen) 42: 199–206.

Corr. author: R. Sass, Biology Dept, Rice Univ., Houston, TX 77251, USA.

Introduction

Wetland rice cultivation is considered to be among the most important sources of atmospheric methane (CH_4), on a par with natural wetlands (Schütz et al. 1990b). Although atmospheric CH_4 concentration is increasing at c. 1% yr^{-1}, it is an open question as to how much of the increase is due to increased emissions from flooded rice fields or should be attributed to other sources or processes. However, it is recognized that to meet the projected rice supply of increased populations, it is estimated that the world's annual rough rice production must increase from today's 473 mill tons to 564 mill tons by year 2000 and 781 mill tons by year 2020, a 65% increase (1.7% yr^{-1}). In south Asia rice production has to be doubled by the year 2020 (Table 1). Because arable land is highly limited in major rice growing countries, the increased production has to be achieved mainly by intensifying cropping (i.e. two or three crops per year) rather than expanding the area of rice cultivation. The irrigated rice areas will continue to dominate production. Irrigated rice land now comprises about half the total harvested area but contributes more than two-thirds the total grain production (Table 2). With present agronomic practices, this will lead to increased CH_4 emissions.

Methane emission from agricultural wetlands is a direct result of soil bacterial activity. It is influenced by temperature, water regime, root exudates, organic residues, plant physiology and soil physical, chemical and biological properties (Conrad 1989). It is the net result of opposing bacterial processes, production and consumption or oxidation, that occur in different microenvironments; it is also the net result of two opposing plant-biophysical processes, transport of oxygen (O_2) to the rhizosphere and transport of methane out of the rhizosphere, inside the plant. These processes may well show independent responses to the same environmental forcing function. Thus, emission measurements must be obtained in conjunction with and correlated with measurements of the opposing processes.

Daily variation of CH_4 emission has been related to diel soil temperature changes (Schütz et al. 1989). While diel CH_4 flux variations have been observed in some studies, they have not appeared universally, perhaps because of influences of other factors such as weather, plant activity and field management. In Italy, maximum daily CH_4 emission rates were observed during late afternoon when temperatures in the soil reached maximum values (Schütz et al. 1989). Methane emission rates in Chinese wetland rice exhibited two different diel patterns for the early and the late rice vegetation period (Wang et al. 1990). During the early vegetation period, two maximum emission rates occurred, one at noon and one during the night. In the late vegetation period there was only a nocturnal maximum in the diel course of CH_4 emission.

Variation in CH_4 emission during the growing season has been correlated with soil temperature in some studies, but not in others. Growth-seasonal CH_4 flux measurements from non-tropical rice fields show a general correlation with temperature, but respond also to seasonal trends in plant development. Three seasonal maxima were found in Italy (Schütz et al. 1989), the first one shortly after flooding, the second one during the reproductive phase of the rice plants, and the third in the late vegetative period. Three seasonal maxima also were observed in Japan (Fig.1), although cultivation practice had a strong influence on methane emission (Yagi and Minami 1989). In Texas, the early maximum was absent

Table 1. Estimated requirement of rough rice production, mill ton, in Asia for the years 2000 and 2020 (IRRI 1990).

	1989	2000	2020	Increase over 1989 %
East Asia	202,264	223,375	267,069	32.0
South East Asia	102,972	120,700	177,036	71.9
South Asia	125,129	157,438	240,134	91.9
Total in Asia	430,365	501,513	684,239	59.0
World	473,489	564,012	781,354	65.0

(Sass et al. 1990), and in China, both two-maximum and three-maximum growth seasonal patterns have been observed (Wang et al. 1990).

For other trace gases, there is insufficient evidence to indicate that flooded rice fields constitute a greater source than other land-use systems, but increased rice production can be expected to increase their emission also, and it might well be that a consequence of changing rice management practices to reduce CH_4 emissions would be to increase emissions of others gases, such as nitrous oxide (N_2O).

Our aims then are to determine how much CH_4 and other climatically revelent trace gases are emitted from rice fields now, to predict how much will be emitted in the future given no change in present agronomic practices, and to develop new management practices which will reduce emissions, yet will meet the need for increased rice production.

Objectives

The objectives are: to determine if rice cultivation is an important cause of the increase in atmospheric concentrations of CH_4 and other trace gases; to understand the causes of the variability in CH_4 emission from rice cultivation, region to region and over time; and to identify management practices for rice cultivation in the future which will stabilize or reduce the emission of CH_4 as grain production is increased to meet the demands of an expanding world population.

Research plans

To achieve these goals, we propose to establish several sites in different, representative rice-cultural systems where core investigations would be conducted to establish the magnitude of the CH_4 emission and the interaction of CH_4 production, consumption, and transport processes in the field. We also will establish a number of ancillary field and laboratory investigations to improve measurement techniques and to determine the effects of soil, plant, and management factors. Since >90% of rice is cultivated in Asia (Tables 3 and 4), the core sites should represent the different Asian rice ecosystems.

Locations nominated for possible core investigations according to climatic region and harvested rice area are listed in Table 5. In the first five years, we will focus on five of them in irrigated and rainfed rice. They will be established in collaboration with national research institutions in the vicinity of national research stations. Concomitantly, empirical, process and predictive models will be developed to coordinate, direct and integrate the trace gas emission studies. Finally, we recommend a plan to train and involve Asian students and scientists in the studies.

We focus on the following specific activities:

Activity 1: Survey of methane exchange under different agricultural management regimes within the major rice growing regions of the world

Objectives

The objectives are: to establish uniform methods of measurement for the field determination of the emission of CH_4 and other trace gases and develop new and improved measurement technologies; and to accurately assess and compare the daily, seasonal, and annual CH_4 emission from rice cultivation at representative sites.

Table 2. Harvested area, yield, and rough rice production in 37 major rice-producing less developed countries, by ecosystem, 1985 (IRRI 1988)

Ecosystem	Area		Yield (t ha^{-1})	Production	
	Mill ha	%		Mill t	%
Irrigated	67	49	4.7	313	72
Rainfed lowland	40	29	2.1	84	19
Upland	18	13	1.1	81	5
Deepwater/tidal wetland	13	9	1.5	19	4
Total	138	100	3.2[a]	437	100

[a]Weighted average.

Table 3. Regional distribution of world rough rice production and trade of milled rice (IRRI 1990).

	,000 T	%
East Asia	202,264	42.7
South East Asia	102,972	21.7
South Asia	125,129	26.4
Latin America	18,386	3.9
Africa	9,789	2.1
Others	14,949	3.2
Total	473,489	100.0
Trade of milled rice	12,366	

Research plan

Almost all our knowledge of emission from flooded rice fields comes from a few investigations employing static flux chamber measurements (e.g. Schütz and Seiler 1989), backed up by some isotopic evidence involving ^{13}C and ^{2}H (Wahlen et al. 1987, 1989). While we envisage that chambers will continue to be the main measurement tools for the study of many of the questions raised here (i.e., fertilizer practices, management practices, comparisons between cultivars), there is certainly a need for corroborative measurements which integrate over larger space and time scales and which do not disturb the natural environment. The relative advantages and disadvantages of chambers and alternative measurement techniques have been much discussed (Fowler and Deyzer 1989, Mosier 1989, Wesely et al. 1989). Micrometeorological measurements represent one such alternative, when implemented with sufficiently sensitive gas sensors which are now becoming available. There are great expectations for eddy-correlation techniques employing tunable diode or gas (He–Ne) lasers. The first such sensors that are now being used for measurement of CH_4 fluxes in high latitude ecosystems and in aircraft flights should become more generally available in another year. Our objective would be to develop eddy-correlation systems suitable for measurement of CH_4 emissions from flooded rice fields as well as other relevant exchanges, i.e., those of water vapor and carbon dioxide (CO_2). This would be deployed in at least two of the core sites where the requirement for large flat areas could be met. Field-scale emission values obtained with these systems will provide comparisons with chamber measurements and provide additional data on water vapor loss and CO_2 uptake.

Profile techniques employing infra-red gas analysis to measure vertical gradients of CH_4 in the air above the rice plants are also being developed (Denmead and Freney pers. comm. 1990), and these would be employed alongside the eddy-correlation equipment in the field investigations.

The development and operation of these micrometeorological techniques require specialized equipment and operational expertise. The use of these techniques should be vested in one or two appropriate research groups. Likewise, flux chamber development is an art and, as far as possible, one proven type of chamber should be used for all the investigations so that intercomparisons are possible. Again, chamber development should be vested in one or two appropriate groups.

To determine clearly the detailed patterns of CH_4 emission, measurements need to be made within daily and seasonal time scales. For the proper interpretation of the data obtained, methane emission studies at each location should be accompanied by measurements of air, water and soil temperatures, soil Eh, flooding intervals, water depth and additions, soil pore-water gas concentrations, fertilization schedule, soil chemical, physical and biological parameters, above and below-ground biomass, and CH_4 production levels.

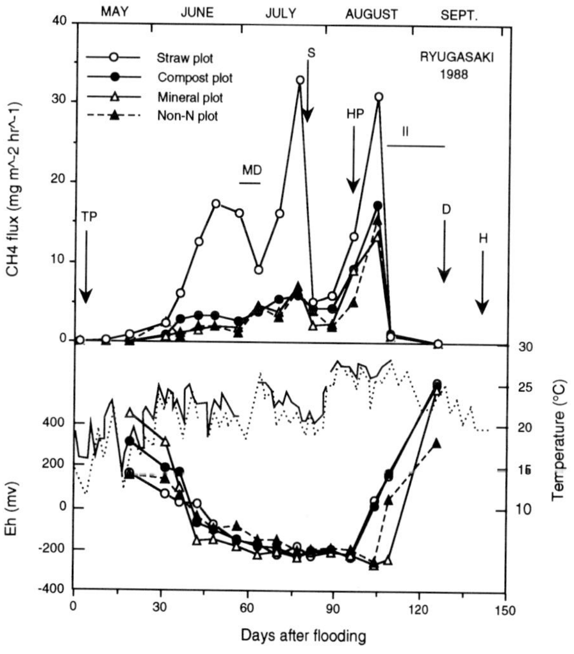

Fig. 1. Seasonal variation of methane fluxes, daily mean soil (solid line) and air (dotted line), temperature, and soil Eh in Ryugasaki paddy field, Japan. The soil temperature and the Eh were measured at 5 cm depth. The arrows indicate the agricultural practices: TP, transplanting; MD, midsummer drainage; S, supplement application of mineral fertilizer; II, Intermittent irrigation; D, drainage; H, harvest; and HP, heading period of rice plants (Yagi and Minami 1989).

Table 4. Harvested rice area by region and rice ecosystem, 1988 (IRRI 1990).

Region	Rice ecosystem (,000 ha) Irrigated	Rainfed	Deepwater	Upland	Total
East Asia	34,253	1,908	0	824	36,985
South East Asia	12,763	15,058	4,263	3,946	36,030
South Asia	18,062	21,450	8,062	7,046	54,620
Latin America	2,634	483	0	4,794	7,911
Africa	1,007	1,315	152	2,986	5,460
Others	3,081	0	0	0	3,081
World	71,800	40,214	12,477	19,596	144,087

Activity 2: Process level studies affecting methane and other trace gas exchange

Subactivity 2.1: Effects of agricultural practices and soil properties on trace gas emission from paddy soils

Differences in agricultural practices can have major effects on annual CH_4 emissions. Application of rice straw significantly increases CH_4 emission rates (Schütz et al. 1989). The effects of compost application on CH_4 emission have been found to be negligible at low amounts (Yagi and Minami 1989) but strongly enhanced emission at high amounts (Schütz et al. 1989). Tillage operations vary according to water availability, soil texture, topography, rice culture, and mechanical resources available. Hand and animal powered tillage are still common in most Asian countries and wet tillage is the preferred land preparation, resulting in earlier seasonal anoxic conditions than those found in soils that have better draining characteristics. Significant variations in CH_4 emission have been observed under different water management treatments (Minami 1992). For example, different rates of water percolation in lysimeter experiments (Fig. 2) strongly effect the seasonal variation of methane flux (Yagi et al. 1990). Both fertilization practices and water management affect trace gas emissions; consequently flux measurements need to be conducted in areas that represent the spectrum of these agronomic practices around the world.

The role of soil differences in CH_4 emission from rice fields is open to question. Koyama (1964) found no significant difference among nine Japanese rice soils,

Table 5. Possible sites for core investigations.

Ecosystem	Region	Country	Site
Irrigated rice	S. E. Asia	Phillippines	Int. Rice Res. Inst., Los Bañoa, Laguna
			Maligaya Rice Res. Cent., Muñoz, Nueva Ecija
	S. Asia	India	Bihar State, Uttar Pradesh State
			Cent. Soil Salinity Res. St, Karnal
Rainfed rice	S. E. Asia	Thailand	Ubon Rice Exp. Stn
		Indonesia	S. Sumatra
	S. Asia	India	Bihar State, Uttar Pradesh State
Deep water rice	S. E. Asia	Thailand	Pratchinburi Rice Exp. Stn
	S. Asia	India	Shinsurah Rice Exp. Stn
		Bangladesh	Ragbari Rice Ecp. Stn
Tidal wetland rice	S. E. Asia	Phillippines	Bani, Pangasinan
		Indonesia	Banjarmasin Rice Exp. Stn, Kalimantan
	S. Asia	India	Salinity Res. Stn, Port Canning, W. Bengal

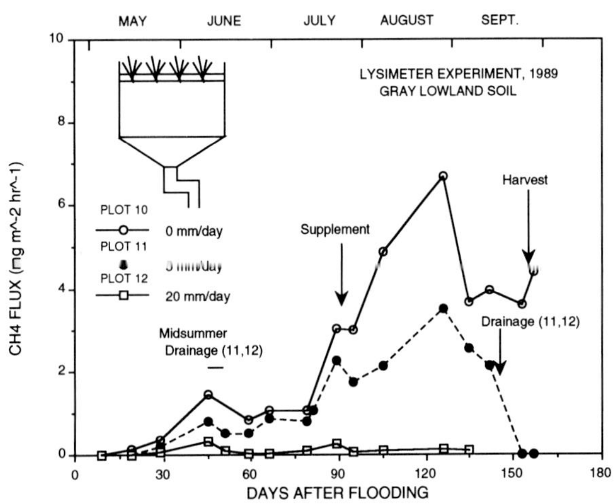

Fig. 2. Effects of water percolation on seasonal variation of methane fluxes from paddy fields (Yagi et al. 1990).

whereas a study in Texas (Sass et al. 1990) shows a three-fold emission difference between two rice fields on different soil types, even though they were treated similarly and were in close proximity. Kawaguchi and Kyuma (1977) found that 40% of the tropical rice soils they studied had at least 45% clay. Soils with such high clay content have a poor structure and are hard when dry. Soils with a greater proportion of sand will have different water movement characteristics and may have different seasonal CH_4 emission characteristics. To better understand the influence of soil properties on CH_4 emission, comparative studies must be made in soils representative of the major rice growing regions of the world, i.e. China and India.

Objective

The objective is: to assess present and projected agricultural practices and various soil types on trace gas emissions.

Research plan

It is proposed to examine the effect of agronomic practice and site on trace gas production and emission. This will be done on sites which are representative of the main rice growing areas (Table 4). In some cases the practices are linked to soil type. However, this is not universal. Further comparative experiments proposed in Subactivities 2.2 and 2.3 will help interpret results obtained from these studies.

Subactivity 2.2: Methane oxidation processes and transport of trace gases

The emission pathways of CH_4 which are generated in flooded rice soils are: diffusion into the floodwater, loss through ebullition, and transport through the aerenchyma tissue of rice (DeBout et al. 1978, Schütz et al. 1990b). Transport through the plant is currently thought to be the major pathway. From the few data available it appears that up to 90% of the CH_4 emitted from flooded rice fields is transported via the plant into the atmosphere (Schütz et al. 1990a). The development of aerenchyma facilitating this transport is induced by flooding. The amount of CH_4 emitted to the atmosphere is also controlled by the oxidation state at the root surface. The oxidation state at the root rhizosphere is largely dependent on the amount of O_2 moved from the leaf to the roots. Differences of aerenchymal structure and root oxidation power between rice cultivars and interrelationships to soil properties or management practices are poorly understood. Oxygen diffusing into the rhizosphere, provides the aerobic environment for CH_4 oxidation. Thus, transport processes in the rice plant provide means of positive and negative control for CH_4 emission from flooded rice fields. However, information on CH_4 and O_2 transport on a diel or seasonal scale for important cultural varieties is lacking (Schütz et al. 1990a). For other trace gases like N_2O or ammonia (NH_3), the path of efflux (via plants, diffusion, ebullition) is unknown.

Objectives

The objectives are: to determine the effects of oxidation processes on the emission of CH_4 and other trace gases from flooded rice fields; and to determine the role of plant transport anatomy in the emission of CH_4 and other trace gases from flooded rice fields.

Research plan

The following experiments are proposed to establish the contribution of transport in rice plants to the emission of CH_4 and other trace gases from flooded rice fields, and to investigate the possibility of using this pathway to reduce trace gas emission from flooded rice fields.

The rates of CH_4 and other trace gas emission from flooded fields will be determined in the presence and absence of plants. Different rice varieties and different soils will be studied. The results will be compared with trace gas emissions in experiments where plants are separated from the water surface, and gas bubbles collected separately. In this way, effluxes via plants, ebullition or diffusion can be distinguished. Experiments will be performed at different times of plant development and will be correlated with anatomical studies on aerenchyma development in the root and the shoot. The role of stomata in providing resistance to CH_4 efflux will be determined by comparing the degree of stomatal opening with CH_4 fluxes. Oxygen transport will be quantified by the use of micro-electrodes and by applying pulses of inert trace gases to the leaves (e.g. ethane). Experiments will be performed mainly in the field.

The transport of O_2 into the rhizosphere affects oxidation processes in the environment and thus these processes may vary on diel and seasonal time scales. Important parameters for CH_4 oxidation include: efficiency of O_2 transport by the plant; diel and seasonal variability of plant physiological activities and anatomical properties; reduced compounds in the soil, e.g. Fe^{+2}, Mn^{+2}, organics, S^{-2}, NH_4^+, which compete for soil O_2; microbial O_2 demand, pH and redox potential; soil physical parameters; faunal activity; O_2 production by organisms in floodwater and at the soil-water interface; and availability of nutrients for methanotrophic bacteria.

Currently, the following techniques are available to study oxidation processes in rice fields: mass balance of produced and emitted trace gases; stable isotopic methods to measure the isotopic fractionation that occurs during oxidation and reduction; and methane oxidation potentials (by laboratory measurements).

Experiments will be designed to: develop methods to quantify CH_4 oxidation in the field; study the effects of field variables (nitrogen, organic carbon, etc.) on CH_4 oxidation; and integrate the results with field flux measurement programs.

Subactivity 2.3: The influence of different cultivars and sources of organic matter on trace gas production and emission from flooded rice fields

The production of trace gases in flooded rice fields is driven by organic matter which is generated by living plants (exudates, roots), senescent rice plants, and the biotic community in the flood water (Bouwman 1990). The degree to which these various sources can be rapidly oxidized follows the pattern of: (living plants) > (organic matter produced photosynthetically in the water column) > (senescent plant material), and is highly dependent on the structures and functions of the associated microbial community (Conrad 1989).

The amount of below-ground plant organic matter in the form of mucigels, shredded root surfaces, etc. is partially controlled by: varietal differences (aboveground:below-ground allocation); soil properties which will affect rooting depths; stage of vegetative growth. Inputs of senescent plant material will depend on the amount, timing and rate of turnover of incorporated material. The contribution of carbon supplied from the biotic community in the water column will depend on the community structure and the proportion of this organic material settling out on the soil surface.

Objective

The objectives are: to assess the influences of rice cultivars and various sources of organic matter on trace gas production and emission from flooded rice fields; and to evaluate the origin of CH_4 carbon as predicated on the variety of rice; soil characteristics; and propagation strategies, i.e., depth and structure of the flood water community.

Research plan

The approach to assessing the impact of these various sources of organic matter in field experiments will be through manipulation of the sources and evaluating the effects on trace gas production and emission.

Microplots will be established at each of the sites where major emission studies will be conducted. Treatments will include: 1) bare soil control; 2) planted soil; 3) planted soil plus 1×litter; 4) planted soil plus 2×litter; 5) soil plus 1×litter; 6) soil plus 2×litter. Fertilizer will be applied at the rate applied in the main site trials. These treatments will be repeated with the overlying water being shaded to remove the major sources of primary and secondary floodwater production. Methane production and emission will be examined at six times during the growing season. Results of these experiments will provide input into process related models developed (Activity 3) to predict effects of these various management practices on yield and trace gas production.

The information gained in field experiments will only provide information on input and output levels. For future management of trace gas producing microbial communities considerably more detailed information is needed requiring laboratory experiments. These studies will simulate the conditions established in field studies and provide the opportunity to examine the spatial and temporal dynamics of carbon and electron flow within these systems. These dynamics describe the terminal trace gases evolved from other systems. The kinetics and controls on the rates of turnover of polymers to monomers to intermediary metabolites to trace gases, will be studied in order to accomplish these goals.

Activity 3: Empirical, process, and predictive models of methane and other trace gas emission from flooded rice fields

Objectives

The objectives are: to provide a retrospective analysis of the role of rice production on increasing atmospheric concentrations of CH_4 and other trace gases; to assist in understanding complex processes of production, oxidation and transport which control trace gas exchange from rice fields; to provide direction, integration and coordination of field research activities; and to predict the influence on trace gas exchange of environmental, physiological, and agronomic changes that may occur in rice production.

Research plan

Three different sets of models concerning the exchange of CH_4 and other trace gases from flooded rice will be developed. These models will provide a retrospective examination of the role of rice production in increasing atmospheric CH_4 and N_2O concentration; assist in understanding the complex processes of trace gas production, CH_4 oxidation, N_2O reduction, and gas transport which control net gas emission or uptake in the system; and predict the influence of environmental, physiological and agronomic changes that may occur in rice cultivation on gas exchange.

Model 1: An empirical model to demonstrate past and present roles of rice production in the increase of trace gas concentrations in the atmosphere.

A first estimate will be made to extrapolate the few available gas emission measurements to current global conditions (Aselmann and Conken 1989). Based upon records of the past and present quantity of rice cultivation, these emission measurements can be used to estimate past, current, and future trace gas emission in the different cropping systems and evaluate their importance in the global context. This estimate will extend earlier extrapolations and will not necessarily be restricted to an areal basis, but will include statistics of other variables (management, soil types, etc.) that will be shown to control CH_4 emissions.

Model 2: Process-model coupling of production, oxidation, reduction and transport of trace gases.

This model should describe the effects of rice plant ageing (physiological and anatomical) on trace gas emission on a daily and seasonal basis. The changes in plant development and diel physiological changes that affect microbial activity are interactive with other factors affecting production, oxidation, reduction and transport of gases. The model will describe the interrelationship between the carbon and nitrogen cycles, and the effects of plant varieties, soil type, climate, and agricultural management (irrigation, tillage, fertilization) on gas exchange.

Model 3: A global predictive model.

This model will incorporate the above process model with a geographic information system (GIS) to provide predictions of the effects of changes in rice cultivation on trace gas production, and to extrapolate the emission measurements from the core sites to much larger spatial and temporal scales. The model would link remotely sensed data on simple variables such as biomass, leaf area index, temperature and water depth to emission rates.

Activity 4: Cooperative training effort between scientists of initiating countries and scientists and students from core site countries

Objective

The objective is: to promote continued research cooperation between participating countries.

Research plan

As described above, 90% of rice cultivation is in Asia. Our core sites will be established in these countries. In order to foster cooperation and continuation of research post IGBP, we recommend an involvement of students and scientists from participating countries, that will not be initiating the research efforts. Students and scientists from participating countries would be trained in cooperating countries and would then be in a position to continue the studies involving trace gas emissions after the initial studies through IGBP were completed in their countries. This program would also aid in local liaison for the initial studies carried out under IGBP.

We envision graduate and postdoctoral students, selected by their countries, spending time in collaborating institutions before the regional experiments are initiated. These scientists would be trained in the principles of the measurements and processes and familiarized with the instrumentation. They would become members of the investigative team and would participate fully in the field measurements or experiments. With time they would be in a position to take over leadership roles and minimize the involvement of the initial IGBP investigators. The same individuals would return post-field season to be involved directly in data reduction and synthesis. The result would be an increase in the pool of scientific researchers who could participate fully in future studies on trace gas emissions within their own countries and, hopefully, supported by their own countries.

References

Aselmann, I. and Conken, P. F. 1989. Freshwater wetlands: global distribution of natural wetlands and rice paddies: their net primary productivity, seasonality and possible methane emission. – J. Atmos. Chem. 8: 302–358.

Bouwman, A. F. 1990. Soils and the greenhouse effect. – Wiley, Chichester.

Conrad, R. 1989. Control of methane production in terrestrial ecosystems. – In: Andreae, M. O. and Schimel, D. S. (eds), Exchange of trace gases between terrestrial ecosystems and the atmosphere. Wiley, Chichester, pp. 39–58.

DeBout, F. A. M., Lee, K. K. and Bouldin, D. F. 1978. Bacterial oxidation of methane in a rice paddy. – Ecol. Bull. (Stockholm) 26: 21–26.

Fowler, D. and Deyzer, J. H. 1989. Micrometeorological techniques for the measurement of trace gas exchange. – In: Andreae, M. O. and Schimel, D. S. (eds), Exchange of trace gases between terrestrial ecosystems and the atmosphere. Wiley, Chichester, pp. 189–208.

IRRI (Int. Rice Res. Inst.). 1988. World rice statistics, 1987. – IRRI, Los Baños, Philippines.

– 1990. World rice facts, 1989. IRRI, – Los Baños, Philippines.

Kawaguchi, K. and Kyuma, K. 1977. Paddy soils in tropical Asia: their material, nature and fertility. – The Univ. Press of Hawaii, Honolulu, Hawaii.

Koyama, T. 1964. Biogeochemical studies on lake sediments and paddy soils and the production of atmospheric methane and hydrogen. – In: Miyake,Y. and Koyama, T. (eds),

Recent researches in the fields of hydrosphere, atmosphere, and nuclear geochemistry. Muruzen, Tokyo, pp. 143–177.

Mosier, A. R. 1989. Chamber and isotope techniques. – In: Andreae, M. O. and Schimel, D. S. (eds), Exchange of trace gases between terrestrial ecosystems and the atmosphere. Wiley, Chichester, pp 175–189.

Sass, R. L., Fisher, F. M., Harcombe, P. A. and Turner, F. T. 1990. Methane production and emission in agricultural wetlands. – Global Biogeochem. Cycles 4: 47–68.

Schütz, H. and Seiler, W. 1989. Methane flux measurements: Methods and Results. – In: Andreae, M. O. and Schimel, D. S. (eds), Exchange of trace gases between terrestrial ecosystems and the atmosphere. Wiley, Chichester, pp. 209–228.

– , H., Holzapfel-Pschorn, A., Conrad, R., Rennenberg, H. and Seiler, W. 1989. A 3-year continuous record on the influence of daytime, season and fertilizer treatment on methane emission rates from an Italian rice paddy. – J. Geophys. Res. 94: 16405–16416.

– , H., Schröder, P. and Rennenberg, H. 1990a. Role of plants in regulating the methane flux to the atmosphere. – In: Charky, T., Holland, B. and Mooney, H. (eds), Environmental and metabolic controls on trace gas emissions by plants. – Academic Press, New York, (in press).

– , Seiler, W. and Rennenberg, H. 1990b. Soil and land use related sources and sinks of methane (CH_4) in the context of the global methane budget – In: Bouwman, A. F. (ed.), Soils and the greenhouse effect. Wiley, Chichester, pp 269–285.

Wahlen, M., Tanaka, N., Henry, R. and Yoshinari, T. 1987. ^{13}C, D and ^{14}C in methane. – Trans. Am. Geophys. U. 68: 1220.

– , Tanaka, N., Henry, R., Deck, B., Seglen, J., Vogel, J. S., Southon, J., Shemesh, A., Fairbanks, R. and Broecker, W. 1989. Carbon-14 in methane sources and in atmospheric methane: the contribution from fossil carbon. – Science 245: 286–290.

Wang , M.-X., Aigno, D., Renzing, S., Schütz, H., Rennenberg, H. and Seiler, W. 1990. CH_4 emission from a Chinese rice paddy field, Hangzhou, China. – Acta Meteorologica Sinica, (in press).

Wesely, M. L., Lenschow, D. H. and Denmead, O. T. 1989. Flux measurement techniques. – In: Lenschow, D. H. and Hicks, B. B. (eds), Global tropospheric chemistry: chemical fluxes in the global atmosphere. Nat. Cent. for Atmospheric Res., Boulder, pp. 31–46.

Yagi, K. and Minami, K. 1989. Effects of organic matter applications on emission from some Japanese paddy fields. – Soil Sci. Plant Nutr. 36: 599–610.

Yagi, K., Minami, K. and Ogawa, Y. 1990. Effects of water percolation on methane emission from paddy fields. – Res. Rep. Div. Environ. Planning, Nat. Inst. of Agro-Environ. Sci. 6: 105–112.